Advanced Series in Agricultural Sciences 11

John R. Parks

# A Theory of Feeding and Growth of Animals

With 123 Figures

Springer-Verlag
Berlin Heidelberg New York 1982

John Rabon Parks, MSc
Department of Animal Husbandry
The University of Sydney
Sydney 2006, N.S.W., Australia

ISBN 3-540-11122-0 Springer-Verlag Berlin Heidelberg New York
ISBN 0-387-11122-0 Springer-Verlag New York Heidelberg Berlin

Library of Congress Cataloging in Publication Data. Parks, John R. (John Rabon), 1908– . A theory of feeding and growth of animals. (Advanced series in agricultural sciences; 11). Bibliography: p. Includes index. 1. Animal nutrition – Mathematical models. Livestock – Growth – Mathematical models. 3. Animal nutrition – Statistical methods. 4. Livestock – Growth – Statistical methods. I. Title. II. Series. SF95.P24. 636.08′52. 82-5535. AACR2.



Printed in Germany.



Typesetting, printing, and bookbinding: Brühlsche Universitätsdruckerei, Giessen
2131/3130-543210

*For Ava H., Ava J., and Gerald*

# Foreword

Geoffrey R. Dolby, PhD

One of the principal characteristics of a scientific theory is that it be falsifiable. It must contain predictions about the real world which can be put to experimental test. Another very important characteristic of a good theory is that it should take full cognisance of the literature of the discipline in which it is embedded, and that it should be able to explain, at least as well as its competitors, those experimental results which workers in the discipline accept without dispute.

Readers of John Parks' book will be left in no doubt that his theory of the feeding and growth of animals meets both of the above criteria. The author's knowledge of the literature of animal science and the seriousness of his attempt to incorporate the results of much previous work into the framework of the present theory result in a rich and imaginative integration of diverse material concerned with the growth and feeding of animals through time, a theory which is made more precise through the judicious use of mathematics. The presentation is such that the key concepts are introduced gradually and readers not accustomed to a mathematical treatment will find that they can appreciate the ideas without undue trauma. The key concepts are clearly illustrated by means of a generous set of figures.

The crux of the theory comprises three differential Eqs. (7.1–7.3), two of which apply to ad libitum feeding and the third to controlled feeding. These are so well motivated by discussion of diverse experimental findings that they do not emerge in final form until Chapter 7. Standard methods exist for solving these equations [see, for example, Articles 14 and 18 of Piaggio's *Differential Equations* (Bell & Sons, London 1958); or at least Chapters 9–12 of Batschelet's *Introduction to Mathematics for Life Scientists* (Springer, Berlin 1975)] and the reader who has had a course in calculus should have little difficulty in quickly acquiring the skills needed to feel at home with the theory. Mathematically, the theory could be deduced starting from these three differential equations and their initial conditions. This would have given a more concise formulation but the reader would have been left wondering about the origin of the crucial equations.

The theory is designed to predict what John Parks calls the "trace" of the animal, that is the triplet $(W, F, t)$ where $W$ is the liveweight and $F$ the cumulative food consumed at time $t$. In the case of ad libitum feeding, solution of the differential equations yields two curves of a form familiar to biologists, namely, the three parameter exponential curve

$$y = P - Q\,e^{-Rx}$$

in which $P$, $Q$, and $R$ are positive constants. Graphically, this equation represents a curve in which $y$ rises exponentially from an initial value of $P-Q$ (at $x=0$) toward an upper asymptote at $P$. The parameter $R$ determines the decreasing "steepness" of the approach to the asymptote. The first such curve [Eq. (2.2)] which occurs in the solution relates $W$ to $F$ with $P$, $P-Q$ and $R$ having values of $A$, $W_0$, and $B$ respectively. This implies that for all animals, the graph of live weight as a function of cumulative food follows the above form, rising from an initial weight $W_0$ to a mature weight $A$. The larger the value of $B$, the more rapidly is the asymptote approached.

The second such curve [Eq. (2.3)] which emerges as a result of solving the differential equations is one connecting the rate of ad libitum feeding $q^*(=dF/dt)$ with the time $t$. Here the daily food, $q^*(t)$, rises exponentially from an initial value $D$ towards an asymptotic level $C$, and the parameter $R$ in this case is denoted by $1/t^*$ in order to relate the theory to previous work by Brody (1945). In Brody's work the same curve arises [Eqs. (1.4, 1.5)], this time as part of a $(W, t)$ relationship. Brody's $t^*$ is the point at which the curve crosses the time axis.

The differential equation governing the trace under controlled feeding has no such simple solutions. In this case, the feeding regime $q(t)$ is given rather than predicted, but solution of the differential equation allows us to predict live weight as a function of time.

The basic parameters characterising the animal and its environment are $A$, $(AB)$, $T_0$, $t^*$ where $T_0$ is the ratio $A/C$ of the mature live weight to the mature rate of feeding. In Chapters 8 and 9 the author turns his attention to the problem of how these parameters may be influenced by the composition of the diet and in Chapter 10 he discusses the application of the theory to questions of quantitative genetics of growth, and in Chapter 11 he remarks on the relation of the theory to the energy balance that all open systems must obey.

Is the theory falsifiable? As the author says (Chapter 7, p. 113): "There is then a need to plan and perform an experiment which will subject the theory to such stringent conditions that the experimental results can be used to declare the theory useless for further research, useful if modified in some rational manner, or useful as presently formulated". After describing such an Experiment, he says, with refreshing candour (Chapter 7, p. 126): "Here I have concluded that the first prediction passed the critical test and the second prediction failed the test...".

There are major problems in falsifying any theory. First, the predictions depend on the values of the parameters for the animal and environment in question. One can measure the parameters for similar animals under similar conditions, but hardly for the experimental animal itself. Therefore any discrepancy between observed and predicted weight may be attributed either to a failure in the theory or to a failure in the specification of the parameters, due to natural variation among animals. Thus a single experiment can hardly be decisive for the theory, even given careful statistical consideration of the various distributions involved.

The theory has the potential to motivate a great variety of ingenious experiments designed to falsify it. The effort which has gone into its formulation surely merits an equally vigorous response on the part of experimentalists. Whether it survives or falls is not very material (I know of no empirical scientific theory that has survived unscathed indefinitely). The value of John Parks' achievement is that it has provided an invaluable stimulus to the production of further theories dealing with this fascinating area. During the course of attempts to refute it, new discoveries will be made. This is how science progresses.

St. Lucia, Queensland, January 1981

CSIRO Division of Mathematics
and Statistics

# Preface

During 1964, while working on an operations research problem for the Agriculture Division of the Monsanto Company, St. Louis, Missouri, I was also assigned to a study of the experimental data produced by the Animal Science Group from short term studies of animal responses to chemical feed additives. The experiments were strictly designed according to the principles of analysis of variance. My association with these men was amiable, resulting in my learning much from them about some of the particulars of how animals fed and grew as well as helping them in their contacts with the Computer Centre.

I felt some intellectual dissatisfaction with the short term experimental approach to nutrition and began to imagine the animal as a black box with input and output which must satisfy the balance equations of mass and energy flow rates. This approach automatically required study of as much long term feeding and growth data on as many animal species as I could find. Prior to launching such an ambitious program, I found Thompson's *Growth and Form*, Brody's *Bioenergetics and Growth*, Kleiber's *Fire of Life* and Sommerhoff's *Analytical Biology* instructive and suggestive. An underlying motive for this self imposed labour was the possibility of helping the Animal Science Group to extract more information from their experimental data than it seemed to me they were getting.

Gathering tables of good data on long term feeding and growth of some domestic species proved to be a less difficult task than I had imagined. During the study of these data and the discussion of them by the research men who produced the data, an idea of a theory of feeding and growth of animals began to take on substantial form.

I cannot say the idea came either by intuition or induction. However, I can say I knew that if a theory could be explicitly stated it should be possible to subject it to critical experimental tests of consequences drawn from the theory.

During the period from 1964 to my retirement in 1969 my ideas led to publication of a new set of mathematical feeding and growth functions and the effects of diet composition and other environmental variables on the parameters of these functions. During this period the concept of this book began to take shape.

After retirement, the late Professor H. L. Lucas, Jr., gave me work and study space in his Biomathematics Program at the University of North Carolina at Raleigh. Here I had critical and rewarding discussions of my approach to feeding and growing animals with him, his faculty members, and postgraduate students. The first drafts of Chapters 1 and 2 were written and critically discussed. I began notes on Chapters 4, 5, 8, 9 and 11.

I had already become doubtful of regarding growth as the realisation of a stochastic process in the time domain and discussions with Professor H. R. van der

Vaart led me to treat growth in the cumulative food consumed domain as a Markov process; Chapter 3 resulted from this study.

In November 1971 I heard Dr. St. Clair Taylor, of the Agriculture Research Council's Animal Breeding Research Organisation, Edinburgh, discuss his experiments on long term controlled feeding and growth of Ayrshire twin cattle. The A.R.C's Underwood Fund made it possible for me to join Dr. Taylor as a visiting scientist at A.B.R.O. in April 1973. Up to this time my studies concerned only the growth of animals fed nutritious diets ad libitum. I saw here a key piece of experimental work which led me to study the growth of animals under feeding regimes other than ad libitum.

Chapter 6 came out of these studies, where I found a first order differential equation of weight versus time, the solutions of which adequately described the experimental growth data when an initial condition was given and the controlled feeding regime was a preselected, well defined function of time.

I then combined this differential equation with those derivable from my ad libitum feeding and growth functions to form a theory. From this theory I drew two predictions which could be put to stringent experimental test, thereby putting the theory at risk. Discussions of the theory and predictions with Mr. G.C. Emmans and Dr. B.J. Wilson led to a long term experiment, in the A.R.C. Poultry Research Centre facilities, aimed at specifically falsifying the predictions. Chapter 7 is the result of this work.

During my stay at A.B.R.O. I worked on the entire manuscript, getting it ready for typing. Dr. Wilson was most helpful in getting the first typescript finished. In July 1975 I left A.B.R.O. to come to Sydney, Australia.

Professor T.J. Robinson, Head of the Department of Animal Husbandry at the University of Sydney, permitted me to work with Doctors J.S.F. Barker and F.W. Nicholas of the Animal Genetics Section in getting Chapter 10, on the application of the theory to the quantitative genetics of animal growth, into an acceptable form. Here the final typescript of this book was produced.

The objective of this book, as perceived some years earlier, is to give the animal scientist a strong feeling for the power of calculus (not higher than first order differential equations, and simple function theory) and analytical geometry in planning both long and short term animal feeding and growth experiments, and interpreting the data in such a manner that the results are applicable to planning experiments in contiguous areas of scientific and economic interest.

The University of Sydney,
February 1982

JOHN R. PARKS

# Acknowledgements

I owe much to Mr. F. Mitchell, Jr., an administrator in the Agriculture Division of the Monsanto Company, who gave much needed support to my early efforts. Many discussions with Mr. A.W. Dickinson and Mr. C.J. McCoy, respectively mathematician-statistician and chemical engineer, and Drs. W. Dudley and K. Maddy, nutritionists, of the Monsanto Company, pointed out the advantages and disadvantages of my engineering physics approach to animals as input output devices. Professor H. W. Norton, of the Animal Science Department of the University of Illinois at Urbana, has followed my work almost from the beginning and offered invaluable critical advice. I appreciate the acquiescence of the publishers of the American Journal of physiology and the Journal of Theoretical Biology to my republishing some of my work in more integrated form, and I hope I have acknowledged in the text all the permissions of authors and publishers of papers and books to quote data and results. I treasure the many discussions with Dr. B.J. Wilson, of the A.R.C. Poultry Research Centre, and Mr. G.C. Emmans, of the South East of Scotland College of Agriculture, on science in agriculture and the relevance of theory to experiment. Dr. Wilson has read the manuscript and offered many valuable suggestions, work for which I am most grateful. I am thankful to Dr. John King, present Director of A.B.R.O., for permitting me the same status as Dr. Donald gave me. I am also indebted to Professor T.J. Robinson, of the Department of Animal Husbandry at the University of Sydney, for accepting me as a visiting scientist and giving me the same privileges that I had enjoyed at the University of North Carolina and A.B.R.O.

I am much obliged to Professor R.M. Butterfield, Head of the Department of Veterinary Anatomy at the University of Sydney, for his generous aid and support in getting this book into final form for publication.

I am indebted to Mrs. Susan Cox of the University of North Carolina at Raleigh, Mrs. Roma Beresford of A.B.R.O. and Miss Pam Armstrong of the University of Edinburgh, who patiently typed and retyped the manuscript until a prepublication typescript appeared. I am most grateful to Mr. John Stapleton who, acting as my secretary and typist, was of great help in getting a final form of the typescript ready for the publisher. I am also grateful to Professor Robinson's secretary, Mrs. Jan Rowe, for her generously given help, and finally, I wish to thank Mr. John Roberts, Cartographer, Department of Geography, at the University of Sydney, for the many illustrations needed in this book.

These people, and many others to whom I am indebted, have influenced the development of this book in one way or another, but any errors of philosophy, concept, fact and mathematical development of the subject matter are strictly my responsibility.

# Contents

## Chapter 5 The Geometry of Ad Libitum Growth Curves



## Chapter 6 Growth Response to Controlled Feeding



## Chapter 7 The Theory

## Chapter 8 A General Euclidean Vector Representation of Mixtures



## Chapter 9 The Effects of Diet Composition on the Growth Parameters



## Chapter 10 The Growth Parameters and the Genetics of Growth and Feeding



## Chapter 11 Energy, Feeding, and Growth

Appendices

# Chapter 1 Introduction

The central theme of this book is a search for elements of determinism in the growth and feeding of animals. Knowledge of these deterministic elements has at least two important uses. First, by elucidating the systematic properties of growth and feeding data, it reduces variability more closely to that due to elements of stochastic variability plus experimental error. Second, it brings into better focus those responses which are caused by conditions imposed on the growing animal.

The complexity, variability, and uniqueness (apparent and real) found in general biology have prompted study of the role of stochastic processes in biology, leaving, until recently, little room for determinism and the implication of mechanism. On the other hand, during the Renaissance the physical sciences began with a strong concept of determinism but of late have considered the probabilistic nature of physical events. It is this difference of approach which emphasises the slow penetration of mathematics, other than statistics, into biology.

This introduction begins with definitions and brief discussions of the terms phenomenology, etiology, and growth, followed by brief historical sketches of efforts to find the growth equation. These sketches serve to emphasise the curious tendency of biologists studying growth to use the universal property of the openness of organisms implicitly. They also serve as background for discussion of explicit use of openness in my approach to growth phenomena.

## 1.1 Phenomenology and Etiology

Webster's New International Dictionary, second edition, gives three definitions of phenomenology. The third fits best the usage in this book; it is "Scientific description of actual phenomena with the avoidance of all interpretation, explanation, and evaluation." Of the two definitions presented for etiology the second is chosen; "The assignment of a cause or reason; as, the etiology of the development of a disease." Etiology then offers scientific interpretation, explanation, and evaluation. In the context of this book phenomenology and etiology embody empiricism and theory respectively.

These definitions strictly separate the science of phenomena from the science of causes. However the history of the practice of science shows that this strict separation is more apparent than real, since it is the interaction of these scientific activities that constitutes the viability of science and technology and contributes to their present exponential growth. It is not necessarily practice before theory or the converse which is important; it is the synergistic effects which count.

In this book the reader will find induction used principally with some measure of deduction from the vast experimental knowledge of growth and feeding of animals in the laboratory and in practice. This is not a choice of empiricism over theory; it is only indicative of our ignorance of a theory of growth. In the history of science empiricism of some particular phenomena has developed into a theory of the phenomena.

Science is based on the empirical findings of regularities within and among classes of phenomena and the explanations of them. Mendeleev's empiricism of the periods among the physical properties of the elements has been explained by the theory of quantum mechanics of the elements. The empiricisms of Copernicus and Kepler regarding the motions of the planets about the sun have been explained by Newton's theory of universal gravity. Darwin, via the theory of evolution and natural selection, sought to explain the classification of presently existing animals. Molecular biologists, through the chemistry of DNA, seek to explain Mendel's empiricism of inheritance and other cellular phenomena. The reader is no doubt aware that there are no final theories or empiricisms. In man's search for integrated knowledge, older theories and empiricisms give way to newer and more comprehensive ones.

The ancients, Aristotle, Ptolemy, and the builders of Stonehenge among them, developed rare uses of phenomenology and uses and misuses of etiology. Centuries later the thoughtful men of the Renaissance began to question the ancient etiologies and to couple experimentation to hypothesis for greater clarity in the study of phenomena. Thus began the development of modern science as a rapid succession of empiricisms and theories of observable events based on well conceived experiments. Historically, phenomenologies have preceded etiologies although the teaching of sciences has de-emphasised phenomenology.

In modern times the theories of physics, astronomy, chemistry, mechanics, etc. have become so fused with fact that the tendency has been to rely heavily on a theoretical approach in a new investigation in the physical sciences. In 1964 Dr. Schwinger brought this into focus by calling for more phenomenology of an engineering nature in the field of the high energy physics of strongly interacting particles, with the hope of breaking the present impasse in theory.

## 1.2 Growth Data and the Growth Equation

The term growth should be defined prior to discussing it, if for no other reason than to clarify what is not being discussed. Among various specialists in biology considerable confusion has developed concerning an adequate or precise definition. This confusion is due to an understandable effort to define the growth of any organism in observable terms but broadly enough to include some factors which lie close to the causes of growth. Needham (1964) has discussed this situation in some detail. Professor Pran Vohra prefers change of body ash as the measure of growth (personal communication, 1974).

The growth of organisms, like the growth of crystals, has an obvious character. The entity increases in size, number or mass as time goes on. In considering this as a definition of growth, one is aware of the dictum "beware of the obvious." The complexities of cell division and differentiation, organ development and membrane function as revealed by physiologists, nutritionists, biochemists, and biophysicists

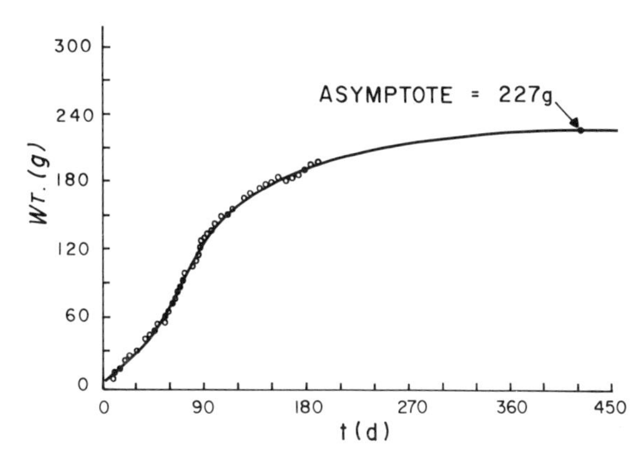


Fig. 1.1. An illustration of the S–shape of the growth curve of female rats. *Open circles* are data points. *Solid curve* is calculated by using a modified form of the logistics equation. [Reproduced from Fig. 16.58 of Brody (1945) by permission of Rheinhold Publishing Company]

certainly give us pause. Consequently there is a tendency among bioscientists to wait and work for a complete microdescription of growth before accounting for the simple and obvious macroscopic phenomenological aspects. All these complex entities and functions participating in the growth of the whole organisms are in the realm of the theory of growth. For this reason the simple definition of growth, as the change of size, live weight or biomass with time or some other variable, will be taken as basic in an empirical description of growth independent of any theory. Choice of this simple definition does not preclude concern for the compatability of the phenomenology with the yet to be revealed etiology of the growth of these complex systems.

There is a tendency here to forget or neglect the energetic aspect of growth. A more complete picture of the phenomenology must contain not only the mass transactions but also the internal mass and energy transactions of the animal during growth. This book deals with both sets of phenomena separately, but it is hoped that as the subject matter develops the reader will see the reasonableness of considering the mass transformations first.

It must be emphasised that here the animal is considered a mobile, self feeding, low pressure, quasi-constant, low temperature macro-assembly of microcatalytic chemical reactions, which transforms the matter and energy of the input chemicals (food) into energy to be dissipated to the environment as heat and work, stored as live weight and packaged as products such as eggs, milk or young. The mechanisms that govern or produce these transformations are important and of great interest but need not be considered in the context of the phenomenology of growth. As a consequence the processes of metabolism, nutrition, and genetics will be discussed only briefly if at all.

Even though the growth of animals is variable and changeable, the biomass of freely fed individuals and populations of animals follows a quite well defined course as they age. Figures 1.1 and 1.2 show that rats and chickens (the open circles are data points) have courses of growth which share the same characteristics. Initially the rate of growth is low but increasing. At age $t'$ the growth rate is maximum (about 60 days for rats and about 84 days for the cockerels and pullets) and then slowly declines to zero when the animal achieves its mature weight. This type of growth curve has been called either the S-shape or sigmoid curve; the names are interchangeable. However, this curve is not universal. There are living organisms

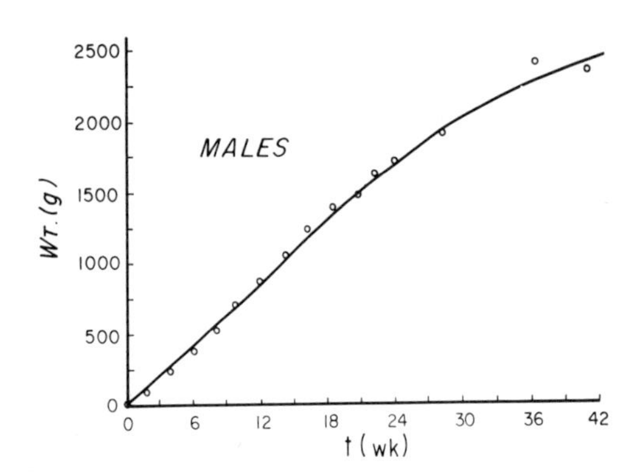


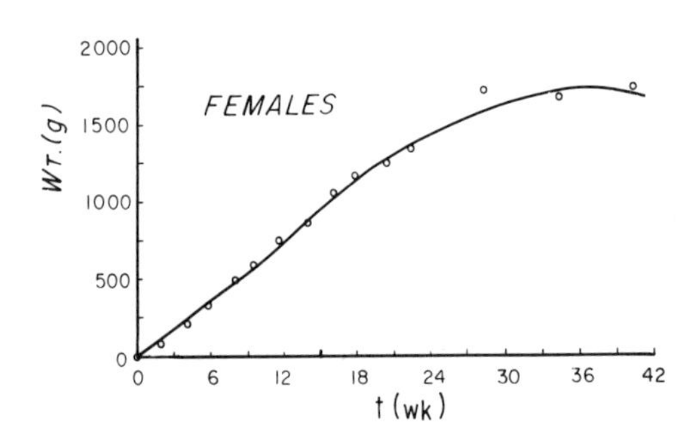


Fig. 1.2. Illustrations of the S–shape growth curves of male and female chickens. *Open circles* are data points. *Solid curves* are calculated from 4th degree polynomial growth functions fitted to the data using linear regression

whose growth possesses neither a t′ nor a mature size or weight. But the sigmoid curve is sufficiently prevalent among animals to warrant study with the objective of developing the equation for this type of growth curve.

Consider what is meant by the phrase equation of growth. The experimental data consists of ordered, number pairs $(t, W)$ where $t$ marks the time at which the animal was weighed and $W$ is the live weight. Generally such data are reported in tables with $W$ ordered according to time $t$. Mathematically these ordered pairs represent a function designed by $W=\phi(t)$; i.e., to every time $t$ there is a corresponding live weight $W$. This is not to say that in general $\phi(t)$ can be so specified that the $W$ corresponding to any time $t$ can be calculated exactly. All the above equation states is that with each $t$ there is associated a unique $W$. Thus $W=\phi(t)$ is an equation of growth.

In creating a table or graph the experimenter displays these ordered pairs, thus providing a means of estimating values of $W$ for times $t$ at which measurements were not made. This is supported by the fact that growth tables and graphs are often used for interpolating to find a value of $W$ at some desired time which lies between two tabular times. Hence tables or graphs are used as though they do in fact represent a growth equation of the type $W=\phi(t)$.

Finding a mathematical form for $\phi(t)$, which can be used to calculate values of $W$ comparable to the tabular values, is highly useful with computers in which it is frequently more efficient to store an equation than a table of data. However, this is only a small part of the challenge of finding a physically and biologically acceptable mathematical representation of $\phi(t)$.

For a given table of data, the equation $W=\phi(t)$ which represents the sigmoid growth curve, must contain a set of numbers $(a_1, a_2, \ldots, a_n)$ as shown in Eq. (1.1). These numbers, which have the generic name parameters, are constant for any table of data but must be changed to a new set of values for a different table.

$$W=\phi(a_1, a_2, \ldots, a_n, t). \tag{1.1}$$

The tables may be different individuals or populations of the same or different species, as shown by the data points in Figs. 1.1 and 1.2. At this point the reader can see the power of an equation of growth in which the mathematical form of the function $\phi$ is invariant and only the parameters change across species and classes of animals under various environmental conditions. All the differences among the species and conditions would be reflected in the different values of the parameter set $(a_i)$, $i=1,2,\ldots,n$. If the parameters happen to have acceptable biological and physical meaning, then subsuming into the parameters the effects on growth of diet, ambient conditions, work, controlled food intake, genetic selection etc. would make possible the unification of much of the present research on growing animals and would suggest new experiments. Such unification would certainly be helpful in the interaction of the empirical and theoretical natures of growth. From this discussion one may understand that the equation of growth is not an idle dream but a truly scientific goal.

Applied mathematics is an indispensable tool in the search for the growth equation. Statistics and the analysis of variance are very useful devices. However, the worker using growing animals in any experiment must also know the calculus, differential equations, and probability, as well as have a working knowledge of how the physicists have used these techniques in solving real world problems with and without the computer (Batschelet 1975).

## 1.3 Present Mathematical Models of Growth in Perspective

The net result of experimentation is a table of data from which the experimenter seeks to draw valid conclusions. This table of data is the historical record of an effort to gain knowledge, insight and understanding of phenomena. It is historical in the sense that the data remain at the end of the experiment while the apparatus is dismantled and experimental objects are disposed of. The experimenter may then use various techniques such as graphing, averaging, and ranking or analysis of variance on the data to point up his expectations and find interesting regions for future experimentation.

Whatever reasons may have existed for performing the experiment, someone (the experimeter, co-worker or other scientist) will enquire about the mathematical structure of the table of data if for no other reason than to reduce, if possible, the volume of numbers to be retained. A low degree polynomial may describe the data sufficiently well that only the coefficients and the statistical measures of fit need be retained. Mitchell et al. (1931) found that 4th degree polynomials described the growth data of cockerels and pullets very well (solid curves in Fig. 1.2). Each polynomial has five coefficients to describe 15 data points representing a three to one reduction of numbers to be retained. Another reason for the enquiry about mathematical structure may be the desire for a mathematical representation of the data for purposes of interpolation or extrapolation to those experimental regions where data were not taken. Too often, however, tables of data are treated as though they possess no mathematical structure, except that extractable by analysis of variance techniques. It is unusual for data to have no mathematical structure.

A table of data, mathematically structureless but containing statistical information, can be visualised by having n rows and m columns where each of the mn cells contains an integer selected randomly from some defined discrete distribution. If the integers represent grey shades in a range of 1 for black to 16 for white and the distribution is rectangular with the probability of the selection of an integer being 1/16, then the table could represent a black and white structureless picture with an average grey shade of 8.5 (Parks 1965). Such a table of data can best be considered a random number table of the integers of 1 to 16. Even here the table of mn integers can be replaced by the algorithm by which it was created. In this case the algorithm is the discrete probability distribution of the integers 1 to 16. At this point it might be properly asked whether experimental work is not really an attempt to discover the natural algorithms by which data are created by the organism plus experimenter.

The mathematical function, which can be used to calculate data points with acceptable precision, is often referred to as a mathematical model of the data. Occasionally a theory of the causes of the data presents the experimeter with a method for deriving a mathematical model (Lucas 1964). This is more frequently the case in the physical sciences than the biosciences.

More often there is no prefabricated model available and the mathematical structure in the data may then be represented by some polynomial of suitable degree. Such an equation can describe experimental data quite well and should be adequate to produce tables of smoothed data. Such a method has been used to create large numbers of tables of smoothed growth data (Brody 1945). It is well known that extrapolation of polynomials yields nonsense. Mitchell et al. (1931) found that extrapolation of their curves beyond 40 weeks of age showed their birds steadily losing weight, which is known not to be true in experiments longer than 40 weeks. A theoretical model which describes the data can be used in both interpolation and extrapolation, whereas emirical functions of the polynomial type can be used safely only within the range of the experimental data.

Between the extremes of theoretical and empirical cases there are situations in which the experimenter has some prior information concerning the system on which he is working. This information can be used with the data to produce a model which is neither purely theoretical nor purely empirical. A notable example of a part theoretical, part empirical model in biology is the work of Skellam et al. (1958) on the simultaneous growth of larvae and adults during the development of the ant *Myrmica rubra*.

Animal growth data are usually presented as a graph of live weight versus age as shown in Figs. 1.1 and 1.2. Since there is no generally accepted etiology of growth from which the equations of growth can be deduced, the form of the function $W=\phi(a_i, t)$ must be developed empirically, keeping in mind that it should have some of the properties expected by biologists. Many biologists, growth model makers such as von Bertalanffy (1938, 1957), Weiss and Kavanau (1957), Laird et al. (1965, 1967) and others would deny the phrase "no generally accepted etiology of growth," since each form of $\phi(a_i, t)$ used by them was derived from some theoretical considerations. The position taken here is that if a generally accepted theory of growth were available then only one form of $\phi(a_i, t)$ would be logically deducible from it, not the competing forms presently found in the literature. The balance of this chapter should give readers enough background to form their own opinions.

The curves in Figs. 1.1 and 1.2 show two important mathematical properties which an acceptable $\phi(a_i, t)$ must have. First, for $t$ large, $\phi(a_i, t)$ is asymptotic to a mature weight $A$, which is one of the parameters $a_i$. This makes it possible to work with the degree of maturity ($u = W/A$) which an animal has reached at any age $t$. In this way the growth of animals can be scaled to lie between 0 and 1, or 0 and 100 if percent maturity is used. Brody (1945, Chap. 16) has effectively used percent maturity in comparing growth curves between species having different mature live weights. Second, the curve shows that there is an age $t'$ and a weight $W'$ at which the growth rate changes from curving upward to curving downward. As a consequence $\phi(a_i, t)$ has an inflection point at $t'$. At $(t', W')$ the growth rate $dW/dt$ is maximum. Biologists refer to growth near the maximum growth rate as the growth spurt. Frisch and Revelle (1969) have used numerical analysis of human growth data to determine not only the time $t'$ of occurrence of the growth spurt but also the time of onset for comparing the growth of children of various races. There is a general feeling that some radical physiological changes are occurring during the growth spurt. For example Brody chose $t'$ to mark sexual maturation where the animal internally changes from the vegetative phase to the procreative phase. Despite gross physiological changes from infancy to maturity experimental growth data generally appear to show that growth proceeds smoothly through age $t'$ as shown in Figs. 1.1 and 1.2.

The observed physiological changes have not yet been used to distinguish the following mathematical properties that a possible growth function may have:

1. Growth is discontinuous in time and is expressed as an accumulation of positive and negative finite increments of live weight.
2. Growth is continuous, but possesses no continuous rates of change.
3. Growth is continuous and possesses a continuous first order rate of change ($dW/dt$), but no continuous higher order rates of change.
4. Growth is continuous and possesses continuous rates of change of all orders.

In Case 1 each of the data points must be considered unique; the table of data points contains no information about relationships among the data points; interpolation and extrapolation would be forbidden. Limited point by point analysis of variance would be permissable with main effects ignored. The concept of a growth function is unacceptable.

In Case 2 the limitations on considering interpolation, extrapolation, and continuous main effects would be removed and a growth function would be possible but with characteristics displayed by the functions of Bolzano (1834) and Weierstrass (1872); namely the functions are continuous everywhere but possess rates of change nowhere. The concept of growth spurt would have no meaning.

[For details about this kind of continuous function see Boyer (1949), pp. 269, 284, 285].

Case 3 would permit the concepts of growth function and growth spurt and would allow the methods of numerical analysis for interpolation and extrapolation between the data points, but it would forbid computing higher order derivatives for the determination of the age or onset of the growth spurt.

Case 4 opens up the possibility of the existence of a growth equation $\phi(a_i, t)$ and the use of regression techniques to distill the information in the set of data points to the parameter set $(a_i)$. Time will have been removed and the techniques of analysis of variance can now be used on the sets of parameters $a_i$ to study the effects of environmental and nonenvironmental factors on the growth of animals.

I have chosen Case 4 for two reasons, namely (a) there appears to be no evidence to the contrary and (b) it is intuitively the most promising of the cases. Because the maximum growth rate occurs at the time $t'$, at which $dW/dt$ is a maximum and $d^2W/dt^2 = 0$, $t'$ is a root of Eq. (1.2).

$$d\phi^2(a_i, t)/dt^2 = 0. \tag{1.2}$$

The maximum value of $dW/dt$ is found by substituting this value of $t$ in the right hand side of Eq. (1.3) and performing the indicated arithmetic operations.

$$dW/dt = d\phi(a_i, t)/dt. \tag{1.3}$$

Since it is likely that the parameters $(a_i)$ are sensitive to the experimental controls placed on growing animals, Eqs. (1.2) and (1.3) imply that the occurrence and intensity of the growth spurt will also be sensitive to these factors.

Excepting Brody (1945), the authors of various growth equations discussed in the rest of this Section accepted Case 4. Brody (1945, Chap. 16) seemed convinced that Case 4 applies at all ages of the animal except at $t'$. His analysis of growth data of many species appeared to show the slope of the growth curve just to the left of $t'$ is different from the slope just to the right (Fig. 1.3). He also found that $t'$ was sufficiently near to, if not coincident with, the age of puberty, that he could conclude that the etiology of the discontinuous growth rate at $t'$ was the physiological change of the animal from being vegetative to being procreative. He (Brody 1945, Chaps. 16.4, p. 514, 16.5, p. 524) then described growth as "self accelerating" before and "self inhibiting or decelerating" after $t'$, and proposed the following mathematical model embodying these features, namely

$$W = W_0 \exp(ct), 0 \leqq t \leqq t', \tag{1.4}$$

$$W = A\{1\text{-}\exp[-k(t-t^*)]\}, t' \leqq t. \tag{1.5}$$

Here $W_0$ is the initial live weight of the animal; $c$ is exponential growth constant in the growth acceleration phase and has the interesting property that $\ln 2/c$ is the time taken for $W_0$ to increase to $2W_0$; $A$ is the mature live weight; $k$ is the exponential growth rate decay constant in the deceleration phase with the property that $\ln 2/k$ is the time taken to reduce $(A-W)$ to $(A-W)/2$, for $t > t'$; and $t^*$ is a translation of Eq. (1.5) along the time axis to complete the description of growth. Brody's $t^*$ and $k$ are symbols used throughout this book with the meanings he attached to them. Figure 1.3 is a general illustration of Brody's model.

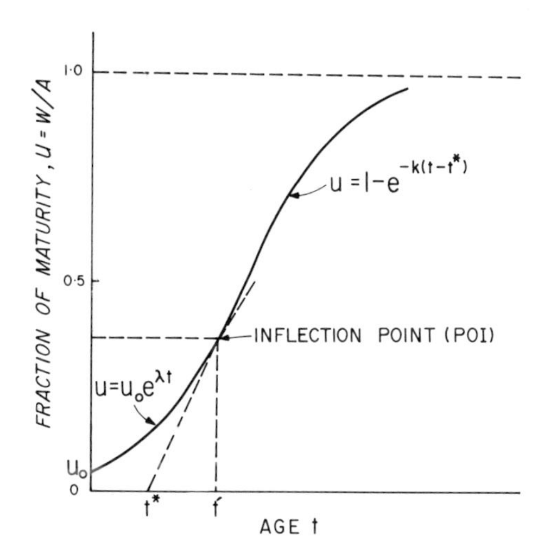


Fig. 1.3. Brody's idea of the continuous growth curve with a discontinous slope at the inflection point. Below the inflection point $W/A$ increases exponentially (vegetative phase). Above the inflection point $W/A$ follows the law of diminishing return (procreative phase) translated along the time axis by $t^*$ units of age

Even though Brody's model has an inflection point at $t'$ a growth spurt does not exist, since $dW/dt$ at $t'$ is discontinuous. However the growth spurt can be given existence mathematically by requiring that the slope of the growth curve be continuous at $t'$, or that $dW/dt$ at $t'$ from Eq. (1.4) be equal to $dW/dt$ at $t'$ from Eq. (1.5). This means that the parameters $W_0$ and $c$ are related to $A, k$, and $t^*$ as shown in Eq. (1.6).

$$W_0 c\exp(ct') = Ak\exp[-k(t'-t^*)]\ . \tag{1.6}$$

Equations (1.4) and (1.5) imply that suddenly at $t'$, the growth of the animal is constrained by the mature live weight $A$, whereas Eq. (1.6) implies that the constraining influence of $A$ would apply back to birth and that the growth during the interval 0 to $t'$ would not be an unconstrained exponential increase of live weight. Equation (1.6) together with Eqs. (1.4) and (1.5) put Brody's concept of $\phi(a_i, t)$ into case 3 (above) since $d\phi^2(a_i, t)/dt^2$ is discontinuous at $t = t'$ and therefore does not exist.

The least squares fit of Eq. (1.5) to growth data from many animals in the age range $t > t'$ so impressed Brody that he described the parameters $A, k$, and $t^*$ as genetic "constants" and created an extensive table listing the values (Brody 1945, Table 16.1, Chap. 16) for a large variety of animals from mice to steers. He also saw Eq. (1.5) in a more general light. If, for $t > t'$, the age $t$ of an animal is transformed to normalised age $T$ by Eq. (1.7) (using the values of $k$ and $t^*$ appropriate to that animal), then Brody showed that using Eq. (1.8) he could put the growth data for all animals from the mouse to the steer on the same graph of degree of maturity ($u = W/A$) versus normalised age $T$.

$$T = k(t - t^*), \tag{1.7}$$

$$u = 1 - \exp(-T), 0 < T. \tag{1.8}$$

Figure 1.4 shows the plot of Brody's data for man, the rat and the cow. The coincidence of the growth data from such widely differing species for $T > 0$ is remarkable. Here it is seen where the growth of different animals has the same char-

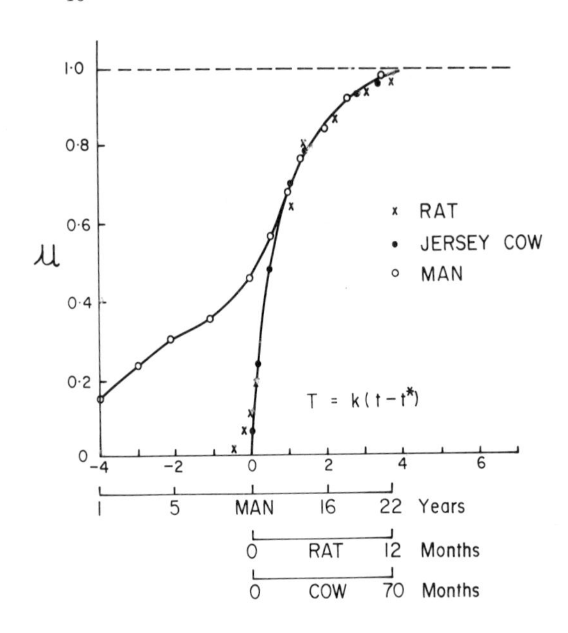


Fig. 1.4. Superposition of growth data of the rat, cow, and man using Brody's normalised age $T$ and fraction of maturity $u$. [Data reproduced from Fig. 16.7 of Brody (1945) with permission of Reinhold Publishing Company]

acteristics and where it differs. The figure illustrates the differences of the growth of these species for normalised ages $T<0$.

The evidence just presented lends support to the idea that a third mathematical property of Eq. (1.1) is its asymptotic approach to Brody's Eq. (1.5). The reader should bear in mind that this is an empirical finding and may not be of biological importance. Also it may not be true that the growth acceleration $d^2W/dt^2$ does not exist at $t'$ during the extensive physiological changes associated with puberty. The truth of these matters awaits research on growth models based on experimental results on the causes of growth.

Prior to Brody's stepwise description of growth, the question of the existence of the derivatives $\phi(a_i, t)$ did not arise. After the impact of his work decayed, growth model builders appeared to again believe that the extensive phases of physiological change from infancy to maturity flowed uniformly into each other without discontinuities in any of the rates of change of the growth, i.e., Case 4 became implicitly accepted. The logistic growth equation is among the growth functions which were in competition with Brody's mathematical description of growth.

Before discussing the logistic growth equation in some detail, and mentioning some work of more recent origin as special cases of Eq. (1.1), it is instructive to briefly consider the background of their development. Prior to publication of Robertson's (1908) famous growth equation, now known as the logistic equation, physicists and chemists were already thoroughly acquainted with first order differential equations of the type

$$dy/dx = f(y). \tag{1.9}$$

Here $x$ is an independent variable such as time when Eq. (1.9) refers to a rate of some kinetic phenomenon, and $y$ is a dependent variable such as displacement, chemical concentration, etc. Equations of this form state that the rate of change of $y$ with respect to the independent variable $x$, at any value of $x$, is uniquely related

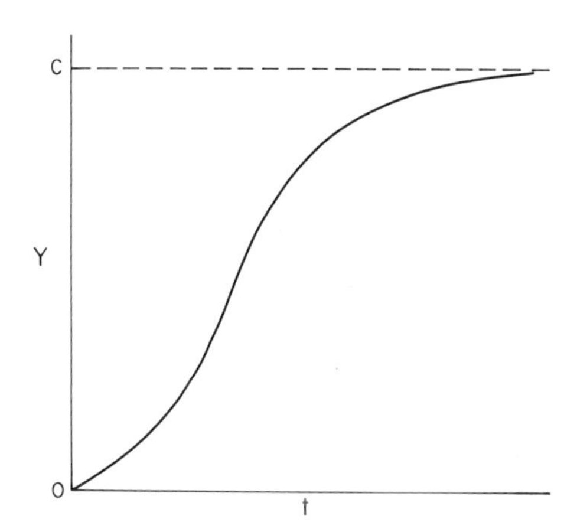


Fig. 1.5. The variation of acid concentration $Y$ versus time $t$ during hydrolysis of an ester showing the sigmoid shape expected of growth curves. Note the similarity to the growth curves in Figs. 1.1–1.3

to the value of $y$ at that $x$. This type of equation was so useful in discovering new laws of physics and physical chemistry during the 18th and 19th centuries that mathematically inclined biologists could not have avoided having some knowledge of them. The concept of the essential molecular nature of living matter had gained adherents and the concept that living systems were incessantly active chemical factories had also developed. Thus the stage was set for the development of rate of growth equations based on various forms of $f(y)$ where $x$ is time and $y$ is live weight $W$. An example is Robertson's use of Ostwald's work as a basis for his work on growth.

Ostwald (1883) showed that the hydrolysis of organic esters in dilute water solution were acid-catalyzed reactions. He found that the rate of appearance of acid was given by Eq. (1.10) (Boudart 1968),

$$dy/dt = (c-y)(k_1 + k_2 y). \tag{1.10}$$

Here $y$ is the concentration of acid at time $t$; $c$ is the initial concentration of ester at $t=0$; $k_1$ is the very low monomolecular rate constant of hydrolysis of ester in the absence of acid; and $k_2$ is the biomolecular rate constant of catalysis of the hydrolysis of the ester by the acid molecules generated by the hydrolysis. Oswald integrated this equation with initial condition $y=0$ at $t=0$, and obtained Eq. (1.11) relating growth of the acid concentration in the solution to time.

$$y = \frac{c\{1-\exp[-(k_1+k_2c)t]\}}{1+(k_2c/k_1)\exp[-(k_1+k_2c)t]}. \tag{1.11}$$

Figure 1.5 illustrates how acid concentration grows with time after the ester has been added to water at concentration $c$. Using this model Ostwald showed that for a given ester the rate constants, $k_1$ and $k_2$, were independent of $c$. The reader can see how this model will aid experimental study of the depencence of $k_1$ and $k_2$ on the temperature of the solution or on some other independent variable, pressure for example.

Successes such as these drew attention to equations in the form of Eq. (1.9) and led scientists to think that if they could adequately describe their data with a particular form of Eq. (1.9) they would be on their way to the discovery of new laws of nature. Nelder (1962) argued that Eq. (1.9) is more likely to lead to natural law than the expression $dy/dx = g(x)$. He felt that in comparing experiments on the same system under two different conditions, more fundamental information is to be had by comparing the slopes, $(dy/dx)_1$ and $(dy/dx)_2$, at the same value of $y$ than by comparing them at the same value of $x$. If $y$ is live weight and $x$ is time, this approach can be rephrased as: it is better to compare growth rates of animals isogravimetrically than to compare them isochronologically. A third alternative will be found in Chap. 5.2.

Willy Feller (1940, pp. 59–61) had already questioned concepts like Nelder's by pointing out that when comparing the two experiments one should find out what happens to the parameters of Eq. (1.9) when the experimental conditions are changed. Ostwald's experiments with autocatalytic chemical reactions are pertinent examples of this experimental approach.

Figure 1.5 shows the curve of acid concentration, $y$, versus time, to be sigmoid, similar to the growth curves of many animals. Further, Eq. (1.11) can be shown to have the three mathematical properties expected of growth curves, i.e.,

1. $y$ asymptotically approaches a maximum, $c$, because as $t$ becomes greater $\exp[-(k_1 + k_2 c)t]$ approaches zero.
2. It possesses an inflection point. Setting the derivative of Eq. (1.10) to zero yields $y = (k_2 c - k_1)/2k_2$. The time ($t'$) of occurrence of the inflection point can then be found by substituting this value for $y$ into Eq. (1.11) and solving for $t$.
3. It approaches Brody's function Eq. (1.15) asymptotically. A time $t_m$ can be found such that for $t > t_m$ Eq. (1.11) asymptotically approaches

$$y = c\langle 1 - \exp\{-(k_1 + k_2 c)[t - \ln(k_2 c/k_1)/(k_1 + k_2 c)]\}\rangle .$$

This is Brody's Eq. (1.5) with $k = k_1 + k_2 c$, $t^* = \ln(k_2 c/k_1)/(k_1 + k_2 c)$, and $A = c$.

Robertson (1908) was so impressed with the strong similarity between Fig. 1.5 on the one hand and Figs. 1.1, 1.2 and growth curves in general on the other, that he embarked on a 15 year study of the chemical basis of growth based entirely on the autocatalytic differential Eq. (1.10), with the exception that he set the monomolecular rate constant $k_1$ to zero and used $W$ and $A$ in place of $y$ and $c$ respectively, leading to Eq. (1.12).

$$dW/dt = kW(A - W). \tag{1.12}$$

The solution of which, with initial conditions $W = W_0$ at $t = 0$, is

$$W = A/\{1 + \exp[-kA(t - t')]\} , \tag{1.13}$$

where $t' = \ln(A - W_0)/kA$ and $W' = A/2$.

This function is a special case of Ostwald's autocatalytic Eq. (1.11). It has the same three mathematical properties with the exception that the live weight at the inflection point and the time of occurrence of the inflection point are more simply expressed as $W' = A/2$ and $t'$ respectively. However it should be noticed that Eq. (1.12) cannot represent a real autocatalytic reaction. Without a monomolecular reaction to generate catalyst initially, the reaction cannot get started. Robertson (1923) overcame this difficulty by assuming catalyst is always present in living matter. Another difficulty is the multitude of chemical reactions constantly going on in the living organism; some produce building material for growth and others produce thermal energy as end product. Not all of these could be autocatalytic in the chemical kinetic sense. However, it had already been shown that the dynamic character of a sequence of reactions has the character of the reaction with the lowest rate constant. Robertson used this fact to support his concept of a "master reaction" of growth. He concluded that the master reaction must be autocatalytic because the autocatalytic function (Eq. 1.13) describes the data so well, as seen in his own words (Robertson 1923, p. 11): "Nevertheless the facts obviously compel us to conclude that the whole of this great complexity of events waits upon and is set in motion by a process which, taken by itself, is of such a character as to admit of representation by the autocatalytic formula." Robertson was so convinced of the truth of this assertion that he set about drawing conclusions from the application of his autocatalytic growth equation to weight versus age data for plants and animals ranging from the unicellular to highly organised, multicellular organisms like the sunflower and man.

The success of Robertson's work led other scientists to use his basic ideas in demography (Pearl and Reed 1923), ecology (Lotka 1956), and genetics (Mendel 1965, Krause et al. 1967). By this time Robertson's Eq. (1.13) had become known as the logistic function and probably is at present the most widely used function in the general biological investigation of growth. Nair (1964) wrote a paper discussing the logistic function as a regression function and the attendant statistical problems. Blumberg (1968) has discussed a generalisation of the logistic growth rate Eq. (1.12).

Several critiques of the scientific basis of Robertson's autocatalytic function have been published (Snell 1929, Feller 1940, Morgan's criticism of Crozier's acceptance of Robertson's thesis, Lindegren 1966, van der Vaart 1968). Feller's criticism is telling. He was emphatic in his belief that the statistical excellence of the fit of a function to growth data is not sufficient justification for drawing conclusions concerning the "true nature" of biological growth controlling factors. He proposed as alternatives to Robertson's autocatalytic function two Eqs. (1.14) and (1.15) which have the mathematical properties expected of growth equations. He showed that they statistically fit growth data as well as or better than the logistic Eq. (1.13). The first is

$$W=(2A/\pi)\arctan\{\exp[k(t-t')\ln 10]\}, \tag{1.14}$$

the differential equation of which is

$$dW/dt=(Ak\ln 10/\pi)\sin(\pi W/A), \tag{1.14a}$$

and the second is

$$W=(A/2)\{1+\phi[k(t-t')]\}, \tag{1.15}$$

where

$$\phi(x)=(2/\sqrt{\pi})\int_0^x \exp(-s^2)ds. \tag{1.15a}$$

The differential equation of Eq. (1.15) cannot be expressed in elementary functions of $W$.

It is clear that the fit of Eq. (1.14) to growth data cannot be used to deduce anything about the biology of growth. However, it is interesting to note that since $\phi(x)$

Table 1.1. Some growth equations of animals as output devices only

| Author | $W=f(t)$, $W_0=f(0)$ $f(t)$ | $dW/dt=F(W)$ |
|---|---|---|
| Gompertz (1825) | $A\exp\{-\exp[-k(t-t')]\}$<br>$t'=(1/k)\ln(\ln W_0/A)$ | $kW\ln A/W$ |
| Robertson (1908)<br>(logistic equation) | $A/\{1+\exp[-k(t-t')]\}$<br>$t'=(1/k)\ln[A-W_0)/W_0]$ | $kW(A-W)$ |
| Brody (1937) | $W_0\exp(\lambda t)$ $0<t<t'$<br>$A\{1-\exp[-k(t-t^*)]\}$, $t'<t$ | $\lambda W$, $0\leq t\leq t'$<br>$k(A-W)$, $t'<t$ |
| Bertalanffy (1938) | $\{N/n-\exp[-(1-m)n(t-t')]\}^{1/(1-m)}$ | $NW^m-nW$ |
| Feller (1940) | I. $(2A/\pi)\arctan\{\exp[k(t-t')\ln 10]\}$<br>II. $(A/2)\{1+\phi[k(t-t')]\}$<br>$\phi(x)=(2/\sqrt{\pi})\int_0^x\exp(-s^2)ds$ | $(Ak/\pi)\ln 10\sin(\pi W/A)$<br>not expressable in elementary functions of $W$ |
| Weiss and Kavanau (1957) | $A[F(t)+D(t)]$<br>$G$=generative matter<br>$D$=differentiated matter<br>$I$ =inhibitor | $dG/dt=H[1-a(I/V)^n]$ $-k[1-(I/V)^n]G-\lambda G$<br>$dI/dt=gG-iI$<br>$dD/dt=k[1-(I/V)^n]G+\lambda G-mD$ |
| Fitzhugh (1976) | $A[1\mp b\exp(-kt)]^M$;<br>$b=\pm[1-(W_0/A)^{1/M}]$<br>upper sign applies when $M\geq 1$,<br>lower sign applies when $M<0$, | $-MkW[(W/A)^{-1/M}-1]$ |
| Richards (1959) | $m=(M-1/M)-1$<br>Gompertz, $M\to\infty$;<br>Logistics, $M=-1$; Brody, $(t^*<t)$<br>$M=1$; Bertalanffy, $M=3$ | |
| Laird (1965) | $a\exp\langle(1/k)\{1-\exp[-k(t-t')]\}\rangle$ | $ZW$, $dZ/dt=-kZ$, $A=a\exp(1/k)$ |
| Parks (1965) | $A[1+a\exp(-bt)+c\exp(-dt)]$ | $gd^2W/dt^2+hdW/dt+iW+n=0$ |

is the probability integral, Eq. (1.15) could, in Fellers words "...easily be pressed to provide a probabilistic background or "explanation" to phenomena of growth, if it be found that it expresses them fairly well."

At the time Feller was developing his thesis, Backman (1939) was publishing his work on growth equations using the probability integral, $\phi(x)$, i.e., Eq. (1.15a). The implication of this work was that growth takes place in cycles with each cycle involving a different form of $\phi(x)$. The preceding remarks of Feller are also relevant to recent work (Frisch and Revelle 1969) on human growth expressed as four cycles, intrauterine, primordial, fundamental, and pubescent using Grubb's (1942) ideas. The stated preference for this approach (David Frisch 1970, private communication) is based on the simple chemical interpretation of the Gaussian integral almost exactly as anticipated by Feller some 30 years previously, and is therefore subject to Feller's criticism of the use of the logistic function in discovery of biological laws of growth.

Table 1.1 is a list of some growth functions and their differential equations frequently found in the literature. It is not by any means exhaustive: it serves only as a focal point for discussion of growth equations deduced from presumed knowledge of internal growth controlling factors. Other lists of families of nonlinear functions suitable for examination of various kinds of growth data exist (Nelder

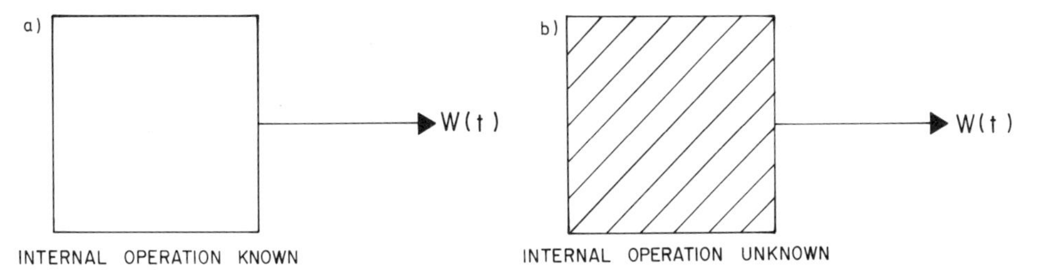


Fig. 1.6 (a, b). Schematic representations of systems with output only, (a) white when internal operation is known; (b) black when internal operation is unknown

1962 a, Grosenbaugh 1965). The growth functions in these latter lists are, as stated by the authors, purely empirical and not based on any knowledge of the internal operations of any growing system whether biological or nonbiological. Kowalski and Guire (1974) have listed 225 published papers dealing with various mathematical approaches to $(t, W)$ tables of data; even this is but a small part of the literature devoted to study of tables of live weight versus age data.

Six of the functions in Table 1.1, namely Gompertz, Robertson, Brody, Bertalanffy, Weiss and Kavanau, and Laird, are generally favoured by biologists because they are based upon some knowledge of the internal mass and energy transactions of the growing animal or of some easily conceived biochemistry of the animal. In the context of this and the following sections I will attempt to put these functions, and functions like them, into perspective.

## 1.4 Animals as White Boxes with Output Only

For the sake of clarity systems analysts picture a system as one box (Fig. 1.6 a) or a set of interconnected boxes. One box, coloured white, suffices for a system with known internal operations. One box, coloured black (Fig. 1.6 b), suffices to picture a system, simple or complex, with internal workings which are unknown or poorly perceived. Between these extremes are grey boxes: the degree of grey depends on the incompleteness of knowledge of the internal dynamics of the system.

The history of the development of a particular area of science shows it to begin as a set of black boxes slowly whitened by painstaking thought and experimental research with not a few false leads and blind alleys. In the prescientific era of chemistry, chemical systems were black boxes and remained black in spite of the efforts of the alchemists to whiten them. But from the time of Dalton's atomic hypothesis (1766) scientific efforts have made chemical systems become even whiter systems of known atoms and molecules with known masses and energy transactions. These systems are not perfectly white, some regions are darker than others. Any scientist working in a black region of some chemical system is guided by the whiteness, i.e., by what is already known, of adjacent regions.

Generally speaking, systems are complex and must be depicted by a set of interconnected boxes each of which represents some recognisable function necessary to the operation of the system. The output of each box are inputs for other boxes.

When the set of boxes is integrated through their inputs and outputs, a box can be drawn around them to represent the system as a whole. Since few systems are independent of their environments, inputs, and outputs of the system box must be depicted. Here the reader will see that the biologist, engineer, and ecologist can be regarded in a general way as systems analysts.

The capacity of the modern computer has recently made possible the writing of programs simulating the growth of animals on the input of food as indicated in the preceding paragraph (Faichney et al. 1976). The set of interconnected boxes represents the various organs, e.g., the brain, liver, muscles etc., and the ways in which they are interconnected are represented by sets of differential equations deduced from slaughter and dissection data versus age of the animal. The resulting set of equations are so numerous that a modern computer is needed to solve them, to give the live weight, i.e., the total weights of the "boxes," as a function of age (Baldwin and Black 1979).

I know of only one type of system, representable as a box with output only, which began as a truly black box and whitened by painstaking research. This system is the science of sequences of radioactive atoms which began with Becquerel's discovery of penetrating rays from a uranium potassium sulphate compound in 1896 (Romer 1964). Pierre and Marie Curie and A. Laborde became directly involved through close association with Becquerel. With Rutherford they set upon a systematic study of the chemistry and physics of uranium in an effort to understand the inner workings of this substance (P. Curie 1908; M. Curie 1903).

These scientists were astonished that these materials could constantly give off energy without any observable input of energy in any form. P. Curie and A. Laborde found, by calorimetric techniques, that 1 gram atom of radium spontaneously generated thermal power at the rate of 22.5 kcal/h. They reported: "The hypothesis of a continuous modification of the atom is not the only one compatible with the development of heat by radium. This development of heat may still be explained by supposing that the radium makes use of an external energy of unknown nature." They were not ready to accept radium as a black box with output only. Originally it was thought radium had a constant power output. Later it was shown to have a very long half life (1629 years), justifying the idea of constancy. The concept of the half life of a radioactive substance had already been developed by Rutherford's use of Eq. (1.16) to follow the reduction of the intensity of the radioactivity of some chemical fraction of uranium or thorium.

$$I = I_0 \exp(-gt). \tag{1.16}$$

Here $I_0$ is the initial intensity of radioactivity; $t$ is the time; $I$ is the intensity at time $t$ and $g$ is the decay constant such that $\ln 2/g$ is the half life of the radioactivity.

What had begun in 1896 as a black box (with output only?) became in 1905 a light grey box (definitely with output only); it was accepted that the power generated by radioactivity had no external source. The box tended towards whiteness because the internal mass and energy transactions were becoming better known. Presently the schemes by which atoms of uranium and thorium decay to stable atoms of lead are very well known.

Reference to growing animals as black or white boxes with output only, certainly seems out of place in the face of the widely accepted openness of organic systems. However, if one considers live weight of a growing animal as an output with no reference to the food intake and considers the brief history of the search for "the equation of growth" presented in previous sections of this book, one will see that the box is black with output only where the suggested functions for $W(t)$ are called empirical by the originators (Richards 1959, Nelder 1962, Grosenbaugh 1965). The box is white with output only in those cases in which the originator bases his equation on some widely accepted idea concerning the internal operation of the animal during growth.

Bertalanffy (1938, 1957) whitened the box by asserting that the rate of growth of an animal is "obviously" the difference between the anabolic and catabolic rates of the tissues of the animal. Equation (1.17) is the quantitative representation of this idea.

$$dW/dt = pW^m - qW. \tag{1.17}$$

He presented biological reasons why the anabolic rate is proportional to the mth power of the live weight $W$ and why the catabolic rate is proportional to the first power of $W$ rather than some other power, say $n$. Integration of Eq. (1.17) yielded an equation (see Table 1.1) which he and others used to study the growth of many animals including fish. Having associated anabolism with metabolism, which is related to the oxygen consumption and heat production of the animal, he used $m$ as a metabolic index. Metabolic type I animals have respirations proportional to their surface area, i.e., $m=2/3$; metabolic type II animals have respirations proportional to their live weights, i.e., $m=1$; metabolic type III animals have respiration which are neither proportional to their surface areas nor to their live weight, i.e. $2/3<m<1$. Based on this type of reasoning Bertalanffy claimed he could predict the growth type of an animal from its metabolic type. It is enlightening to see that respiration data can lead to a Bertalanffy equation of growth, but determining $m$ from growth data may not correctly determine an animal's metabolic type (Richards 1959).

Weiss and Kavanau (1957), following the widely accepted concept that growth is the difference between biomass produced and retained and biomass destroyed or otherwise lost, gave a detailed analysis of the metabolic processes that Bertalanffy had lumped into anabolism and catabolism. They invoked the following basic biological concepts. The biomass of an organic system consists of two different components, generative and differentiated cellular materials. The increase of generative biomass is catalysed by some kinds of molecules characteristic of each cell type. Each type of cell also produces specific and freely diffusable molecules antagonistic to the catalytic molecules, i.e., inhibitors. The regulation of growth occurs automatically by a negative feedback mechanism which increases the concentration of the inhibitors which progressively blocks the action of the catalytic molecules.

Weiss and Kavanau converted this type of the etiology of the growth process into a set of differential equations (see Table 1.1) with "biologically meaningful" parameters. The set of parameters include five rate constants $k_1$, $k_2$, $k_3$, $k_4$, $k_5$, a proportionality factor, $p$, between the negative feedback and the concentration of the inhibitor molecules, and a ratio, $b$, of the actual feed back inhibition at terminal equilibrium and the potential of complete inhibition at equilibrium. Using the in-

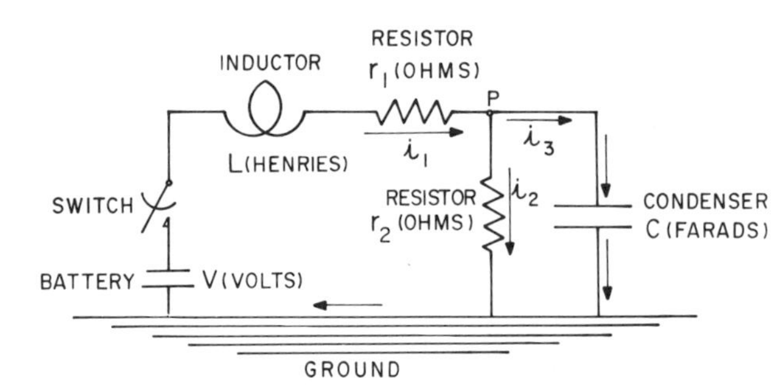


Fig. 1.7. An inductance-resistance-capacitance (*LRC*) circuit which simulates animal growth

tegrated forms of their differential equations they studied the growth data of the chick from conception to young adulthood and obtained "biologically" reasonable values of the seven constants which they then used to reproduce the growth of the chick over 200 or more days as a continuous process. The reader should note that this is an example of circular reasoning often found in biology and in theoretical studies of other complex systems. Experimental data are used to determine the structural parameters of a theoretical function which is then used to reproduce the data. This circular process is often regarded as validation of the theory (Dolby 1982).

They also theoretically explored what would happen to the growth of an organ in an adult when a portion of the organ is excised. Some data in the literature tended to support their conclusions.

The reader sees, in this brief sketch, the apparent profound depth of the theory of the growth process proposed by Weiss and Kavanau in comparison with the etiologies of Robertson and of Bertalanffy. There is little doubt of its great appeal to biologists, especially since it suggests experimental approaches to growth regulation and organ regeneration. However, in a critical assessment of their theory the reader should carefully consider Weiss and Kavanau's own words to the effect that the entities they name, e.g., generative and differentiated matter, catalysts and inhibitors, in practice are not readily resolvable.

Their model, then, is a more detailed description of the growing organism mixed with acceptable chemical concepts including the relatively modern ideas of feedback control in biology (Milsum 1966). It is a good story but it may not be any more real than any of the other models. Even though it is a more realistic attempt to whiten the black box of growth, it is never the less a box with output only, since Weiss and Kavanau nowhere explicitly use nutrient input in their model.

Another illustration of how the description of the biology of growth influences the mathematics of growth is given in the following discussion of an electrical analogue of the growth of an animal (Parks 1964, see Table 1.1). The Parks function for $W(t)$ and its differential equation shown in Table 1.1 were derived from the inductance-resistance-capacitance circuit shown diagramatically in Fig. 1.7. The circuit elements are a battery of voltage $V$ in series with a switch, an inductor of $L$ henries of self inductance, and a resistor of $r_1$ ohms connected to ground through a resistor of $r_2$ ohms and a capacitor of $C$ farads in parallel. The circuit is completed through the ground. When the switch is closed at time zero the voltage across the capacitor will build up in time storing a charge $q$ on it. The imaginative

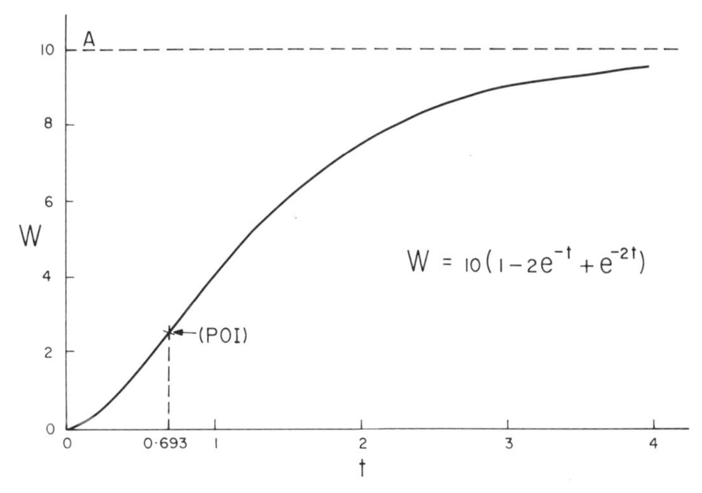


Fig. 1.8. A sigmoid growth curve obtained from the electrical analogue of a growing animal shown in Fig. 1.7

biologist can see the following analogues: accumulation of charge on the capacitor is like storing live weight during growth; the maximum charge accumulated in infinite time is like the mature live weight; current $i_1$ is like food intake; at junction $P$ the splitting of $i_1$ into currents $i_2$ and $i_3$ is like apportioning the feed intake, $i_1$, between maintenance, $i_2$, and growth, $i_3$; resistor $r_2$ is like the thermal resistance (skin, fat, fur, feathers, etc.) between the energy stores as live weight and the environment; the inductor and resistor $r_1$ act continuously to adjust current $i_1$ up to its maximum with time just as the animal adjusts its feed intake (appetite) upward to its mature food intake as it ages. Finally the battery voltage $V$ is like the animal's urge to eat.

Using the above analogies, solution of the differential equation for this circuit is the growth function $W(t)$, shown in Table 1.1, and is repeated here for convenience.

$$W(t) = A[1 + a\exp(-bt) + c\exp(-dt)].$$

If $L$, $r_1$, $r_2$, $C$, and $V$ are so chosen that $A = 10$, $a = -2$, $b = 1$, $c = 1$, and $d = 2$, $W(t)$ will have the form generally expected of the growth curve of an animal (Fig. 1.8). The curve possesses the expected three mathematical properties discussed in Chap. 1.3; e.g., it is sigmoid; it has an inflection point at $t' = \ln 2$ units; it has an asymptotic mature live weight, $A = 10$ units; it is asymptotic to Brody's curve, i.e., Eq. (1.5), since for large $t$, $W(t) \doteq 10\{1\text{-}\exp[-1(t\text{-}\ln 2)]\}$. Here Brody's $k$ and $t^*$ are respectively unity and $\ln 2$.

Although this treatment of an electrical analogue of a growing animal is crude and possibly repugnant to biologists, it does two things no other biologically acceptable treatment of growth has done. First, it uses the openness of organisms explicitly by making food intake analogous to current $i_1$. Second, it anticipates how the live weight $W(t)$ will vary with time during a starvation regime.

If, at some time during the charging of the capacitor (growth) the switch is opened, thereby reducing $i_1$ (food intake) to zero, the charge on the capacitor (live weight $W$) will leak to earth through resistor $r_2$ according to the following equation:

$$W = W_0 \exp[-t/(r_2 C)].$$

Here, $W_0$ is analogous to the charge on the capacitor at the time the switch was opened; $t$ is the time lapsed since the switch was opened simulating starvation. The product of $r_2$ ohms (thermal resistance to flow of energy to the environment) and $C$ farads (capacity of the animal to store energy) has the units of time in seconds, hence $1/(r_2 C)$ can be called the decay constant $k$ and the preceding equation rewritten as:

$$W = W_0 \exp(-kt).$$

Kleiber (1961, pp. 29–38) used this equation to discuss the weight changes during starvation of Succi, a professional human faster, and of a dog named Oscar. Robertson (1923, p. 238) reported the application of the above equation by A. G. Mayer to the starvation of the coelenterate *Cassiopea xamachana* (a species of jellyfish). These cases are discussed in detail in Chap. 6.

Despite these "successes" I refuse to emphasise this electrical analogue of growth. I am convinced that the consideration of the animal as a black box with output only which must be whitened by some plausible biological hypotheses prior to development of a growth equation, must be foregone for a different approach. A serious defect in all these growth equations is the lack of explicit expression of the food or energy input to the growing animals. There is a strong reliance on the assumption that the energy and the raw materials of growth are never limiting, therefore negating the need for explicit expression of the source of the nutrients. Although the authors of these equations knew that organisms are open system they have nevertheless described growth as they would a machine, with output only. It is possible the great amount of $(t, W)$ data in the animal science literature has forced growth model makers to take the positions discussed in this section. This is most certainly true in quantitative botany where nutrient uptake may be virtually impossible to measure and the empirical approach may be the only one available. Crossing the nebulous boundary between the phenomenology and the etiology of growth cannot be done by associating any output function with any collection of physiological, biophysical, and biochemical facts thought to be controlling growth.

## 1.5 Animals as Black Boxes with Input and Output

Figure 1.9 depicts a black system with input $F(t)$ and output $W(t)$. Here all knowledge of the internal structure, macro and micro, and operation of animals is set aside for the time being. The first order of business becomes comparing what is known about the input $F(t)$ with the output $W(t)$ of individuals or populations of various species of animals under various known conditions of growth. This is strictly an empirical approach to growth which may be unsatisfactory to biologists who know something of the internal biology of growing animals from conception to natural death. To them the animal is not a black box. They may consider it fool-

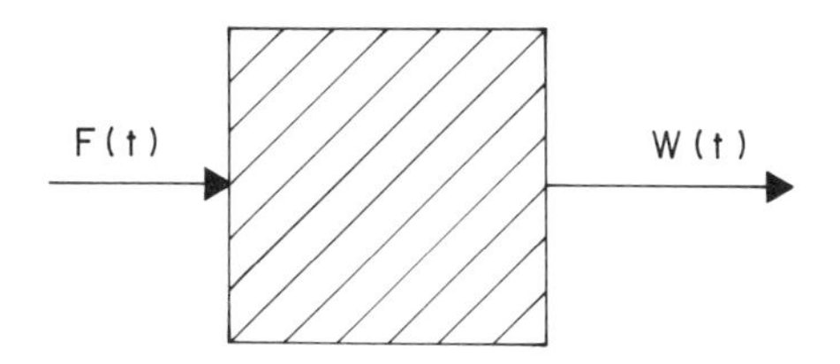


Fig. 1.9. Schematic representation of a system of unknown internal operation with known input $F(t)$ and output $W(t)$

hardy to ignore the scientific treasure house of knowledge of internal activities of growing, developing animals. However, Fig. 1.9 serves to focus attention on the input function $F(t)$ as being just as important as the output $W(t)$ which has been almost exclusively studied since Robertson's work in 1908. The promise of this book is to show 1) the relation between the cumulative food $F(t)$ consumed by an animal in time $t$ and the gain in live weight, $W - W_0$, and 2) the relation of rate of growth $dW/dt$ to the appetite ($dF/dt$), i.e., the rate of change of the cumulative food consumed $F$.

Consider the differential equation

$$dW/dt = (dW/dF)(dF/dt). \tag{1.18}$$

This equation means the growth rate is the product of the growth efficiency, $dW/dF$, and the food intake, $dF/dt$. $dW/dF$ is called the true growth efficiency because it is the ratio of the differential change in output $dW$ and the differential change in input $dF$ (Batschelet 1975, p. 249).

The reader should consider Eq. (1.18) with care, because there is the possibility of a double infinity of ways by which treatments or conditions can affect the growth rate. The conditions imposed experimentally can affect either the true growth efficiency or the appetite or both. There are numerous papers on nutrition where the response is taken to be growth rate or specific growth rate, i.e. $(dW/dt)/W$, only, ignoring the possibilities of how the treatments affect the growth efficiency and/or which are related to the chosen response through Eq. (1.18) (Robbins et al. 1979, Gibney et al. 1979, Grau and Almquist 1946). Nutritionists and geneticists might use more caution in considering the underlying nature of the characters they choose in nutrition and selection exercises (see Chaps. 9 and 10).

Equation (1.18) also has a special mathematical meaning in the context of growth and feeding of animals. There is a theorem stating that if a continuous positive function $Y$ is the product of a decreasing positive function and an increasing positive function, then the function $Y$ will have a maximum value at some value on the range of its independent variable. It is well known that the growth efficiency ($dW/dF$) of an animal is positive and decreases and its appetite ($dF/dt$) is positive and increases as it ages; therefore by the above theorem, the growth rate ($dW/dt$) must have a maximum value at some age between infancy and young adulthood. Consequently the live weight, $W(t)$, must have an inflection point (POI) at the age where $dW/dt$ is maximum (Fig. 1.10). The reader should note the implication that the existence of a (POI) may be a mathematical consequence and unrelated to any physiological change of development at that age.

Equation (1.18) also serves to focus attention on the derivatives on the right hand side, namely $dW/dF$ and $dF/dt$ individually.

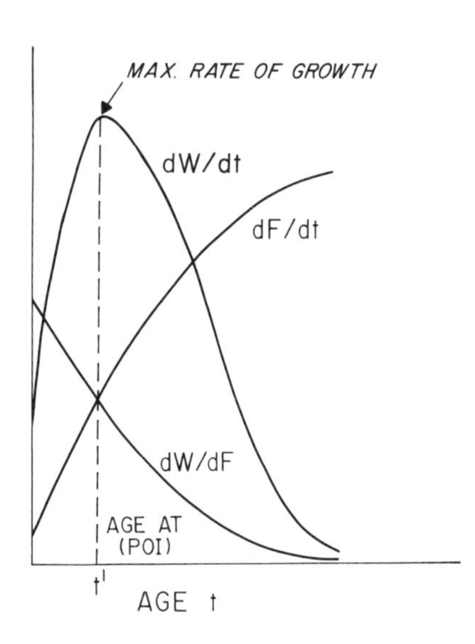


Fig. 1.10. An illustration of how expressing the growth rate as the product of the efficiency and the food intake leads naturally to a point of inflection (*POI*) in the growth curve

In any experimental feeding regime a set of data points $(t, F, W)$ will be obtained. From these a graph of $W$ versus $F$ data displays the form of $W(F)$ for the particular experimental feeding regime. Study of such a graph can imply an expression of $W(F)$ in terms of elementary functions of $F$, thus making it possible to study $dW/dF$ analytically as either a function of $F$ or a function of $W$. Similarly the graph of $F$ versus $t$ could suggest the function $F(t)$ from which the food intake, $dF/dt$, might be obtained analytically.

## 1.6 Conclusions

This chapter has laid the foundation for the study of the growth of animals as input-output systems, guided by the basic ideas of Kleiber and Brody. Chapter 2 will be devoted to finding mathematical functions for the ordered pairs of data, namely $(t, dF/dt)$, $(t, F)$, $(F, W)$, and $(t, W)$ taken from $(t, F, W)$ data of long term experiments under ad libitum feeding regimes of as many kinds of animals as I could glean from the literature.

Undoubtedly there will be systematic deviations of the experimental $(t, F, W)$ data points from the curves computed from the functions found to be good approximations of the true growth and feeding functions (Figs. 1.1 and 1.2). To meet the objections to these systematic variations I propose considering the growing animal as being in a sequence of equilibrium states and accepting the systematic deviations as indications that the growing animal, in reality, is in a sequence of steady states which are homeostatically constrained by internal control mechanisms to follow generally some "true" course of growth, describable by mathematical approximations (Chap. 7.4.3).

This is the major assumption on which are based the answers to the questions of the relation of growth to environmental and nonenvironmental conditions discussed in the following chapters. If this assumption and the growth and feeding functions are found useful, then the systematic deviations of growth and feeding data from the computed values may be considered relevant to the physiological control of metabolism and appetite.

## References


Backman GV (1939) Die organische Zeit. Gleerup, CWK Lund

Baldwin RL, Black JL (1979) A computer model for evaluating effects of nutritional and physiological status on the growth of mammalian organs and tissues. Anim Res Lab Tech Pap No 6, CSIRO, Australia

Batschelet E (1975) Introduction to mathematics for life scientists, 2nd edn. Springer, Berlin Heidelberg New York

Bertalanffy L von (1938) A quantitative theory of organic growth [inquiries on growth laws (2)]. Hum Biol 10:181–213

Bertalanffy L von (1957) Quantitative laws in metabolism and growth. Q Rev Biol 32:218–231

Blumberg AA (1968) Logistic growth rate functions. J Theor Biol 21:42–44

Bolzano B (1934) Paradoxien des Unendlichen. In: Wissenschaftliche Klassiker in Faksimile-Drucken, vol II, Berlin

Boudart M (1968) Kinetics of chemical processes. Prentice, New Jersey

Boyer CB (1949) The history of the calculus. Dover, New York

Brody S (1945) Bioenergetics and growth. First published: Reinhold, New York (Reprinted: Hafner Press, New York, 1974)

Curie M (1904) Radioactive substances, 2nd edn. Chem News 1904 (Reprinted from Chem News 88:1903)

Curie P (1908) Oeuvres de Pierre Curie. Gauthier, Paris

Dolby GR (1982) The role of statistics in the methodology of the life sciences. Biometrics (accepted for publ August 1981; to be publ 1982)

Faichney GJ, Black JL, Graham N McC (1976) Computer simulation of the feed requirements of shorn sheep. Proc N Z Soc Anim Prod 36:161–169

Feller W (1940) On the logistic law of growth and its empirical verification in biology. Acta Biotheor 5:51–66

Fitzhugh HA (1976) Analysis of growth curves and strategies for altering their shape. J Anim Sci 42:1036–1051

Frisch R, Revelle R (1969) The height and weight of adolescent boys and girls at the time of peak velocity of growth in height and weight. Hum Biol 41:536–559

Gibney MJ, Dunne A, Kinsella IA (1979) The use of the Mitscherlich equation to describe amino acid response curves in the growing chick. Nutr Rep Int 20:501–510

Grau CR, Almquist HJ (1946) The utilisation of the sulfur amino acids by the chick. J Nutr 26:630–640

Grosenbaugh LR (1965) Generalisation and reparameterisation of some sigmoid or other nonlinear functions. Biometrics 21:708–714

Grubb R (1942) Acta Paediatr Scand 30:67–93

Kleiber M (1961) The fire of life. Wiley, New York

Kowalski CJ, Guire KE (1974) Longitudinal data analysis. Growth 38(2):131–169

Krause GF, Siegel PB, Hurst DC (1967) A probability structure for growth curves. Biometrics 23:217–225

Laird AK, Howard A (1967) Growth curves in inbred mice. Nature (London) 25:786–788

Laird AK, Tyler SA, Barton AD (1965) Dynamics of normal growth. Growth 29:233–248

Lindegren CC (1966) The cold war in biology. Planarian Press, Ann Arbour Mich

Lotka AJ (1956) Elements of mathematical biology. Dover Publ, New York

Lucas HL (1964) Stochastic elements in biological models: their source and significance. In: Gurland JT (ed) Stochastic models in medicine and biology. Univ Wisconsin, Madison

Mendel JL (1965) Statistical analysis of nonlinear response functions with application to genetics of poultry growth. PhD Thesis, Univ Michigan

Milsum JH (1966) Biological control of systems analysis. McGraw-Hill, New York

Mitchell HH, Card LE, Hamilton TS (1931) A technical study of the growth of White Leghorn chickens. Bull 367 Ill Agric Exp Stn

Nair CR (1964) A new class of designs. Am Stat Assoc J 59:817–833
Needham AE (1964) The growth process in animals. Van Nostrand, New Jersey
Nelder JA (1962a) Critique of Dr Best's paper. In: Lucas HR Jr (ed) The Cullowhee conference on training in biomathematics. Typing Service, Raleigh
Nelder JA (1962b) An alternative form of a generalised logistic equation. Biometrics 18:614–616
Ostwald W (1883) Studien zur chemischen Dynamik: Zweite Abhandlung: Die Einwirkung der Säuren auf Methylacetat. J Prakt Chem 28:449–495
Parks JR (1964) An electrical analogue of a growing animal. Unpublished
Parks JR (1965) Prediction and entropy of half-tone pictures. Behav Sci 10:436–445
Pearl R, Reed LJ (1923) On the mathematical theory of population growth. Metron Vol III, Nl:6–19
Richards FJ (1959) A flexible growth function for empirical use. J Exp Bot 10:290–300
Robbins KR, Norton HW, Baker DH (1979) Estimation of nutrient requirements for growth data. J Nutr 109(10):1710–1714
Robertson TB (1908) On the normal rate of growth of an individual and its biochemical significance. Arch Entwicklungsmech Org 25:581–614
Robertson TB (1923) The chemical basis of growth and senescence. Lippincott, Philadelphia
Romer A (1964) The discovery of radioactivity and transmutation. Dover, New York
Skellam JG, Brain MV, Proctor JR (1958) The simultaneous growth of interacting systems. Acta Biotheor 13:131–144
Snell GD (1929) An inherent defect in the theory that growth rate is controlled by an autocatalytic process. Proc Natl Acad Sci USA 15:274–281
van der Vaart HR (1968) The autocatalytic growth model: critical analysis of the conceptional framework. Acta Biotheor 13:133–142
Weierstrass K (1872) Mathematische Werke. Akademie, Berlin
Weiss P, Kavanau JL (1957) Model of growth and growth control. J Gen Physiol 41:41–47

# Chapter 2 Ad Libitum Feeding and Growth Functions

Chapter 1 discussed all growth functions so far developed for animals as systems with output only, empirical in nature despite the appeals to various theories of growth. The animal is best considered as a black box with input and output not whitened by any theories of growth. It was shown that the rate of growth ($dW/dt$) of an animal is the product of the true growth efficiency ($dW/dF$) and the food intake ($dF/dt$) whether an animal is fed ad libitum or on a controlled feeding regime. Since the animal is considered a black box with input $F(t)$ and output $W(t)$, only experimental data can reveal acceptable functions $W(F)$ and $F(t)$ so that the true growth efficiency ($dW/dF$) and the food intake ($dF/dt$) can be calculated. These functions will be acceptable if the $W(t)$ function possesses the three mathematical properties discussed in Chap. 1 and if the parameters of the $W(F)$ and $F(t)$ functions are interpretable in physical terms.

## 2.1 Live Weight as a Function of Cumulative Food Consumed

Spillman (1924) gave a detailed description of how he developed a form of $W(F)$. Prior to this time there were many published tables of agricultural data which expressed the response of $Y(n)$ of many products to levels of $n$ units of some material. In reference to animals $Y(n)$ was the mean weight, $W$, in pounds or kilograms per individual and $n$ was the cumulative food consumed, $F$, expressed in some unit such as 100 pounds or kilograms depending on how $W$ was expressed. In working through these tables, where the $Y(n)$'s were not only weights of animals versus $n$ units of cumulative food, but also were yields of products, like cotton, cabbage etc., per acre versus units of fertilizer, irrigation water etc. applied per acre, Spillman found a simple relation between successive values of $Y(i)$, $i=1,2,...$, for a given table of data. Arithmetically he found the ratios $[Y(i+1)-Y(i)]/[Y(i)-Y(i-1)]$ to be very nearly a constant less than unity which he designated $R$.

Designating the differences in the ratio as $\Delta Y(i+1)$ and $\Delta Y(i)$, Spillman realised that the successive differences could be calculated from the recursive equation

$$\Delta Y(i+1)=\Delta Y(i)R, \text{ i.e.}$$

$$\Delta Y(2)=\Delta Y(1)R, \Delta Y(3)=\Delta Y(2)R \text{ etc. (Fig. 2.1).}$$

Combining these relations he wrote $Y(n)$ as the sum of the successive differences, namely

$$\begin{aligned} Y(n) &= \Delta Y(1)+\Delta Y(2)+\Delta Y(3)+\ldots+\Delta Y(n), \\ &= \Delta Y(1)(1+R+R^2+\ldots+R^{n-1}). \end{aligned}$$

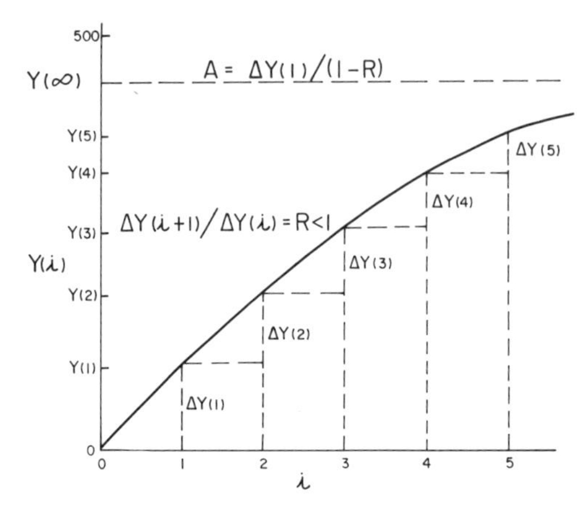


Fig. 2.1. A geometric illustration of Spillman's algebraic analyses of tables of weights, $Y(i)$, of animals versus $i$ units of food consumed and other tables of agricultural data which lead him to the equation of diminishing increments

Recognising the term on the right as a geometric series Spillman could write this equation as

$$Y(n)=[\Delta Y(1)/(1-R)](1-R^n).$$

Since $R<1$, $Y(n)\rightarrow\Delta Y(1)/(1-R)$ as $n\rightarrow\infty$, i.e. $\Delta Y(1)/(1-R)$ is the asymptotic upper limit of $Y(n)$ as $n$ becomes larger. Spillman designated this upper bound by $A$ so he could write

$$Y(n)=A(1-R^n), \tag{2.1}$$

as a general empirical description of all the agricultural tabular data he had at hand.

Figure 2.1 illustrates Spillman's arithmetic leading to Eq. (2.1) as his form of the law of diminishing returns. It is very interesting to note that the change of $Y$ on the first unit, namely $\Delta Y(1)$ foretells the asymptotic value of $Y(n)$ given by $A$.

Our interest focuses on Spillman's study of the growth of animals on the cumulative food they consumed, where $Y$ represents the animals weight $W$ and $n$ is a measure of the cumulative food consumed $F$ to reach that weight. He showed how he found values of $A$ and $R$ from Haecker's (1920) data on steers and Henry's (1898) data on hogs. The fit of Eq. (2.1) was sufficiently good in both cases that he could calculate the weights within 1% of the tabular values.

Equation (2.1) can be put in more modern notation for animal growth by replacing $R$ by its equal $\exp[-\ln(1/R)]$ and letting the unit of food be 1 kg so that $n$ becomes $F$. The equation then becomes

$$W(F)=A\{1-\exp[-\ln(1/R)F]\}.$$

But here $\ln(1/R)$ is clumsy notation and, as Spillman realised, an initial weight $W_0$, i.e. the weight when $F=0$, is missing. Including $W_0$ and designating $\ln(1/R)$ as $B$, Eq. (2.1) can be rewritten as

$$W(F)=(A-W_0)[1-\exp(-BF)]+W_0. \tag{2.2}$$

In the spirit of Chap. 1, Spillman developed Eq. (2.1) purely as an empirical expression of the "mathematical structure" he found in published growth tables. Going from the growth and feeding data of steers to the data of capons, he found the

Table 2.1. The mean weight per chick in each of four lots at hatching and biweekly thereafter. (Data of Jull and Titus 1928)

| Age (weeks) | Mean weight (g per chick) | | | |
|---|---|---|---|---|
| | Lot 1 (females) | Lot 2 (females) | Lot 3 (males) | Lot 4 (males) |
| 0 | 34.54 | 34.19 | 34.97 | 34.77 |
| 2 | 90.82 | 79.76 | 77.12 | 82.27 |
| 4 | 215.31 | 198.10 | 185.44 | 255.59 |
| 6 | 387.25 | 355.76 | 366.77 | 399.43 |
| 8 | 618.13 | 582.71 | 608.10 | 656.59 |
| 10 | 835.94 | 770.85 | 773.17 | 924.23 |
| 12 | 985.45 | 936.95 | 1091.10 | 1225.66 |
| 14 | 1334.55 | 1236.59 | 1505.87 | 1629.74 |
| 16 | 1541.29 | 1480.37 | 1772.95 | 1933.03 |
| 18 | 1651.77 | 1574.15 | 2046.03 | 2157.37 |
| 20 | 1860.16 | 1785.43 | 2386.79 | 2435.53 |
| 22 | 2026.93 | 2012.07 | 2736.41 | 2744.86 |
| 24 | 2211.45 | 2170.97 | 2906.28 | 2926.89 |

form of Eq. (2.1) did not change, only the values of $A$ and $R$ changed. This was an excellent curve fitting achievement.

H.W. Titus, M.A. Jull, and W.A. Hendriks who were coworkers of Spillman in the United States Department of Agriculture decided to test Eq. (2.1) on some experimental growing and feeding data. Titus and Jull (1928) reported the results of a 24-week experiment in which two lots of female and two of male chickens were freely fed. The expressed purpose of this experiment was "... to provide data, obtained under controlled conditions, to test the application of the law of diminishing increment, Eq. (2.1), as applied to growing chickens."

The experimental animals were 170 Barred Plymouth Rock females and Rhode Island Red males. A sex-linked character permitted separation of the chicks into lots of 84 females and 86 males. The females were divided into lots 1 and 2 of 40 and 44 chicks respectively, and the males were divided into replicate lots 3 and 4 of 43 chicks each. Thus they tested Spillman's hypothesis for males and females separately. The lots were fed the same feed and in "like manner" over the 24 weeks.

The data are summarised in Tables 2.1 and 2.2 in Appendix A. Table 2.1 shows mean live weight per bird in each lot versus age, and Table 2.2 displays the mean biweekly food intakes per bird. Over the experimental period 2 birds died in lot 1, none in lot 2, 4 in lot 3 and 2 in lot 4. In taking data for mean live weight and food consumed per bird, these deaths were taken into account.

Jull and Titus tested Spillman's hypothesis on each lot by following his recipe. They found the fit of Eq. (2.1) satisfactory but concluded that $A$ and $R$ can be considered as empirical constants only having no physical meaning. However, in the light of a year long experiment on chickens which permitted them to attain their mature weights, they changed their minds to considering the parameters to have real physical significance (Titus et al. 1934). Equation (2.2) appears to be more than a mere empirical formula for describing the relationship between the weight

Table 2.2. Mean food consumption per chick for each two week period. (Data of Jull and Titus 1928)

| Age (weeks) | Mean feed consumption (g per chick) | | | |
|---|---|---|---|---|
| | Lot 1 (females) | Lot 2 (females) | Lot 3 (males) | Lot 4 (males) |
| 2 | 127.14 | 132.76 | 111.82 | 136.08 |
| 4 | 271.16 | 246.16 | 236.29 | 257.39 |
| 6 | 511.28 | 472.40 | 489.99 | 536.40 |
| 8 | 688.19 | 677.62 | 788.26 | 814.14 |
| 10 | 846.94 | 824.76 | 863.49 | 1116.54 |
| 12 | 1176.41 | 1045.48 | 1219.03 | 1287.96 |
| 14 | 1256.89 | 1080.88 | 1322.22 | 1406.14 |
| 16 | 1251.04 | 1063.18 | 1248.02 | 1309.45 |
| 18 | 1316.88 | 1231.34 | 1639.91 | 1647.26 |
| 20 | 1534.90 | 1205.89 | 1502.70 | 1715.89 |
| 22 | 1559.77 | 1416.09 | 1721.33 | 1745.11 |
| 24 | 1613.18 | 1261.21 | 1625.37 | 1764.11 |
| Total | 12153.78 | 10657.77 | 12822.43 | 13736.47 |

$W$ of an animal and the cumulative food consumed by it to attain that weight and Eq. (2.2) will be accepted as a fundamental equation in the remainder of this book.

It is immediately apparent that, if a form of the function $F(t)$ could be established from study of experimental cumulative food consumed versus age data, then substitution of $F(t)$ in Eq. (2.2) would yield a $W(t)$ which could be explored for meeting the three mathematical properties expected of a growth function discussed in Chap. 1, namely (1) having an asymptotic maximum; (2) possessing a point of inflection (POI); (3) being asymptotic to Brody's $W(t)$ function [Eq. (1.5)] for $t > t^*$.

## 2.2 The Ad Libitum Feeding Function

Spillman, Titus, Jull, and Hendricks did not take the opportunity to study the $(t, F)$ data of the animals (steers, hogs, and chickens) at their disposal.

Figures 2.2 and 2.3 show respectively how the steers and lot 4 chickens consumed food as they aged. Both figures show that $F$ rises curvilinearly and appears to approach a straight line asymptotically, the slope, $dF/dt$, of which is the mature food intake required by the animals to maintain their mature weights $A$. Titus (1928) reported graphing the $(t, F)$ data for White Pekin ducklings showing the asymptotic approach to the straight line $F = 3.23(t - 3.6)$. Here 3.23 lbs of food per week is mature food intake; $t$ is age in weeks and the 3.6 weeks appears to be analogous to Brody's $t^*$.

The feeding characteristics of animals have long been known to be important in growth experiments designed to explore the effects of dietary factors. Hopkins (1912) was foremost in this field and gave great impetus to the scientific search for dietary growth factors such as vitamins and coenzymes. He reported his observations of the food intake of rats given a diet with and without the addition of

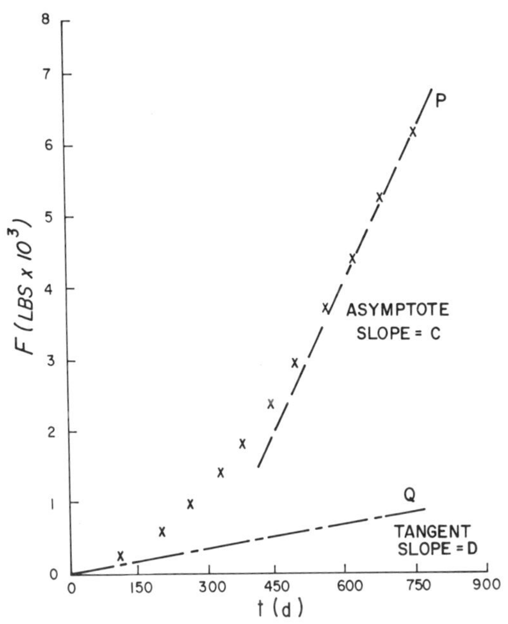


Fig. 2.2. Cumulative food consumed, $F$, versus age, $t$, of Haecker's steers. Also indicated are the tangent, $Q$, to the $F(t)$ curve at $t=0$ (birth) and the asymptote, $P$, the $F(t)$ curve gradually approaches for ages beyond 500 days

a few millilitres of milk per day but did no further mathematical analysis of his results. After Hopkins, other workers designed growth experiments in which food intake was claimed to be important but they failed to evaluate its importance quantitatively (see, for example, Drummond and Marrian 1926, Mitchell and Beadles 1930). Harte et al. (1948) made a good start but obscured their basic data by reporting the ad libitum food intake as "voluntary caloric intake per square decimeter of body surface per day." Well before 1948, the surface law (Rameaux and Sarrus 1837) had become accepted as a "law" in bioenergetics, so that many experimental data of feeding have been confused by relating intake to "metabolic mass, $W^{2/3}$"

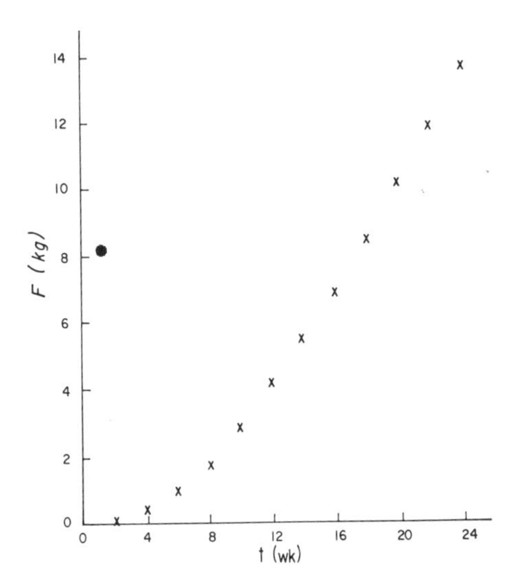


Fig. 2.3. Cumulative food consumed, $F$, versus age, $t$, of the lot four male chickens of Titus et al. (Table A-2 in Appendix A). Note the similarities of the disposition of these data points to those for steers in Fig. 2.2

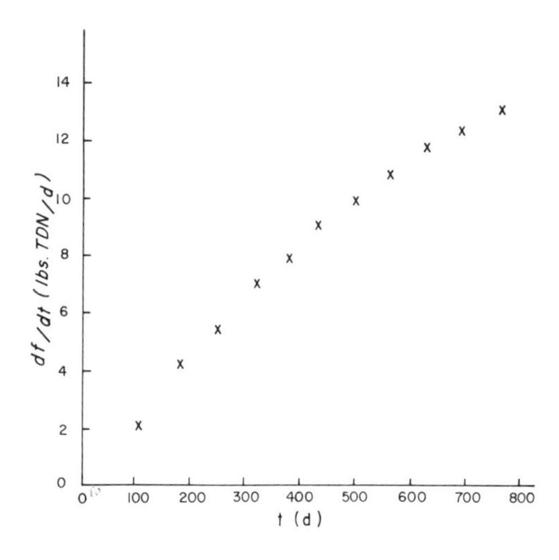


Fig. 2.4. Food intake of Haecker's steers versus age, $t$. Note the way the food intake initially rises rapidly followed by a slowing towards what might be a maximum asymptote. The disposition of these data points is similar to the curve in Fig. 2.1

which is approximately proportional to the surface area of an animal of live weight $W$ (Brody 1945, Chap. 13; Kleiber 1961, Part IV).

There have been very few explicit statements of the input $F(t)$ to the growing organism from which to glean some quantitative ideas about $F(t)$ in the light of the mathematical properties shown in Figs. 2.2 and 2.4. Wamser (1915), a student of Mitscherlich, studied the possibilities of expressing the specific ad libitum feed intake $[(dF/dt)/W]$ of an animal as an exponential function of time or age. He offered a plausible physiological reason for his formulation in an effort to whiten the black box but lacked sufficient long term data to give it meaning. Rashevsky (1939), in his study of the mathematical biophysics of the growth of a cell, assumed that the cell is immersed in a nutritive fluid and that the nutrients diffuse through the cell wall at a rate per unit area, $dc/dt$, proportional to the difference between the nutrient concentration, $c$, outside the cell and the concentration, $c_1$, just inside the cell wall; that is $dc/dt = h(c - c_1)$. Rashevsky then whitened the box by postulating how the nutrients were used internally to increase the size of the cell. Van der Vaart (1968) uses the same idea for input of "food" and whitened the black box by postulating an autocatalytic mechanism for conversion of food to "biomass" which diffuses back to the environment. The findings of these authors are not very helpful in formulating the food intake function $dF/dt$ or the cumulative food consumed function $F(t)$ in any manner suitable to describe the common mathematical properties of the $F(t)$ data shown in Figs. 2.2 and 2.3. The only procedure available is a direct approach to the data as such.

The first hint of a possible food intake function, i.e., $dF/dt$ versus age, came from Fig. 2.4, which is a graph of the daily food intake of Haecker's steers in units of pounds of total digestible nutrients (TDN) per day versus age $t$ in days. The graphs of the biweekly food consumed per chick versus age data for the four lots of chicks by Titus et al. showed the same geometrical features of increasing with diminishing slope and apparently becoming asymptotic to a maximum food intake analogous to Fig. 2.1. In these graphs, especially in Fig. 2.4, the disposition of the data points suggests the law of diminishing return for the increase of appetite, $dF/dt$, as the animal ages.

It seemed reasonable to try

$$dF/dt = (C - D)[1 - \exp(-t/t^*)] + D \quad (2.3)$$

as a possible formula for the relation of $dF/dt$ to the age $t$. Here $C$ is the maximum food intake the animal can achieve as it ages; $D$ is the initial food intake at birth, hatching or weaning, when $t$ is assigned the value of zero. The geometrical meanings of $D$ and $C$ are illustrated in Fig. 2.2 where $D$ is the slope of the tangent, $Q$, to the $F(t)$ curve at $t=0$ and $C$ is the slope of the asymptote of $F(t)$ when $t$ is large. The time constant in the exponential was taken to be Brody's $t^*$ because following the consequences of Eq. (2.3) indicated the choice was reasonable.

A possible form of the cumulative food consumed function, $F(t)$, is obtained by integrating Eq. (2.3) with the initial condition $F(0)=0$. Equation (2.4) is the result of the integration after some algebraic adjustment.

$$F(t) = C\{t - t^*(1 - D/C)[1 - \exp(-t/t^*)]\}. \quad (2.4)$$

Equations (2.3) and (2.4), which are tentative descriptions of the way animals feed as they age, can be used to explore how reasonable they are. For example suppose measurement of weight $W$ and cumulative food $F$ begin at birth, i.e., age $t=0$. At birth, or hatching, it can be assumed that the initial food intake, $D$, is small enough compared to its mature food intake, $C$, that it can be taken as zero and Eqs. (2.3) and (2.4) become respectively

$$dF/dt = C[1 - \exp(-t/t^*)],$$

$$F(t) = C\{t - t^*[1 - \exp(-t/t^*)]\}. \quad (2.5)$$

These equations show that as $t$ gets larger than $t^*$

$$dF/dt \to C \text{ and } F(t) \to C(t - t^*).$$

In other words as the animal enters young adulthood, its appetite asymptotically approaches its mature appetite, and the cumulative food it has consumed up to age $t$ asymptotically approaches the straight line $C(t-t^*)$ as suggested by the data points in Figs. 2.2 and 2.3 respectively for ages beyond 500 days for the steers and 15 weeks for the lot 4 chickens. These deductions find support in the observation made by Titus that the cumulative food consumed by the White Pekin duckling approached the straight line $F(t)=3.23(t-3.6)$. Here we may conclude that the 3.23 pounds of food per week is the mature food intake of these animals and 3.6 weeks is the $t^*$ for them. A further indication that $t^*$, as used here is Brody's, is shown in

$$W(t) = A\{1 - \exp[-(BC)(t - t^*)]\}.$$

Here $F(t)=C(t-t^*)$ is substituted in Eq. (2.2) and $W_0$ at birth or hatching is small enough to be negligible. Replacing the product $BC$ by $k$ makes this equation identical with Brody's equation, namely Eq. (1.5) discussed in Chap. 1.3. This relationship of Brody's $k$, which has been called the maturing rate, to the product of $B$ and $C$, is a new finding. If Eqs. (2.2), (2.3), and (2.4) are deemed suitable descriptors of how animals feed and grow, the substitution of Eq. (2.4) in Eq. (2.2) gives the live weight as a function of age, $W(t)$, namely

$$W(t) = (A - W_0)[1 - \exp\langle -BC\{t - t^*(1 - D/C)[1 - \exp(-t/t^*)]\}\rangle] + W_0, \quad (2.6)$$

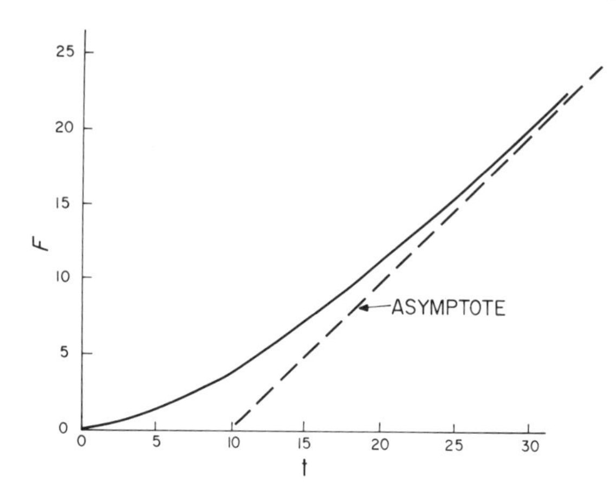


Fig. 2.5. A simulation of cumulative food consumed, $F$, versus age, $t$, calculated using Eq. (2.5) and 0.9 for the mature food intake, $C$, and 10 for Brody's $t^*$. Note the similarity of this curve to the disposition of the data points in Figs. 2.2 and 2.3

which can be expected to be a suitable descriptor of how animals grow as they age under normal conditions with nutritious food freely available. The solid curve in Fig. 2.5 was calculated, using 0.9 for $C$ and 10 for $t^*$ in Eq. (2.5), to show the similarity to the disposition of the data points in Figs. 2.2 and 2.3. And Fig. 2.6 illustrates the calculation of $W(t)$ using $A=6$, $C=0.9$, $t^*=10$, taking $B=0.101$ and $W_0=0$ in Eq. (2.6). Figure 2.6 is similar to Figs. 1.1 and 1.2 and shows that Eq. (2.6) has the three mathematical properties required of a growth function, namely it is S–shaped with a (POI); it is asymptotic to a mature weight; and it is asymptotic to Brody's function which in this case is $W(t)=6\{1-\exp[-0.091\,(t-10)]\}$.

Aside from $W_0$ and $D$, the parameters $A$, $B$, $C$, and $t^*$ characterise the animal's response to the food offered within the management scheme. It was remarked in Chap. 1 that the parameters of an acceptable growth function should have some biophysical meaning. Among the four characteristic parameters, the meanings of

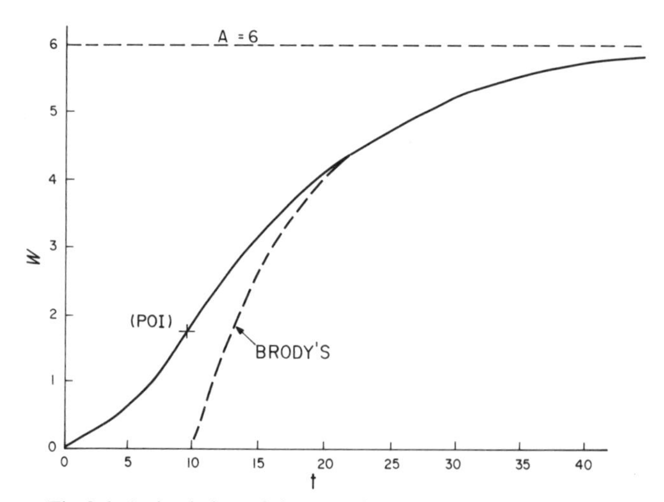


Fig. 2.6. A simulation of the growth curve, $W(t)$, using Eq. (2.6) with 6 for the mature weight, $A$; 0.9 for the mature food intake, $C$; 10 for Brody's $t^*$ and 0.101 for $B$ and zero to $W_0$. This is the response of an animal fed according to the simulation shown in Fig. 2.5. Note how this curve has the three expected properties of real growth curves (Figs. 1.1 and 1.2)

$A$ and $C$ are the mature live weight and the mature food intake required to maintain the mature weight. The meanings of $B$ and $t^*$ are obscure. However in the context of Eqs. (2.2) and (2.3), the way physicists approach the problems of the meanings of the parameters in equations like these can be useful. Considering Eq. (2.2) first, we could ask how much food the animal must consume in order to reach a weight gain $W(t) - W_0$ which is one half the maximum gain achievable $(A - W_0)$. From Eq. (2.2), this food consumption, which we shall designate $F_{1/2}$, is found to be given by

$$F_{1/2} = \ln 2/B.$$

Here $B$ is shown inversely related to the half food $F_{1/2}$. The larger $B$ is, the less the food required to achieve half the gain yet to be made. Similarly the time, which we shall designate $t_{1/2}$, required to increase the gain of appetite to half that achievable is, from Eq. (2.3),

$$t_{1/2} = t^* \ln 2.$$

Here $t^*$ is seen to be directly related to the half time $t_{1/2}$. The larger $t^*$ the more time is required for the animal to halve the sum of its present appetite and the maximum achievable, namely $C$. Solving these relations for $B$ and $t^*$ and substituting them into Eqs. (2.2) and (2.3) respectively, and remembering that $\exp(\ln 2) = 2$, we can write them as

$$W = (A - W_0)(1 - 2^{-(F/F_{1/2})}) + W_0,$$

and

$$dF/dt = (C - D)(1 - 2^{-(t/t_{1/2})}) + D.$$

These equations may help the reader to visualise the physical significance of $B$ and $t^*$. However, in the remainder of the book, Eqs. (2.2), (2.3), and (2.4) will be used to test their usefulness in describing as much long term ad libitum $(t, F, W)$ experimental data on various animals as can be found in the literature. The characteristic parameters will be $A, B, C$, and $t^*$ with parameters $W_0$ and $D$ considered initial conditions. However, in the development of the differential equation of Eq. (2.2) in the following section, it will be seen that $(AB)$, the product of the mature live weight, $A$, and $B$, may be a more suitable parameter than $B$ alone.

## 2.3 The Differential Equations of Ad Libitum Feeding and Growth

Differential equations, which for convenience we shall symbolise by *DE*, express the general ways in which systems respond to natural or artificial treatments. They are independent of initial conditions, which for convenience we shall designate as i.c., but dependent only on quantities I have called characteristic parameters, namely $A, B, C$, and $t^*$ that have been used here and considered independent of time.

Given a function $y(x) = \phi(a_i, x)$, $i = 1, 2 \ldots, n$, the DE, of which $y(x)$ is the solution, is found using the following steps.

First: set $x=0$ in $y(x)$ and note the number, say $j$, of parameters of the set $(a_i)$ which remain. These $j$ parameters are the i.c. This step indicates the order of the desired DE; that is, it will involve the differential coefficients $d^jy/dx^j$, $d^{j-1}y/dx^{j-1}$, ..., $dy/dx$. Second: the j parameters, comprising the i.c., are algebraically eliminated from $\phi(a_i, x)$ and its $j$ derivatives with respect to $x$ leading to the DE, namely

$$d^jy/dx^j + b_1 d^{j-1}y/dx^{j-1} + b_2 dy^{j-2}/dx^{j-2} + \ldots$$

$$+ b_{j-1}dy/dx + b_j y + b_{j+1} = 0.$$

Here the set of factors $(b_m)$, $m=1,2...,j+1$, are functions of the set of characteristic parameters $(a_i)$, $i=1,2,...,n-j$ (Batschelet 1975).

Examples of this process of deriving the DE from a function are given in the following. We need the DE of the $W(F)$ function in Eq. (2.2). Setting $F=0$ gives $W(0)=W_0$ as the i.c. The DE of Eq. (2.2) is therefore first order and taking the first derivative with respect to $F$ gives

$$dW/dF = (A - W_0)B\exp(-BF),$$

thereby completing the first step. The second step requires using Eq. (2.2) with $dW/dF$ to eliminate $W_0$. After some algebraic manipulation the DE is found to be

$$dW/dF + BW - AB = 0. \tag{2.7}$$

These expressions involving the growth efficiency $dW/dF$ show that it decreases as the animal ages, i.e. consumes more food and grows towards its mature size, $A$. Equation (2.7) indicates that the characteristic parameters of the growth of an animal on the food it consumes are two in number, namely $A$ and $B$. However, this equation can be rewritten as

$$dW/dF = (AB)(1 - W/A),$$

and since $W/A$ is Brody's fraction of maturity, $u$, this revision of Eq. (2.7) implies that $(AB)$ is a better growth parameter than $B$ alone. As such $(AB)$ can be designated as the growth efficiency factor. The importance of choosing $(AB)$ rather than $B$ is shown in Chaps. 4 and 9. This choice means Eq. (2.2) is better written as

$$W = (A - W_0)\{1 - \exp[-(AB)F/A]\} + W_0.$$

Equation (2.3) as it stands is not a DE because it contains the i.c., $D$, when $t$ is set to zero. Before proceeding to the DE it is convenient to designate $dF/dt$, given by Eq. (2.3), as $q^*(t)$ because this equation is proposed as a description of the ad libitum feeding function of the animal and is therefore an upper limit of the food intake an animal will voluntarily choose. All other food intake functions of time will be designated as $q(t)$.

Proceeding to eliminate $D$ from Eq. (2.3) as we did in eliminating $W_0$ from Eq. (2.2), the ad libitum feeding DE is found, namely

$$dq^*/dt + q^*/t^* - C/t^* = 0. \tag{2.8}$$

Here it is seen that the characteristic parameters of ad libitum feeding are again two in number, namely $C$ and $t^*$. Equation (2.8) is also a second order DE of the cumulative food consumed $F$ versus $t$. Substituting $dF/dt$ for $q^*(t)$ in Eq. (2.8) yields

$$d^2F/dt^2 + (1/t^*)dF/dt - C/t^* = 0, \tag{2.9}$$

the solution of which requires two i.c., namely $t=0$, $dF/dt=D$ and $F=0$.

Differential Eq. (2.7) coupled with Eq. (2.8) is a general description of how the animal is feeding and growing at any age $t$, and the solution of them with some given i.c. such as $t=0$, $q^*(0)=D$, and $W(0)=W_0$ selects the particular way the animal will feed and grow for ages $t>0$ from the many courses of feeding and growth the animal might have taken on any other specified i.c. Similarly coupling DE's (2.7) and (2.9) the general description of how the animal grows on the food it has consumed up to age $t$, and the solution of them with i.c. $t=0$, $dF/dt=D$, $F(0)=0$, and $W(0)=W_0$ selects the particular way an animal will grow on its cumulative food consumed from the many ways it could grow depending on the specification of the i.c.

Here it is seen why the characteristic parameters of a system plus environment should be distinguished from those parameters which can be shown to be initial conditions only. In this respect grave errors can be made using the popular growth functions discussed by Fitzhugh (1976) as nonlinear regression equations.

McCarthy and Bakker (1979) used the following form of the Gompertz equation,

$$W = A\exp[-b\exp(-kt)]$$

to study bending the postweaning growth curves of mice. Since the mice were weaned and weighed at three weeks of age, the weight, $W_3$, is the initial condition if $t$ is the age less three. In the use of this equation the constant $b$ was regarded as a scaling factor and was determined along with $A$ and $k$ in the regression as independent constants. However, if $t$ is set to zero, then it follows that

$$W_3 = A\exp(-b).$$

Here it is seen that

$$b = \ln(A/W_3),$$

indicating that $b$ is confounded with the characteristic parameter $A$ and the i.c. $W_3$, so that their determinations of $A$ and $k$ were biased. The form of the Gompertz equation McCarthy should have used is

$$W = A(W_3/A)^{\exp(-kt)}.$$

Here the characteristic parameters are clearly distinguished from the i.c. and regression should yield unbiased estimates of $A$ and $k$.

Robbins et al. (1979) used a form of the logistic equation such as

$$y = p + q/[1 + \exp(r + sx)],$$

to study the effect of the level of addition, $x$, of sulfur amino acid to a basal diet on the growth of chicks over an eight day period from age 8 to 16 days posthatching. Here $y$ is the gain $W(x) - W(0)$ so that the curve is forced to pass through $y=0$

when $x=0$, in which case there is a relation between the parameters $p, q$, and $r$, namely

$$p = -q/(1+\exp)(r),$$

indicating that their estimates of what they call the parameters, e.g., $p+q, r$, and $s$, are biased by the relation among parameters $p, q$, and $r$.

Designating $(p+q)$ as $A$, and $s$ as $Ak$ their formula for $y(x)$ can be rewritten as

$$y = AW(0)/\{W(0)+[A-W(0)]\exp(-Akx)\} - W(0),$$

where the parameters characteristic of the logistic equation, namely $A$ and $k$, are clearly distinguished from the i.c. $W(0)$. If this form of the logistic function is used, the estimates of $A$ and $k$, and $W(0)$ if it is considered a regression constant, will be unbiased by any hidden relation such as that between $p, q$, and $r$ shown above.

Robbins et al. and Gibney et al. (1979) make this type of error in using the diminishing returns equation. Robbins et al. have used

$$y = u + v[1-\exp(wx)],$$

and Gibney et al. used

$$y = A - B\exp(-Cx).$$

Here, it is seen that the parameters characteristic of the diminishing returns equation are not distinguished from the initial condition, that is, $y=A-B$ when $x=0$. We have already seen that the diminishing returns equation should be written in the form Eq. (2.2) or Eq. (2.3) which clearly distinguish between the characteristic parameters and the i.c.

Chapter 4 will discuss the application of the solutions of the DE's (2.7) and (2.8) to as much long term experimental feeding and growth data as I could find in the literature and other sources and on as wide a range of species and varieties of animals as was available. Here there will be a clear distinction between the characteristic parameters $A, B, C$, and $t^*$ and the i.c. experimentally imposed. If these functions prove useful in the study of ad libitum feeding and growth, the remarks in Chap. 1.5 regarding the growth rate, $dW/dt$, as the product of the growth efficiency $dW/dF$, and the food intake, $dF/dt$, will find support. Equations (2.7) and (2.8) respectively show that $dW/dF$ decreases and $dF/dt$ increases as age increases and the growth rate will therefore pass through a maximum at some age of the animal between infancy and young adulthood (Fig. 1.10) and the growth curve, $W(t)$, will naturally possess a (POI). However some experimental $(t, W)$ data may not show the (POI) when the age of the animal at the beginning of the experiment is sufficiently large that the i.c., namely $W_0$ and $D$, for solving the DE (2.7) and (2.8) are sizeable fractions of their respective maxima $A$ and $C$.

During these studies I found quite a number of references to growth as a realisation of a stochastic process as a function of age or time. Intuitively I thought Spillman's conjecture that the equation of diminishing increments of weight versus cumulative food consumed [Eq. (2.2)], merited study as the realisation of a stochastic growth process as a function of food consumed thereby providing additional insight. Chapter 3 is the result of this study.

# References

Batschelet E (1975) Introduction to mathematics for the life sciences, 2nd edn. Springer, Berlin Heidelberg New York
Brody S (1945) Bioenergetics and growth. First published: Reinhold, New York. (Reprinted: Hafner Press, New York, 1974)
Drummond JC, Marrian GF (1926) Physiological role of vitamin B; relation of vitamin B to tissue oxidations. Biochem J 20:1225–1229
Fitzhugh HA (1976) Analysis of growth curves and strategies for altering their shape. J Anim Sci 42:1036–1051
Gibney MJ, Dunne A, Kinsella IA (1979) The use of the Mitscherlich equation to describe amino acid response curves in the growing chick. Nutr Rep Int 20:501–510
Haecker RW (1920) Investigations in beef production. Bull 193: Minn Agric Exp Stn
Harte RA, Travers JJ, Sarich P (1948) The effect on rat growth of alternated protein intakes. J Nutr 35(3):287–293
Henry WA (1898) Feeds and feeding. First published: Henry, Madison (10th edn 1910)
Hopkins FG (1912) Feeding experiments illustrating the importance of accessory factors in normal dietaries. J Physiol (London) 44:425–460
Kleiber M (1961) The fire of life. Wiley, New York
McCarthy JC, Bakker H (1979) The effects of selection for different combinations of weights at two ages on the growth curve of mice. Theor Appl Genet 55:57–64
Mitchell HH, Beadles JR (1930) The paired feeding method in nutrition experiments and its application to the problem of cystine deficiencies in food proteins. J Nutr 2:225–243
Rameaux, Sarrus (1837/8) The relation of heat production to linear size and surface area for animals. Bull Acad Med 2:538–561
Rashevsky N (1939) Mathematical biophysics of growth. Bull Math Biophys 1:119–127
Robbins KR, Norton HW, Baker DH (1979) Estimation of nutrient requirements for growth data. J Nutr 109(10):1710–1714
Spillman WJ, Lang E (1924) The law of diminishing increment. World, Yonkers
Titus HW (1928) Growth and the relation between live weight and feed consumption in the case of White Pekin ducklings. Poult Sci 7(6):254–262
Titus HW, Jull MA (1928) The growth of Rhode Island reds and the effect of feeding skim milk on the constants of their growth curves. J Agric Res 36:515–540
Titus HW, Jull MA, Hendricks WA (1934) Growth of chickens as a function of the feed consumed. J Agric Res 48:817–835
van der Vaart HR (1968) The autocatalytic growth model: critical analysis of the conceptional framework. Acta Biotheor 13:133–142
Wamser HP (1915) Die Futteraufnahme wachsender Tiere – eine mathematische Gesetzmäßigkeit. Dissertation, Königsberg

# Chapter 3 A Stochastic Model of Animal Growth

Thus far growth has been envisaged as purely deterministic and its variability as due to instabilities internal and external to the animal. In this chapter, in apparent deviation from the goal of disclosing the deterministic elements in animal growth, the concept of determinism will be viewed in the light of a stochastic approach to growth. This necessitates deeper consideration of the variability exhibited by experimental growth data. The stochastic element of variability refers only to the sequence of pure chance events, inherent in the system itself, which expresses the observed course of development of the system under the imposed experimental conditions as if experimental variability were absent.

Suppose $y$ is a stochastic variable of a given system on the continuous domain $x$; that is to say at each value of $x$ in the range $0 \leqq x \leqq \infty$, $y$ has values distributed, according to some probability distribution function $\Psi(x,y)$, with a mean value $\bar{y}$ and dispersion $\sigma(x)$. If the development of $y$, as $x$ proceeds, is observed experimentally then the experimental error $s(x)$ becomes involved and the total variance observed is the sum $[\sigma(x)]^2 + [s(x)]^2$. A stochastic model of $y$ is found by stating an ideal situation in which the course of development of $y$, as $x$ increases, is dependent only on the distribution of $y$ values given a value for $x$. The model is then an expression of the most frequent development of $y$ given an infinite number of identical systems at $x=0$. Nothing can be said about what the exact value of $y$ will be at a given $x$, but if $\Psi(x,y)$ is known, the odds of $y$ being in the range $y_1 < y < y_2$ at an $x$ can be stated.

The results of solution of a problem of a stochastic process can serve as a basis for discussion of the concept of determinism. If $\sigma(x)$ is inherently so small as to be experimentally undetectable then all the paths of $y$ in the $(x,y)$ plane (for all the systems governed by the supposed stochastic process) will cluster very closely around $\bar{y}$ at any $x$ and thus appear to yield some type of empirical function $f(x)$ within experimental error $s(x)$. This is bound to be the case since $\sigma(x)$ implies a sharp distribution of $y$ at any $x$ and hence the observed development of $y$ as a function of $x$ will more closely follow the average path in the $(x,y)$ plane. Further if $\bar{y}=f(x)$ can be shown to be the solution of a differential equation such as $dy/dx = \phi(y)$, then $\bar{y}=f(x)$ becomes a law of the nature of the observed system and $f(x)$ is considered the deterministic description of the phenomenon. Thus it can be concluded that, in general, the concept of determinism is a special case of the concept of stochasticism where the stochastic variance $\sigma^2(x)$ is so small, relative to the resolution of observing devices, as to be experimentally undetectable. In this sense this chapter does not deviate much from my goal but does present the reader with some new ideas about growth.

The concept of determinism received strong support in the early development of science. This was especially so in astronomy after the works of Kepler, Galileo, and Newton (Moore 1973). The motions of heavenly bodies were found to follow closely the solutions of Newton's differential equations of the mechanics of motion. However, as physics and chemistry developed, it became clearer that the exactness of the law of nature of a particular phenomenon depends on the number ($n$) of particles which cooperate to express the law (Schrodinger 1967, p. 18). In general the departure of the observed phenomena from the law will be of the order of $n^{1/2}$ and the relative precision of the law will be $n^{1/2}/n = 1/n^{1/2}$. In $n = 25$ the relative precision of the law will be 1/5 or 20% and one may be reluctant to say that the experiment confirms the law. However if $n$ is very large, as it is in astronomy or in physical and chemical experiments where $n$ is of the order of Avogadro's number ($n = 0.6025 \times 10^{24}$), the relative precision of the laws ($1/n^{1/2}$) approaches 0 and one may speak of the exactness of the laws of nature in these cases.

These general remarks on determinism versus stochasticism are illustrated in the next section with a specific example of animal growth as the outcome of a stochastic process. Because biological responses are regarded as being largely unpredictable it is important to attempt solution of stochastic models of biological phenomena (Gurland 1964) so that the relative importance of the mean stochastic variance and experimental variance of an experiment may be assessed. In the preceding chapters of this book deterministic functions have been proposed for the ad libitum feeding and growth of animals. However, little was said about the error structure of the functions.

## 3.1 Growth in the Food Consumed Domain as a Markov Process

Many physical and biological processes have been studied as time developing collections of randomly distributed events (Lindsay 1941, Bartholomay 1958a, b, 1959, 1964, 1968, Gurland 1964, Heyde and Heyde 1969, 1971). The results of these studies were the expected time development of the collections and the variances of the developments. The mathematics of Markov processes were found useful in the study of collections of events the underlying causes of which are constant over time and the future state of which are dependent only on the present states.

The many successful applications of Markov processes give strong motivation for viewing animal growth as a Markov process in the time domain (Mendel 1965, Zotina and Zotin 1972, Zotin et al. 1971). In such a view, growth is considered a random accumulation of live weight in time, the processes generating live weight fulfilling the requirements of Markovian mathematics. However, studies of animals as input output devices lead one to conclude that the concept of growth as a stochastic process in the time domain restricts development of a suitable model because the food consumed is not explicitly considered (Parks 1970a, b, c, 1971a, b). Of the food consumed by a growing animal a portion participates in the growth process with the balance used in maintenance. The stochastic model to be discussed need only consider random accumulation of live weight in the food consuming domain, because connection between any two of the three processes, food consumption, growth, and maintenance, implies the third. The model will be of the growth of an animal reared in a strictly regulated environment with a very large (infinite) supply of food freely available. The effects of feeding regimes on growth in the food consumed domain are presented in Chap. 3.6.

Cellular theory suggests that growth of an organism is the accumulation of discrete units of biomass created from ingested food (Conklin 1912, Thompson 1917). It is attractive to assume that the integral number of cells defining the live weight of the adult is somehow coded in the fertilized ovum and that the processes involved in constructing these cells are also specified at conception. These ideas suggest formulation of growth as a Markov process in the food consumed domain defined by the following assumptions:

1. at conception, a number, $a$, of potential biomass units is generally defined for an individual animal;
2. the demise of each potential unit produces a real unit, $b$, of biomass, (a pure birth process). The composition of the unit is not defined here;
3. production of an adult is the conversion of $a$, potential units to $A/b$ real units, where $A$ is the mature weight of the animal;
4. conversion of a potential to a real unit of biomass requires food;
5. if, after consumption of $F$ units of food, the animal has $m$ real units of biomass with $i$ potential units yet to be converted, then $i+m=A/b$. This conservation principle holds for all $F$ in the interval $0 \leqq F \leqq \infty$;
6. all the internal processes of the animal, for consuming food and converting it to live weight and energy of maintenance, are constant and remain the same in the interval $0 \leqq F \leqq \infty$;
7. food consumed is measured, at the portal of entry of the animal, in units which can be subdivided as small as desired;
8. the real units appearing on utilisation of any increment of food $\Delta F$ are countable;
9. a future value of $i$ depends on its present value only and is independent of past values.

These assumptions define the growth of an animal as a Markov process starting at conception with $a$ potential units of live weight at $F=0$ and approaching $A/b$ real units ($i$ approaches 0) as food is cumulatively consumed. A fertilized ovum starting with $a$ potential units will develop to maturity along any one of an infinity of paths of $i$ versus $F$ in the $(F, i)$ plane. Each path is a possible route of development of the animal as it consumes and utilizes food for growth and maintenance. By way of illustration let $f$ be an amount of food ingested for which there can be either conversion of 1 potential unit to a real one (growth) or no conversion (maintenance). Paths 1 and 2 (Fig. 3.1) are examples of such paths, differing in the number of conversions and sequences of conversions and nonconversions. All the possible paths must lie on or between the paths $i=a$, where for each $f$ consumed and utilized there is no conversion (growth efficiency is zero and each $f$ goes to maintenance), and $i=a-nf$ where for each $f$ there is one conversion (growth efficiency $=100\%$).

Such paths as 1 and 2 can cross and recross each other as the growth process develops. At any $F=nf$ there is a distribution of values of $i$ for all the possible paths in the region of the $(F, i)$ plane defined. This distribution, its mean and variance as functions of $F$ will now be developed. Digitizing both the live weight and food consumed was for heuristic purposes only. In the following $F$ will be considered continuous.

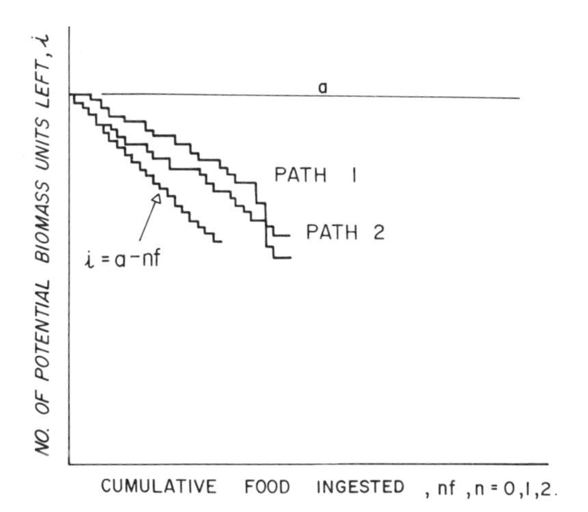


Fig. 3.1. Two possible paths of conversion of potential units, $i$, to real units of live weight expressed as step functions for stochastic growth of an organism on cumulative food ingested, $nf$

## 3.2 The Markov Probability Transition Matrix

The crux of developing a Markov model of the growth of an animal is the specification of the elements of the probability transition matrix. The needed matrix will be an $(a+1)\times(a+1)$ square matrix, with row subscript $i$ and column subscript $j$, $i$, and $j$ running from 0 to $a$. Note that $i$ may start at 0 even though the problem as stated had $i$ starting at $a$, the solution is unaffected. Each element $P_{ij}$ of the matrix is the probability that $i$ will experience a transition to value $j$ as the stochastic process proceeds in the domain. Since $i$ can only remain the same or decrease ($j$ must be $\leqq i$) the matrix elements $P_{ij}=0$ for $i\leqq j\leqq a$, the above diagonal elements are all zero. The transition matrix contains all the information relevant to the stochastic process by which $i$ diminishes and therefore by which $m$ increases to $A/b$ (the animal grows to maturity as it feeds).

Consider the following hypothetical experiment. Suppose a unit of $F$ is divided into $N$ subunits $\phi=1/N$ and the $N$ subunits of food $\phi$ are fed, one at a time, to the animal. Nonappearance or appearance of real live weight units is noted for each $\phi$ consumed. Let $n_0=$ number of $\phi$'s for each of which there was no conversion of potential units, then the basic probability notion of the process of food not going to live weight can then be expressed as $p_0(\phi)+d(\phi)=n_0/N$, where $p_0(\phi)$ is the probability that $i$ will remain the same on consuming 1 sub-unit of food; $d(\phi)$ is the difference between $p_0(\phi)$ of the event and its relative frequency $n_0/N$, and $d(\phi)$ has the property that $d(\phi)/\phi$ approaches 0 as $\phi$ approaches 0. Then $[1-p_0(\phi)+d(\phi)]/\phi$ is the relative frequency of the appearance of real live weight units per unit of food consumed. Suppose the experiment is repeated many times with $N$ increasing indefinitely. It is intuitive that, as $N$ approaches $\infty$, $p_0(\phi)$ approaches 1 but in such a manner that $[1-p_0(\phi)+d(\phi)]/\phi$ has the limit $B>0$, which is the probability of appearance of a real unit of live weight per unit of food consumed. Since $B$ measures the effectiveness of the process for generating live weight from food consumed it is, by assumption 6, constant over the interval $0\leqq F\leqq\infty$. The probability of the appearance of 1 real unit of live weight on utilization of an increment $\Delta F$ of food is then $p_1(\Delta F)=B\Delta F$. If there are $i$ potential live weight units yet to be converted, the probability of $k$ conversions of potential units to real live weight units, i.e. $i$ being reduced to $i-k$, is

$$P_{ii-k} = \binom{i}{k}(p_1(\Delta F))^k(p_0(\Delta F))^{i-k}, \quad \text{or since} \quad p_1 + p_0 = 1$$

$$= \binom{i}{k}(B\Delta F)^k(1 - B\Delta F)^{i-k}.$$

Here $\binom{i}{k}$ is the binomial coefficient.

If $B\Delta F$ is sufficiently small that $(B\Delta F)^{2+}$ is negligible, this probability can be approximated by

$$p_{ii-k} \simeq \binom{i}{k}[(B\Delta F)^k - (i-k)(B\Delta F)^{k+1}].$$

Therefore with $i$ potential units yet to be converted, $k=0$, meaning no conversion, gives $\binom{i}{o}=1$ and the transition probability is

$$p_{ii} \simeq 1 - iB\Delta F.$$

For $k=1$, meaning one conversion has taken place, $\binom{i}{1}=i$ and the transition probability in this case is

$$P_{ii-1} \simeq iB\Delta F,$$

and for $k=2+$, meaning two or more conversions, $\binom{i}{2}=i(i-1)/2$ and the transition

$$P_{ii-2+} = i(i-1)/2[(B\Delta F)^{2+} - (i-2+)(B\Delta F)^{3+}] \simeq 0$$

with the assumption of the small magnitude of $B\Delta F$. By taking $\Delta F$ small enough, say $dF$, the equations for $P_{ii}$, $P_{ii-1}$ and $P_{ii-2+}$ are exact and give the elements $P_{ij}$ of the transition matrix, designated $P(dF)$.

Since the assumptions make the process of growth a linear homogenous stationary Markov process, it is indifferent whether $i$ is taken initially as 0 or $a$. It is convenient to take $i=0$, signifying that the animal is mature. At maturity $i$ cannot increase, so $P_{00}=1$ and $P_{oj}=0$ for $j=1, 2, \ldots, a$. The equation for $P_{ii-2+}$ *shows that* $P_{ij}=0$ for $j<i-1$. Therefore $P(dF)$ is

| $i/j$ | 0 | 1 | 2 | $a$-1 | $a$ |
|---|---|---|---|---|---|
| 0 | 1 | | | | |
| 1 | $BdF$ | 1-$BdF$ | | | |
| 2 | | 2$BdF$ | 1-2$BdF$ | | |
| · | | | | | |
| · | | | | | |
| · | | | | | |
| $a$ | | | | $aBdF$ | 1-$aBdF$ |

$P(dF)=$ (the matrix above)

where all the elements above the diagonal and all those below the subdiagonal are zero.

$P(dF)$ is the first step towards specification of $P(\omega)$, where $\omega$ is any segment of the food consumed scale $0 \leqq F \leqq \infty$. Let $S$ represent the state of the animal, namely the number of potential units yet to be converted. Then $S(0)=i$ and $S(\omega)=j$ represent the states before and after consumption of $\omega$ units of food. Hence the probability, $P_{ij}(\omega)$, of transition from $S(0)$ to $S(\omega)$ is the conditional probability that $S(\omega)=j$ given $S(0)=i$, or

$$p_{ij}(\omega)=p[S(\omega)=j \mid S(0)=i].$$

Therefore matrix $P(\omega)$, the elements of which are $P_{ij}(\omega)$, will contain complete information about all the transition routes from $i$ to $j$ on consumption of $\omega$ units of food anywhere on the interval $0 \leqq F \leqq \infty$. When the distribution of routes is known, the most probable route and its variance can be specified. The $P(\omega)$ matrix is derived in the following section by evaluating the elements of $P_{ij}(\omega)$.

## 3.3 The Matrix Differential Equation of Animal Growth

Since the proposed stochastic process is linear, homogeneous and stationary, the Chapman-Kolmogorov forward and backward equations apply equally, i.e. $P(\omega+dF)=P(\omega)P(dF)$, (forward), and $P(\omega+dF)=P(dF)P(\omega)$, (backward). Thus the immediate future state of the animal at $\omega+dF$ is the same whether the present state is taken at $\omega$ or at $dF$. Using the backward equation, $P(\omega+dF)-P(\omega)$ becomes $[P(dF)-I]P(\omega)$ where $I$ is the identity matrix. The matrix $P(dF)-I$ is the $P(dF)$ matrix with unity subtracted from each diagonal element, thereby yielding matrix $R$ after extracting the factor $BdF$ from $P(dF)-I$:

$$R=\begin{pmatrix} 0 & & & & & \\ 1 & -1 & & & & \\ & 2 & -2 & & & \\ & & \cdot & \cdot & & \\ & & & \cdot & \cdot & \\ & & & & a & -a \end{pmatrix}$$

thus giving the difference equation $P(\omega+dF)-P(\omega)=BRP(\omega)dF$. Dividing this equation by $dF$ and taking the limit as $dF$ approaches 0, yields the matrix differential equation $p'(\omega)=BRP(\omega)$, where $P'(\omega)$ is the derivative of $P(F)$ with respect to $F$ evaluated at the terminus of any segment $\omega$ of $F$ in the food consumed domain. This differential equation describes all the routes of growth of the animal on consuming $\omega$ units of food anywhere in the $F$ domain. Its solution with initial condition $P(0)=I$ leads to the distribution $\psi(\omega,k)$ of states $S(\omega)=k$ starting with $S(0)=i$. Here $i$ has the same meaning as $i$ in the previous section but it also denotes the state of the system before consumption of $\omega$ units of food.

## 3.4 Growth as the Solution of the Differential Equation

Section 3.7 gives details of finding the solution of the above differential equation. The solution is the set of transition probabilities, $p_{ik}(\omega)=\binom{i}{k}[\exp(-B\omega)]^k$ $[1-\exp(-B\omega)]^{i-k}$, where $\binom{i}{k}$ is the binomial coefficient. These transition probabilities state exact transitions from state $i$ to state $k$ on consuming $\omega$ units of food. The probabilities of interest here are those for which $i=a$ and $\omega=F$: $P_{ak}=\binom{a}{k}[\exp(-BF)]^k[1-\exp(-BF)]^{a-k}$. Now the distribution $\psi(F, K)$ defined by $P_{ak}$ is the binomial distribution $(p+q)^N$ in which $N=a$, $p=\exp(-BF)$ and $q=1-\exp(-BF)$, with the mean state $\bar{k}=Np$ on consuming $F$ units of food and variance $\sigma^2(k)=Npq$. Hence the mean and variance of $k$, the number of potential live weight units remaining after cumulative consumption of $F$ units of food, are respectively $\bar{k}=a\exp(-BF)$, and $\sigma^2(k)=a\exp(-BF)\,[1-\exp(-BF)]$.

Application of the conservation principle $\bar{m}=a-\bar{k}$ (item 5, Chap. 3.2), and the relational equations $W=b\bar{m}$, $A=ba$, $\sigma^2(W)=b^2\sigma^2(k)$ respectively, give the expected growth and standard deviation, namely

$$W=A[1-\exp(-BF)]+\sigma(W), \tag{3.1}$$

$$\sigma(W)=\{bA\exp(-BF)\,[1-\exp(-BF)]\}^{\frac{1}{2}}. \tag{3.2}$$

The first term on the right of Eq. (3.1) is the mean route of growing in the $(F, W)$ plane, the second term, $\sigma(W)$, is the inherent variability of growth given by Eq. (3.2).

Spillman (1924) empirically developed Eq. (3.1) from weights of pigs and steers and cumulative food consumed by them. Jull and Titus (1928), Titus et al. (1934), and Parks (1970 a) experimentally verified Spillman's suggestion on fowl and rats, and in Chap. 4 it is applied to species from mice to cattle.

## 3.5 The Inherent Error $\sigma$ ($W$) of Growth

Choosing $b, A$, and $B$ as 1,100 and (ln 2) respectively, Fig. 3.2 shows the expected growth curve [Eq. (3.1)] with the plus and minus one standard deviation curves [Eqs. (3.1) and (3.2)] to define the one standard deviation confidence region. The confidence region is seen to have maximum width $(bA)^{1/2}$ at the value of $F$ for which $\exp(-BF)=½$, and narrows to zero for both $F$ approaching 0 and $F$ approaching $\infty$. This region is expected to encompass about 68% of the $(F, W)$ data points for a single growing animal, for the above assumed values of $b, A$, and $B$, excluding measurement errors. The ratio $b/A=1/100$ is very exaggerated, but the values $A=100$ kg and $B=(\ln 2)$ are fairly realistic.

Cellular theory suggests that $b$ is of the order of mass of a cell, $10^{-12}$ kg (Conklin 1912 a, b). Hence $ba$ is of the order of $10^{-10}$ kg$^2$ for a 100 kg mature weight and $\sigma(W)$ $10^{-5}$ kg. If the cell plays the role of the biomass unit, detection of stochastic variability of $W$ would require measuring $W$ to about 1 part in $10^7$. It can be concluded that stochastic variability in animal experiments may never be of any prac-

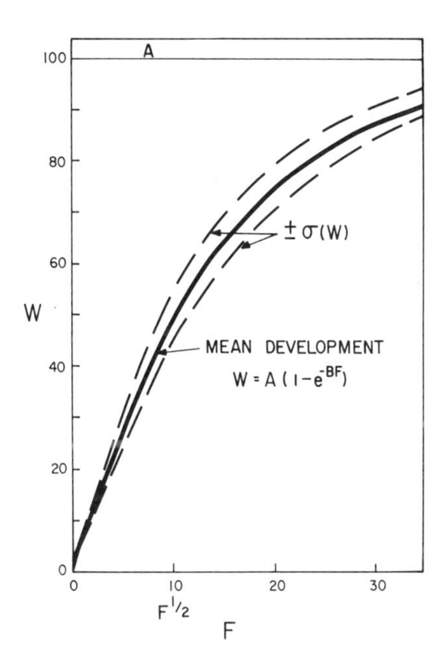


Fig. 3.2. Illustration of the stochastic mean live weight, $W$, as a function of cumulative food consumed, $F$, with the plus and minus one standard deviation curves for stochastic variability

tical importance (Heyde and Heyde 1971). In usual animal experiments the stochastic variability of $W$, namely $\sigma^2(W)$, is most likely not resolvable from measurement plus internal and external environmental variability of the animals live weight. However, animal science would be advanced if an experiment on a suitably chosen animal were designed in an attempt to resolve the stochastic variability. Even though the stochastic theory of growth in the food consumed domain may not be proven by experiment, it supports an old empirical formulation, namely the diminishing returns law.

The stochastic model of growth proposed here is an over simplification. The cells of an animal do not all continue to replicate. Nerve cells, for example, simply grow larger after the ground plan of the CNS has been laid. Conklin (1912 a, b) reported on much work centred around the question: what determines mature size; cell size, cell numbers or both? The conclusion was that there is no general biological law of increase in size by cell multiplication only. Further, the cells in an animal are not all the same size and there are tissues which increase by accretion only, i.e., tissues such as connective tissue, blood and fat deposits. Hence the stochastic model finds no support in assigning the role of biomass unit to the cell.

It is conceivable that the biomass unit is the equivalent of an aggregate of living matter co-operating as a whole to form the biomass unit, implying that $b$, the inherent live weight of the unit of growth, is very large compared to $10^{-12}$ kg. If this is so, it might be possible to show experimentally that the growth of a suitable animal, in a well regulated environment with ample supply of food, does not follow exactly the mean curve,

$$W = A[1-\exp(-BF)], \text{ and}$$

the variance $\sigma^2(W)$ might then be used to estimate $b$ by

$$b = \sigma^2(W)/\{A\exp(-BF)[1-\exp(-BF)]\},$$

and using the estimated value in conjunction with other biological facts to test the validity of the concept. Such an experiment would be a critical experiment for the proposed stochastic model.

The model has served well in at least one respect, namely leading to the examination of the experimental error $s(W)$ to be presented in Chap. 4. In following the consequences of the model, I was forced to consider $s(W)$. It will be shown that the variance of weight in the growth of a population of animals is not as unpredictable as most biologists seem to believe: $s(W)$ appears to be describable, in the food domain, by the same law of diminishing returns as applies to live weight, $W$. Further, it will be shown that the coefficient of variation of the live weight of a population of genotypes estimates the coefficient of variation of mature weights of the individuals in the population.

## 3.6 Role of Food Intake

The model affirms the stability of growth in the food consumed domain when conditions are maintained as uniformly as experimental technology permits.

Time is implicit in the model, growth in the time domain depending heavily on the feeding regime. This is confirmed by Lister and McCance (1967) who fed one group of weanling pigs on a good diet in the conventional manner, and fed a second group the same diet at a constant rate of 90 g/d (approximately maintenance intake) for a year after which the food intake was increased over a period of 1 to 2 weeks and free access to food allowed thereafter to maturity. During the year the undernourished pigs gained slightly, from about 5 to 6 kg.

Lister and McCance found that the mean growth of the group of rehabilitated pigs, in the time domain, is similar in shape to that of the group of normally fed pigs, but displaced by 1 year (Fig. 3.3), though they reach only about 80% of normal mature weight. Carcase analysis indicated that some irreversible physiological changes from normal were associated with the severe undernourishment and probably contributed to stunting of the animals.

Although undernourishment for a year is drastic treatment in the time domain, it does not appear drastic in the food consumed domain (Fig. 3.4). A normally fed pig needs about 1,080 kg of food to reach 90% of its mature weight. Feeding the undernourished pigs 90 g/day for a year (cumulatively 30 kg) maintained their

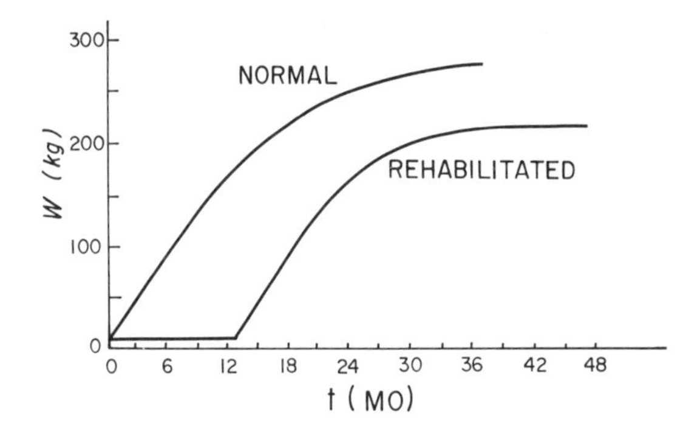


Fig. 3.3. Comparison of the live weights, $W$, as functions of time, $t$, of normally and severely undernourished pigs. The undernourished pigs were kept on a constant food intake for about a year prior to rehabilitation

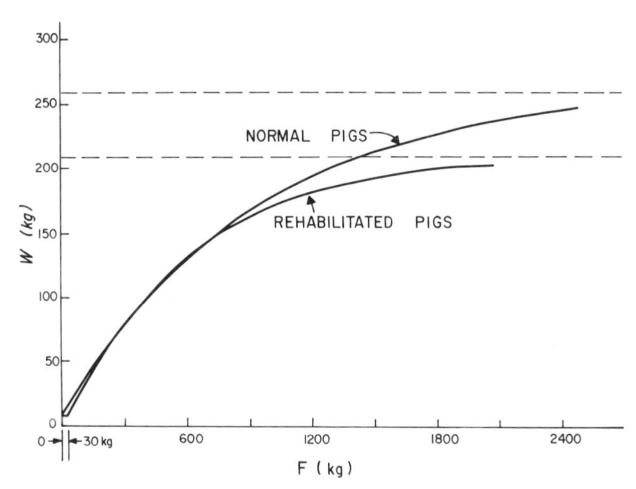


Fig. 3.4. Comparison of the growth of the normal and rehabilitated pigs as functions of cumulative food consumed, $F$. Here is seen the minor distortion of the growth curve of the undernourished pigs compared to the apparently major distortion in the time domain

weight at about 5 kg. On rehabilitation the animals fed and grew similarly to the normal pigs, except for the lower mean mature weight. Plotted on the same graph of $W$ vs $F$ there is minor displacement of the growth curves from each other.

If the undernourished pigs had received 90 g/day for a period of 1 month they would have consumed 2.7 kg cumulatively (a very small displacement) and on rehabilitation would almost follow the curve of normal growth. This has been observed in experiments on male and female rats undernourished for moderate periods of time, less than half the time required to grow from weanlings to young adults (Widdowson and McCance 1963, Widdowson et al. 1964). Figures 3.5, 3.6, and 3.7 show that moderate undernourishment results mostly in displacement of the growth curve in the time domain. Figure 3.5 shows there is less differences between the growth curves of normal and rehabilitated rats when undernourishment is carried on for 3 weeks compared to the effect of 8 weeks undernourishment as shown in Figs. 3.6 and 3.7. The recovery towards normal mature body weight is all but complete in both cases.

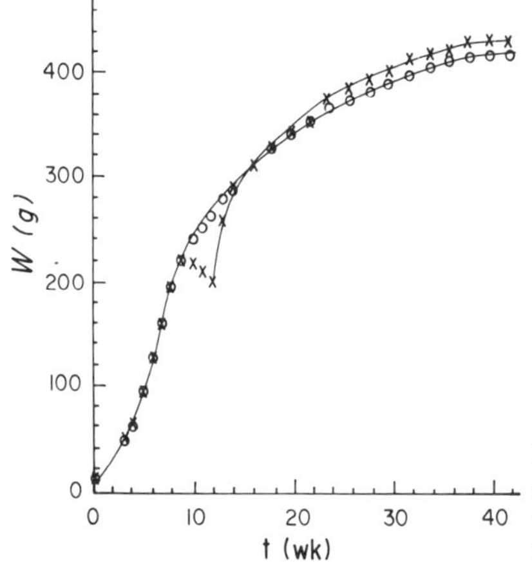


Fig. 3.5. Comparison of the growth of full-fed rats (○) and rats undernourished between 10 and 12 weeks old, age (X). [Reproduced from Fig. 3 of Widdowson and McCance (1963) with permission of the authors]

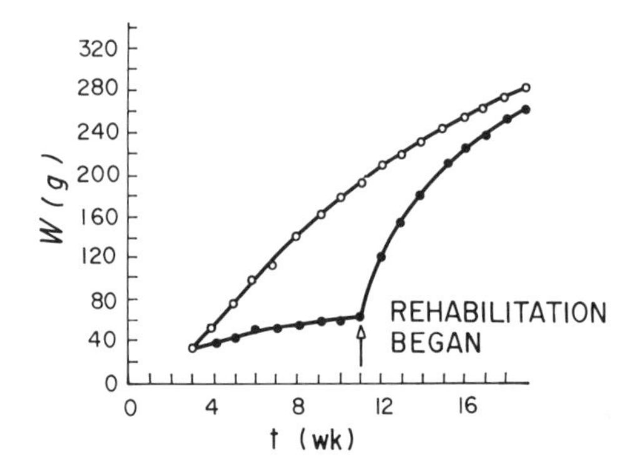


Fig. 3.6. Growth in time of normal (○) and undernourished (●) male rats. [Reproduced from Fig. 1 of Widdowson, Mavor, and McCance (1964) with permission of the authors and the Cambridge University Press]

The physiological changes of the undernourished pigs observed by Lister and McCance (1967) were extensive, thus casting serious doubt on assumption (3), Chap. 3.2, which is the basis of the concept that $B$ is constant over the range $0 \leqq F \leqq \infty$. However, before rejecting the stochastic model on the physiological basis, one should consider the following remarks of Widdowson (1968).

"Although rehabilitation is not strictly within the scope of this paper, it might be mentioned in conclusion that rehabilitation after even such severe undernourishment as was produced in the pigs is not only possible but easy and the animals grow very nearly, if not quite, to the size they would have been had they never been undernourished (Lister and McCance 1967). Their bodily proportions also became normal, as did the composition of their tissues. Careful measurement of the bones caused us to think that the bones of the rehabilitated animals do not grow to quite the same length as those of the well nourished ones, but the astonishing thing is that they very nearly do. Functionally the animals become normal too; they mate and produce good litters and the females lactate normally. There is no evidence of any effect of the year of undernutrition on the first generation, and the young, if fed normally, become full sized adults."

It is regrettable that these scientists did not record the data of cumulative food consumed so that graphs similar to Fig. 3.4 could be displayed. However, it is clear that environmental factors which moderately disturb the growth curve in the time domain will minimally disturb growth in the food consumed domain. The implication is that the law of diminishing returns, $W = A[1 - \exp(-BF)]$, is a very stable function compared to the arbitrariness of feeding regimes. For this reason a search for stochastic models of animal growth in the time domain may be unfruitful. Here

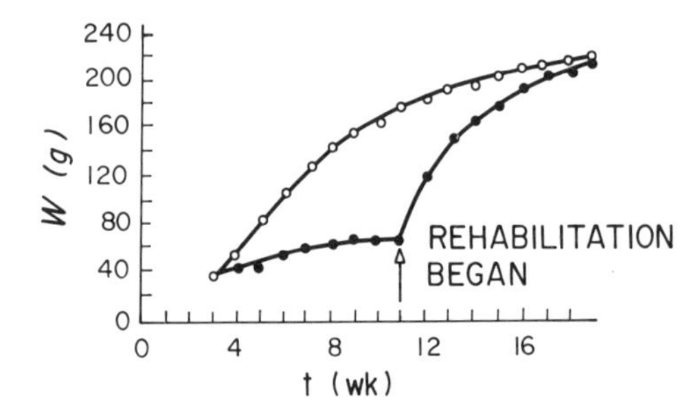


Fig. 3.7. Growth in time of normal (○) and undernourished (●) female rats. [Reproduced from Fig. 2 of Widdowson, Mavor, and McCance (1964) with permission of the authors and the Cambridge University Press]

$B$ has been shown to have a probabilistic meaning, namely it is the probability of a conversion of a potential unit to a real unit of live weight per unit of food consumed. $B$ also has an interesting relationship with Spillman's $R$ and the mature live weight $A$ (Chap. 2.1). There $B$ was set equal to $\ln(1/R)$ and Fig. 2.1 shows $A = \Delta Y(1)/(1-R)$. This results in $R = e^{-B}$ and with $B$ being small $1-R \simeq B$, thereby giving the very interesting result $AB \simeq \Delta Y(1)$.

The product $AB$ is therefore the expected gain in weight on the first unit of food consumed. This statement has profound implications to be discussed in Chap. 4.

## 3.7 Solving the Differential Equation

The differential equation characteristic of the stochastic process of growth in the food consumed domain is $P'(\omega) = BRP(\omega)$, where $\omega$ is any segment of the cumulative food consumed interval, $0 \leqq F \leqq \infty$, beginning at $F_0$ and ending at $F_0 + \omega$. Since it is a first order differential equation in the independent matrix $P(\omega)$, an initial value $P(0)$ is required for the solution. At $F = F_0\,(\omega = 0)$ it is certain that the animal has $i$ potential units yet to be converted, where $0 \leqq i \leqq a$, so that $P(0)$ is the identity matrix. Solution of the equation with this initial condition will yield the probabilities $P_{ij}(0 \leqq j \leqq i)$ as functions of $\omega$, given the animal has $i$ potential units remaining at $\omega = 0$.

The solution of the differential equation uses probability generating functions, p.g.f. ($s$) (Feller 1968, p. 264). The p.g.f. ($s$) is the sequence $1, s, s^2, \ldots, s^n, \ldots$ running as far as needed. Post multiplying the differential equation by the p.g.f. ($s$) as a column vector

$$\{s^{k-1}\},\ k = 1, 2, \ldots, a+1, \text{ gives}$$

$$P'(\omega)\{s^{k-1}\} = BRP(\omega)\{s^{k-1}\}.$$

Representing $P(\omega)\{s^{k-1}\}$ by column vector $\{g(\omega,s)\}$ and $P'(\omega)\{s^{k-1}\}$ by $g'(\omega,s)$, yields Eq. (3.3):

$$g'(\omega,s) = BRg(\omega,s). \tag{3.3}$$

From matrix $R$ (see Chap. 3.3) it is apparent that the nth component of Eq. (3.3) is the recursive equation

$$g'_n(\omega, s) = (n-1)B[g_{n-1}(\omega, s) - g_n(\omega,s)] \tag{3.4}$$

reducing the problem to that of solving $a+1$ linear differential equations with initial conditions $\{g(0,s)\} = I\{s^{k-1}\} = \{s^{k-1}\}$, the $n$th component of which is $g_n(0,s) = s^{n-1}$.

Because $n-1$ appears as a factor in Eq. (3.4) and the first component of $\{g(\omega,s)\}$ is 1, $g_n(\omega,s) = [h(\omega,s)]^{n-1}$ is chosen as a trial solution. Substitution yields

$$(n-1)[h(\omega,s)]^{n-2}h'(\omega,s) = (n-1)B\{[h(\omega,s)]^{n-2} - [h(\omega,s)]^{n-1}\},$$

from which the following differential equation is found,

$$h'(\omega,s) = B[1 - h(\omega,s)],$$

with initial condition $h(0,s)=s$ obtained from $g_n(0,s)=[h(0,s)]^{n-1}$. Integration between limits $[F_0, h(0,s)]$ and $[F_0+\omega, h(\omega,s)]$ gives $h(\omega,s)=1-\exp(-B\omega)+s\exp(-B\omega)$.

Hence the solution of Eq. (3.4) is

$$g_n(\omega, s)=[1-\exp(-B\omega)+s\exp(-B\omega)]^{n-1}$$
$$=\sum_{j=0}^{n-1}\binom{n-1}{j}[1-\exp(-B\omega)]^{n-1-j}[\exp(-B\omega)]^j s^j, \tag{3.5}$$

on expanding the right hand by the binomial theorem. Since $g_n(\omega, s)$ is the $n$th component of the vector $P(\omega)\{s^{k-1}\}$ it follows that $g_n(\omega, s)$ is also given by

$$g_n(\omega, s)=\sum_{j=0}^{n-1} p_{n-1\,j}(\omega)s^j. \tag{3.6}$$

Since $n$, running from 1 to $a+1$, corresponds to $i$, running from 0 to $a$, we may take $i=n-1$ at $F_0$. Equating coefficients of like powers of $s$ in Eq. (3.5) and (3.6), the transition probabilities, $P_{ij}(\omega)$ are

$$P_{ij}(\omega)=\binom{i}{j}[1-\exp(-B\omega)]^{i-j}[\exp(-B\omega)]^j,\ 0\leq j\leq i,$$

which are the coefficients of the binomial distribution of $j$ for given $i$ with mean $\bar{j}=i\exp(-B\omega)$ and variance $\sigma^2(j)=i\exp(-B\omega)[1-\exp(-B\omega)]$ applying on any segment, $F_0$ to $F_0+\omega$ of the food consumed domain. Since at $F_0=F=0$ the animal has $a$ potential units remaining, the required transition probability for development of the animal from $a$ to $k$ potentials units on consuming $F$ units of food is

$$p_{ak}(F)=\binom{a}{k}[1-\exp(-BF)]^{a-k}[\exp(-BF)]^k,$$

with mean $\bar{k}=a\exp(-BF)$, and variance $\sigma^2(k)=a\exp(-BF)[1-\exp(-BF)]$.

Bartholomay (1958a b, 1964, 1968) and others (Doob 1953; Feller 1968) solved differential Eq. (3.3) in studying monomolecular chemical reactions and disintegration of radioactive elements in the time domain. Here the problem is formulated in terms of stochastic growth of an animal in the food consumed domain, yielding a probabilistic meaning for the parameter $B$.

## References

Bartholomay AF (1958a) Stochastic models for chemical reactions I. Theory of the unimolecular reaction process. Bull Math Biophys 20:175–190

Bartholomay AF (1958b) On linear birth and death processes of biology as Markoff chains. Bull Math Biophys 20:97–118

Bartholomay AF (1959) Stochastic models for chemical reactions II. The unimolecular rate constant. Bull Math Biophys 21:363–371

Bartholomay AF (1964) The general catalytic queue process. In: Gurland J (ed) Stochastic models in medicine and biology. Univ Wisconsin, Madison

Bartholomay AF (1968) Some general ideas on deterministic and stochastic models of biological systems. In: Locker A (ed) Quantitative biology of metabolism. Springer, Berlin Heidelberg New York

Conklin EG (1912) Body size and nuclear size. J Exp Zool 12:1–98

Doob JL (1968) Stochastic processes. Wiley, New York
Feller W (1968) An introduction to probability theory and its application, 3 rd edn. Wiley, New York
Gurland J (1964) Stochastic models in medicine and biology. Univ Wisconsin Press, Madison
Heyde CC, Heyde E (1969) A stochastic approach to a one substrate one product enzyme reaction in the initial velocity phase. J Theor Biol 25:159–172
Heyde CC, Heyde E (1971) Stochastic fluctuations in a one product enyzme system: are they ever relevant? J Theor Biol 30:395–404
Jull MA, Titus HW (1928) Growth of chickens in relation to feed consumption. J Agric Res 36:541–550
Lindsay RB (1941) Physical statistics. Wiley, New York
Lister D, McCane RA (1967) Severe undernutrition in growing and adult pigs. Br J Nutr 21:787–799
Mendel JL (1965) Statistical analysis of nonlinear response functions with application to genetics of poultry growth. PhD Thesis, Univ Michigan
Moore P (1973) Watchers of the stars. Michael Joseph Ltd, Milano
Parks JR (1970 a) Growth curves and the physiology of growth. I. Animals. Am J Physiol 219:833–836
Parks JR (1970 b) Growth curves and the physiology of growth. II. Effects of dietary energy. Am J Physiol 219:837–839
Parks JR (1970 c) Growth curves and the physiology of growth. III. Effects of dietary protein. Am J Physiol 219:840–843
Parks JR (1971 a) The effect of ambient temperature on the thermochemical efficiency of growth of cold-acclimating rats. Am J Physiol 220:578–582
Parks JR (1971 b) Growth curves and the physiology of growth. IV. The effect of dietary methionine. Am J Physiol 221:1845–1848
Schrodinger E (1967) What is life? Cambridge Univ Press, Cambridge
Spillman WJ, Lang E (1924) The law of diminishing increment. World, Yonkers
Thompson D (1917) On growth and form. Cambridge Univ Press, Cambridge
Titus HW, Jull MA, Hendricks WA (1934) Growth of chickens as a function of the feed consumed. J Agric Res 48:817–835
Widdowson E (1968) The effect of growth retardation on postnatal development. In: Lodge GA, Lamming GE (eds) Growth and development of mammals. Plenum Press, New York
Widdowson E, McCance RA (1963) The effect of finite periods of undernutrition at different stages on the composition and subsequent development of the rat. Proc R Soc London Ser B 158:329–342
Widdowson E, Mavor WO, McCance RA (1964) The effect of undernutrition and rehabilitation of the development of the reproductive organs of the rat. J Endocrinol 29:119–126
Zotin AA, Zotina RS (1972) Towards a phenomenological theory of growth. J Theor Biol 35:213–225
Zotin AA, Grudnitzky VA, Shagimordanov NS, Terentieva NV (1971) Growth equation with allowances made for interaction of growth and differentiation. Wilhelm Roux' Arch Entwicklungsmech Org 168:169–173

# Chapter 4 Treatment of Ad Libitum Feeding and Growth Data

This chapter outlines the various procedures I have used on various classes of long term growth and feeding data. Because the experiments were done for reasons other than quantitative analysis of the animals' feeding and growth, published data fell into five classes, defined below. The classes determine which of the feeding and growth functions are best suited to determine the values of the growth parameters appropriate to the data. The growth and ad libitum feeding functions are repeated here for convenience.

1. Ad libitum feed rate
   $$dF/dt=(C-D)[1\text{-exp}(-t/t^*)+D \qquad (4.1)$$
2. Cumulative food consumed [integral of Eq. (4.1)]
   $$F=C\{t-t^*(1\text{-}D/C)[1\text{-exp}(-t/t^*)]\} \qquad (4.2)$$
3. Live weight versus cumulative food consumed
   $$W=(A-W_0)\{1\text{-exp}[-(AB)F/A]\}+W_0 \qquad (4.3)$$
4. Live weight versus time [Eq. (4.2) substituted in Eq. (4.3)]
   $$W=(A-W_0)[1\text{-exp}\langle-(AB)C/A\{t-t^*(1\text{-}D/C)[1\text{-exp}(-t/t^*)]\}\rangle]+W_0\,. \qquad (4.4)$$

The reader will have noticed in Eqs. (4.3) and (4.4) the substitution of $(AB)/A$ for $B$. This was done because the number designated by $(AB)$, which from Chap. 3 is the expected gain on the first unit of food consumed, has been found very constant across species for which the mature live weight $A$ lies in the range of about 0.05 to 1,000 kg. Hence, it is more appropriate to consider $(AB)$ as a growth parameter than $B$ alone. Evidence for the near constancy of $(AB)$ across species is collected in Table 4.1.

Readers who are not interested in the methods of reducing data may skip to Chap. 4.2. However I suggest they examine the Figures in Chap. 4.1 to satisfy themselves as to how well the calculated (solid) curves track the examples of the five classes of data.

## 4.1 Nonlinear Regression – Method and Results by Class of Data

It has been remarked in Chap. 2 that Titus, Jull, and Hendricks realized Eq. (4.3) could not be put into a form linear in the parameters $A$ and $(AB)$ for application of the then well-known linear regression techniques. They developed an iterative technique, starting with approximate values of the parameters, for finding values

Table 4.1. Summary of the mean values of *A*, (*AB*), *C* and $t^*$ obtained by methods outlined in Chapter 4.1 assuming all the data collected in part A of the Appendix are from ad libitum feeding experiments. Table number refers to original data in Appendix and Chapter 4.1

| Animal | *A* kg | (*AB*) | *C* kg/wk | $t^*$ wks | Table No. |
|---|---|---|---|---|---|
| Steers | | | | | |
| (Spillman 1924) | 1,000.0 | 0.235 | – | – | – |
| Cattle | | | | | |
| (Brody 1945, p. 51) | | | | | |
| Jersey | 369.0 | 0.277 | 77.0 | 104.0 | A-1 |
| Holstein | 502.0 | 0.262 | 117.0 | 168.0 | A-1 |
| (Monteiro 1974) | | | | | |
| Jersey | 474.0 | 0.176 | 58.0 | 23.0 | A-2 |
| Friesian | 714.0 | 0.190 | 76.0 | 25.0 | A-3 |
| Pigs | | | | | |
| (Headley et al. 1961) | 380.0 | 0.409 | 61.0 | 8.7 | A-5 |
| (Spillman 1924) | 500.0 | 0.247 | – | – | |
| Dog | | | | | |
| (A. Hedammar 1972) | | | | | |
| Male Great Dane | | | | | |
| Ad libitum | 59.0 | 0.50 | 8.4 | 8.4 | A-6 |
| 67% ad libitum | 48.0 | 0.51 | 6.0 | 10.0 | A-7 |
| Chickens | | | | | |
| (Hammond et al. 1938) | | | | | |
| Dietary protein | | | | | |
| Male ad libitum | | | | | |
| 25% | 3.31 | 0.312 | 0.956 | 7.8 | A-8 |
| 23% | 3.26 | 0.385 | 1.14 | 15.4 | A-8 |
| 21% | 3.21 | 0.391 | 1.23 | 18.3 | A-8 |
| 19% | 3.17 | 0.376 | 0.98 | 12.4 | A-8 |
| 17% | 3.20 | 0.354 | 1.06 | 13.8 | A-8 |
| 15% | 3.25 | 0.337 | 0.950 | 11.9 | A-8 |
| 13% | 3.33 | 0.164 | 1.21 | 19.7 | A-8 |
| 70% ad libitum | | | | | |
| 25% | 2.87 | 0.465 | 0.686 | 10.8 | A-9 |
| 23% | 2.74 | 0.434 | 0.686 | 10.8 | A-9 |
| 21% | 2.72 | 0.430 | 0.686 | 10.8 | A-9 |
| 19% | 2.66 | 0.389 | 0.686 | 10.8 | A-9 |
| 17% | 2.73 | 0.401 | 0.686 | 10.8 | A-9 |
| 15% | 2.89 | 0.352 | 0.686 | 10.8 | A-9 |
| 13% | 3.01 | 0.267 | 0.686 | 10.8 | A-9 |
| (Card 1952) | | | | | |
| Male 1930 diet | 4.26 | 0.362 | | | A-10 |
| 1960 diet | 4.24 | 0.644 | | | |
| (Emmans 1974) | | | | | |
| Female broiler | 3.81 | 0.382 | 0.989 | 4.19 | A-4 |
| Rat | | | | | |
| (Parks 1970) | | | | | |
| Male | | | | | |
| Percent relative specific energies (RSE) | | | | | |
| 100% | 0.431 | 0.390 | 0.175 | 2.5 | A-11 |
| 90% | 0.434 | 0.370 | 0.194 | 2.6 | A-11 |

Table 4.1 (continued)

| Animal | *A* kg | (*AB*) | *C* kg/wk | t* wks | Table No. |
|---|---|---|---|---|---|
| 80% | 0.391 | 0.330 | 0.208 | 2.2 | A-11 |
| 70% | 0.383 | 0.294 | 0.236 | 2.9 | A-11 |
| 36% dietary protein | 0.426 | 0.500 | 0.153 | 1.5 | A-12 |
| 25% | 0.439 | 0.497 | 0.153 | 1.4 | A-12 |
| 20% | 0.438 | 0.473 | 0.156 | 1.6 | A-12 |
| 14% | 0.429 | 0.400 | 0.154 | 1.4 | A-12 |
| Individual rats | | | | | |
| (Parks, unpublished) | | | | | |
| Male E31 | 0.338 | 0.458 | 0.132 | 1.4 | A-13 |
| D23 | 0.364 | 0.484 | 0.132 | 3.4 | A-13 |
| A22 | 0.433 | 0.525 | 0.154 | 2.0 | A-13 |
| A24 | 0.451 | 0.414 | 0.168 | 1.8 | A-13 |
| D10 | 0.467 | 0.480 | 0.161 | 1.8 | A-13 |
| (Mayer and Vitale 1957) | | | | | |
| Male 60% dietary protein | 0.431 | 0.609 | 0.126 | 1.6 | A-14 |
| 25% | 0.422 | 0.618 | 0.126 | 1.6 | A-15 |
| 10% | 0.506 | 0.335 | 0.126 | 1.6 | A-14 |
| Mouse | | | | | |
| (Bateman and Slee 1973) | | | | | |
| Male | 0.039 | 0.456 | 0.044 | 0.81 | A-16 |
| Female | 0.032 | 0.362 | 0.039 | 1.0 | A-16 |
| (Timon and Eisen 1970) | | | | | |
| High male | 0.038 | 0.428 | 0.046 | 0.8 | A-17 |
| Control male | 0.033 | 0.410 | 0.044 | 1.1 | A-18 |
| High female | 0.032 | 0.308 | 0.042 | 0.8 | A-19 |
| Control female | 0.028 | 0.301 | 0.044 | 0.7 | A-20 |
| (Rutledge 1971) | | | | | |
| Male | 0.036 | 0.410 | 0.050 | 0.4 | A-21 |

of the parameters which minimize the sum of squares of the residuals. This technique has been further developed, improved and programmed for computers. There are many modern modifications of this more sophisticated version of the method Titus et al. developed for use with a desk calculator and tables of exponential and logarithmic functions. These modifications are referred to generally as least squares nonlinear regression techniques for finding best values of the parameters in a very general class of functions, such as the feeding and growth equations, which are not linear in the parameters. Most computer centres have a version of a nonlinear regression program. The version I have used, not by preference but because it was available, is named NLIN.

For details of the procedure and statistics of nonlinear regression the reader would do well to consult Chap. 10 of Draper and Smith (1967).

NLIN is based on a method first described and programmed by Marquardt (1964) and updated by Middleton, Fulton, and Usanis on the staff of the Computer Center of the University of North Carolina at Raleigh in 1972. The program can provide the user with much output beside the best values of the parameters and the residual standard deviation of fit of the function to the data. The amount of output is under the control of the user who supplies either a Fortran or Pl/1 program of the function to be fitted to the data. Initial estimates of the values of $A$, $(AB)$, $C$, and $t^*$ must also be supplied. Since the values differ by orders of magnitude they should, for best results, be scaled to the same numerical values with the scaling factors entered in the regression equations as constant factors. The regression equation, or equations, selected from Eqs. (4.1) to (4.4) depend on the kind of data under study.

I have placed the published data in five classes for convenience of displaying the forms of Eqs. (4.1) through (4.4) which are applicable to the various kinds of data. Each class will be defined with an example of the kind of data and the appropriate equation to be used and the results of regression. Most of the data are population averages of live weight and food consumed over time; there are a few for individual animals. The tables of data are all compiled in Appendix A.

Class I data are periodic average live weights and food consumptions during the periods of time and hence are representable as tables of data points $(i, f_i, W_i)$. Hence $i$ is the number of the period; $f_i$ is the food consumed during the ith period and $W_i$ is the weight of the animal at the end of the $i$th period. The duration of a period is an integral number, $n$, of units of time; days, weeks, or months.

Since $F_i$ is the cumulative food consumed up to and including the ith period the reader may be tempted to sum the $f_i$ data to get $F_i$ directly. This is a manoeuvre for approximate work with such data, but to get best results a manoeuvre of this type should be avoided since $\sum_{j=1}^{i}(F_j - F_{j-1}) = F_i$ accumulates the experimental errors encountered in each of the periods from 1 to $i$.

Equation (4.2) can be used to develop, as follows, the regression function applicable to $f_i = F_i - F_{i-1}$ data. Since $F_i = Ct_i - Ct^*(1\text{-}D/C) + Ct^*(1\text{-}D/C) \cdot \exp(-t_i/t^*)$, and $-F_{i-1} = -Ct_{i-1} + Ct^*(1\text{-}D/C) - Ct^*(1\text{-}D/C)\exp(-t_{i-1}/t^*)$, adding them term by term yields

$$F_i - F_{i-1} = C\langle (t_i - t_{i-1}) + t^*(1\text{-}D/C)\{1\text{-}\exp[(t_i - t_{i-1})/t^*]\}\exp(t_i/t^*)\rangle .$$

But $F_i - F_{i-1} = f_i$, $t_i - t_{i-1} = n$ and $t_i = ni$, so that Eq. (4.5) becomes the regression equation for finding values of $C$, $t^*$ and $D$ from Class I ad libitum feeding data, where $n$ = number of units of time per period, i.e.

$$f_i = C\{n + t^*(1\text{-}D/C)[1\text{-}\exp(n/t^*)]\exp(-ni/t^*)\} . \tag{4.5}$$

The equation for $W_i$ is obtained by substituting $ni$ for $t$ in Eq. (4.4), to get

$$W_i = (A - W_0)[1\text{-}\exp\langle -[(AB)C/A]\{ni - t^*(1\text{-}D/C)[1\text{-}\exp(-ni/t^*)]\}\rangle] + W_0 . \tag{4.6}$$

Although Eq. (4.5) can be used with the $(i, f_i)$ data to determine $C$, $D$, and $t^*$ the values of which can be subsequently substituted in Eq. (4.6) for use with $(i, W_i)$ data to determine $A$ and $(AB)$, I recommend that Eqs. (4.5) and (4.6) be used together to determine $A$, $(AB)$, $C$, $D$, and $t^*$ in one pass of the $(i, f_i, W_i)$ data through NLIN or any nonlinear regression programm available to the reader. The duty programmer in the Computer Centre can help in developing a program for the determination of these parameters in a single pass through all the $(i, f_i, W_i)$ data.

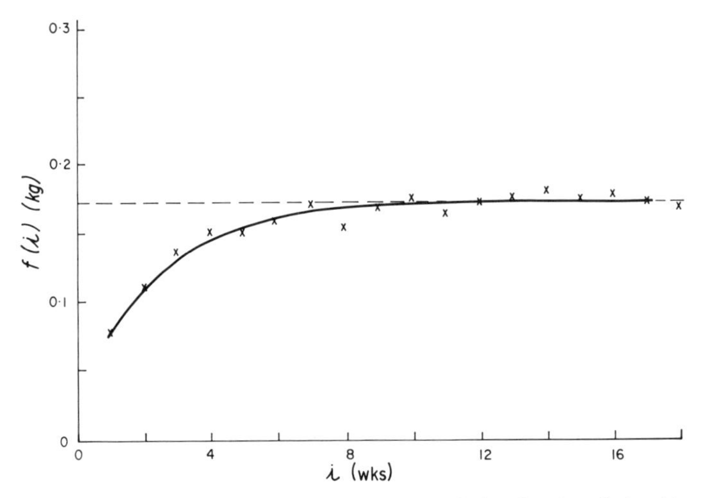


Fig. 4.1. An example of Class I $(i,f_i)$ data. A graph showing the relationship of the calculated curve to the data of average weekly food consumed by Sprague-Dawley male albino rats versus time since weaning

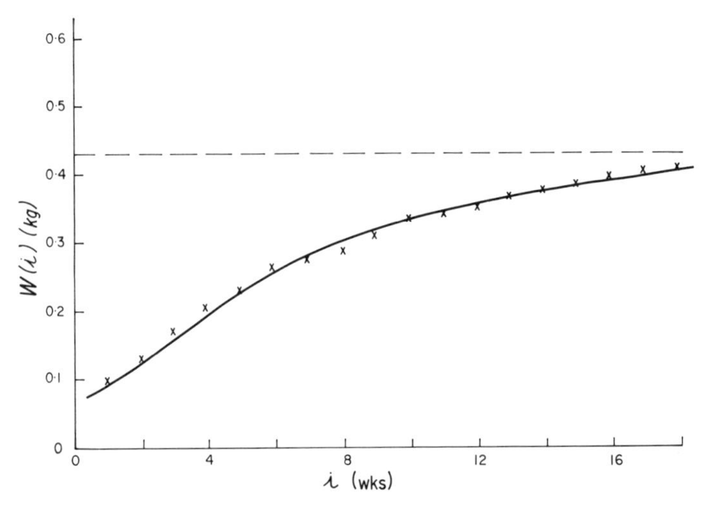


Fig. 4.2. An example of Class I $(i, W_i)$ data. Graph showing how the calculated growth curve tracks the average live weight versus time since weaning of Sprague-Dawley male albino rats

The following is an example of Class I data and the results of regression. Table A-11 (Appendix A, relative energy density, RSE = 100%) contains mean weekly, $n=1$, $(i,f_i, W_i)$ data from a group of 25 Sprague-Dawley albino male rats individually caged and fed for 20 weeks from weaning on a diet consisting solely of a commercial ground meal (LAB-BLOX; Allied Meals, Inc.). These rats were one group of four grown simultaneously on meal blended with various percentages of precipitated silicic acid to reduce the specific metabolisable energy of the diet without changing the bulk density. This experiment is discussed in detail in Chap. 9. The results of the regression of the data in Appendix Table A–11 are $A=$

0.431 kg, $(AB)=0.390$, $C=0.175$ kg/wk, $D=0.056$ kg/wg, and $t^*=2.45$ wk; rsd of fit $=0.008$ kg.

Figure 4.1 displays how well the calculated curve tracks the $(i,f_i)$ data of food consumed weekly. It also illustrates how the data may be used to obtain first estimates of $C, D$, and $t^*$ for entry in the nonlinear regression program. Most important, it shows the deviations of the experimental data from the expected curve thus indicating the weakness of examining biological data solely by periods, i.e., implying phases of growth, rather than using regression techniques to subsume the total variation of the dependent variable over all periods into the parameters of an appropriate function.

Figure 4.2 shows the fit of the calculated growth curve to the mean live weight data of the 25 rats and the systematic undulations of the data points about the calculated curve. Brody (1945, p. 548) has cautioned against placing undue emphasis on the systematic deviations; they may reflect the averaging process rather than some underlying phenomenon of growth. These systematic deviations will be considered in Chap. 7; the important point is that the calculated curve tracks the major variation of live weight during growth and shows the S-shape usually taken as characteristic of growth.

Class II data are $[i, (dF/dt)_i, W_i]$ as shown in Appendix Table A–1. Brody (1945, p. 51) used these data from Jersey cattle along with comparable data from Holsteins to illustrate the monotonic decrease of the gross efficiency of growth of these cattle from about 35% at age 1 month to about 5% at 21 months. The manner in which $dW/dF$ decreases with age is studied in detail in Chap. 9. These data are not mean sequential measurements of food intake and body weight of one group of Jersey cattle; they are composite data considered by Brody consistent with the growth of Jersey and Holstein cattle. However, in Chap. 5 I submit reasons why I think the feed rate $(dF/dt)_i$ data from Jersey cattle and from Holstein cattle are inconsistent with the $W_i$ data for ad libitum fed animals. Tables A–2 and A–3 (Appendix A) contain more recent data on Jersey and Friesian cattle, respectively, obtained at The A.R.C. Animal Breeding Research Organisation, Edinburgh, Scotland. These data are Class I and are sequentially and accurately taken and are consistent with ad libitum feeding and growth as discussed in Chap. 5.3.

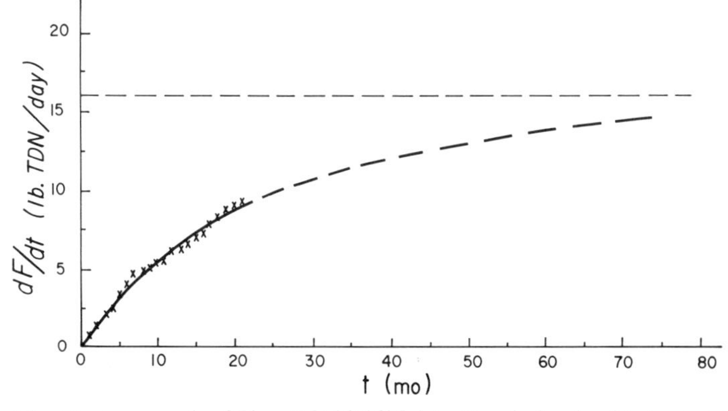


Fig. 4.3. An example of Class II $[i, (dF/dt)_i]$ data. Graph showing fit of calculated curve to data of feed rate versus age for Brody's Jersey cattle

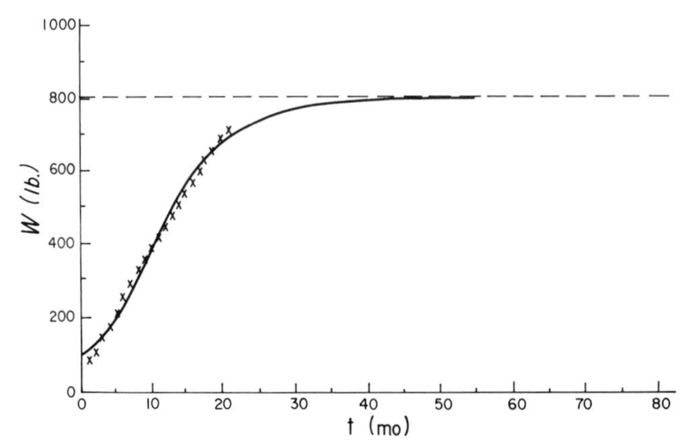


Fig. 4.4. An example of Class II $(i, W_i)$ data. Graph of live weights versus age of Brody's Jersey cattle showing how the calculated curve tracks the data

Class II data also can be put through the nonlinear regression program in one pass in the same manner as Class I data. For Brody's Jersey cattle the results of regression are $A = 812$ lbs, $(AB) = 0.456$, $C = 16.0$ lbs (TDN)/day and $t^* =$ 26 months with the rsd = 18 lbs. Figures 4.3 and 4.4 illustrate how well the calculated curves track the data points $[i, (dF/dt)_i, W_i]$ of the Jersey cattle in Table A–1 (Appendix A).

Class III data are listings of $(F, W)$ data points. Table A–10 is an example. The FORTRAN version of Eq. (4.3) is used in studying data of this type. These data were used by Card (1952, p. 242) in partial support of the improved growth of chicks on present feeds. It is not intended to discuss this matter here, but will be studied in Chap. 9 where effects of diet are analysed.

The data are for male progeny of crossbred Rhode Island Red x Barred Plymouth Rock chickens. The first estimates of $A$, $W_0$, and $(AB)$ are respectively 4 kg, 0.035 kg, and 0.5. The value for $A$ is taken from Brody (1945, Table 16.1, p. 568) and that for $W_0$ is the commonly accepted hatching weight for chickens. Brody's Table 16.1 is useful for obtaining estimates of mature weight $A$ for animals from mice to beef cattle but it supplies no information about the mature feeding characteristics $C$ of these animals. The final values of the growth parameters were $A =$ 4.195 kg, $(AB) = 0.3665$ kg and the rsd = 0.00033 kg. Figure 4.5 shows the very close fit of Eq. (4.3) to the data. Fits of this precision are rarely encountered when raw data are studied. Undoubtedly Card's (1961, p. 242, curve B) data were read from diminishing returns curves of the type found by Titus, Jull, and Hendricks in their studies during the 1930's.

Class IV data are listings of $(t, W)$ data points. Table A–21 (Appendix A) is an example of mean live weight versus age data for some white mice. Graphs of such data have long been considered "the experimental growth curve," as discussed in Chap. 1. Equation (4.4) is used as the regression equation applicable to $(t, W)$ growth data, but it must be rewritten as Eq. (4.7). Since the parameters $C$ and $(AB)$ are no longer separately determinable from the data, their product is defined as another parameter, $a$.

$$W = (A - W_0)[1 - \exp\langle -(a/A)\{t - t^*[1 - \exp(-t/t^*)]\}\rangle] + W_0 . \tag{4.7}$$

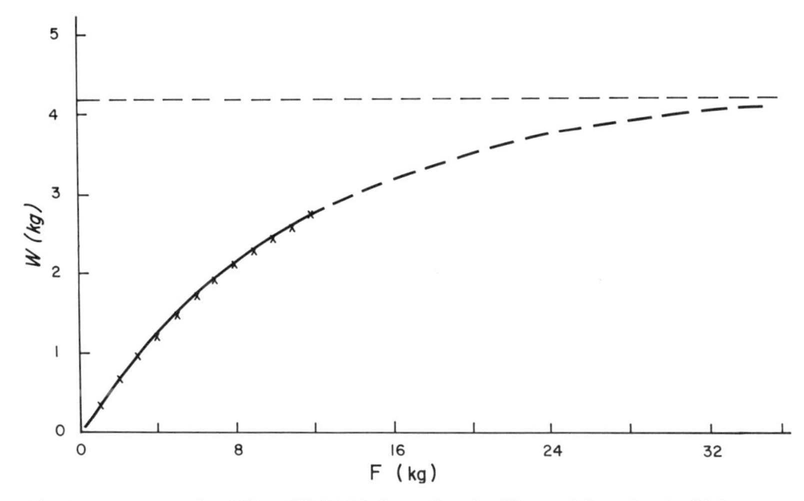


Fig. 4.5. An example of Class III ($F$, $W$) data. Graph of live weights of male chickens versus cumulative food consumed. (Data by Card 1952)

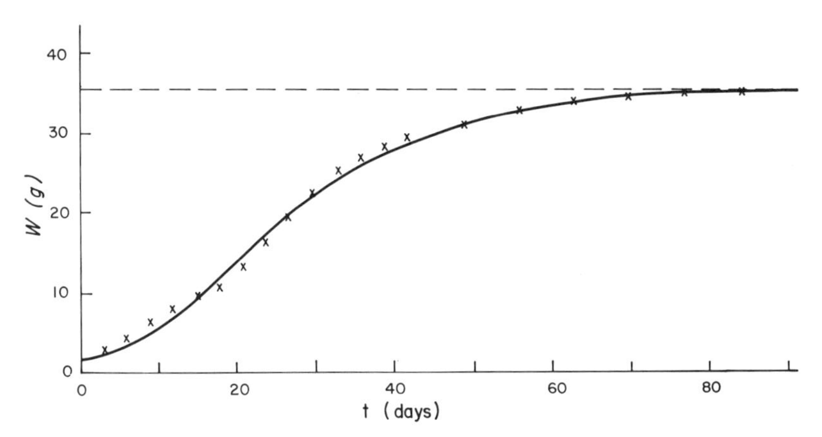


Fig. 4.6. An example of Class IV ($t$, $W$) data. Graph showing how the calculated curve tracks the weight of mice versus age from birth

Using the FORTRAN version of Eq. (4.7) on the data (Appendix A Table A–21) gives the following results, $A = 35.29$ g, $a = 5.0$ g/day, and $t^* = 5.9$ days. Because the ratio $a/A$ is Brody's $k$, its value is calculated as 0.14 days$^{-1}$. Figure 4.6 shows how the calculated curve tracks the data. The fit is not very good. The large positive deviations in the period between 0 and about 15 days of age are probably due to the diet during this period being predominantly mother's milk with more efficient use of the nutrients for growth. The application of Eq. (4.7) to data gathered from birth violates the assumption that the type of food is kept constant during the observations of growth.

It is apparent that the data considered for regression should have begun when the mice were weaned to meet the assumption of food of constant composition. In this case $W_0$ may be fixed at 12.87 g and time = 21 days. The final values of $A, a,$

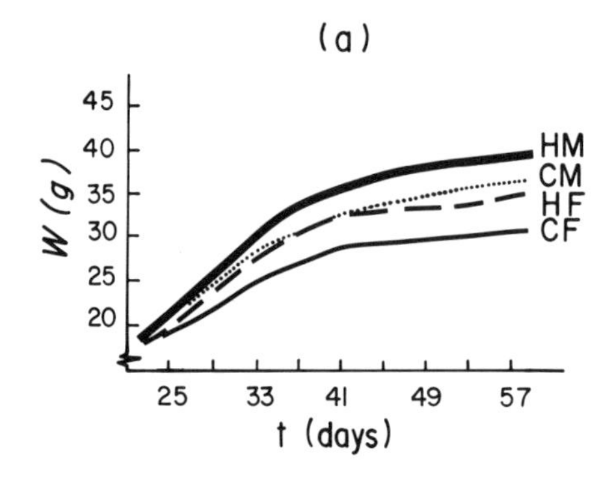


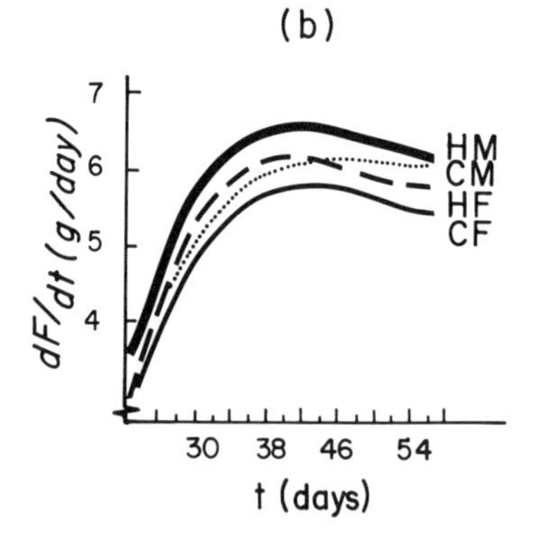


Fig. 4.7(a, b). A typical source of Class V data. Graphs of weight (a) and feet rates (b) versus age of mice since weaning. *HM* and *HF* are males and females of the ninth generation selected for increased gain in 21 days postweaning. *CM* and *CF* are of the ninth generation randomly selected. [Reproduced from Figs. 1 and 2a of Timon and Eisen (1970) with permission of the authors and Genetics Society publisher]

and $t^*$ quoted previously could be the new initial estimates. However, Eq. (4.7) as it stands cannot be used, since it does not have the term $D/C$ as required by Eq. (4.4). At weaning the initial feed rate $D$ is not negligible relative to $C$. Equation (4.4) can be used by regarding $(AB)C/A$ as Brody's $k$ and $(D/C)$ as a parameter to be determined by regression. This is one of the weaknesses of growth data which do not include measures of food consumed, i.e., data consisting of the usual measures of $W$ versus $t$ only.

Class V data are read as accurately as possible from published graphs, such as shown in Fig. 4.7, and are treated in this case as Class II data in the regression analysis. Figure 4.7 is a reproduction of graphs of Timon and Eisen (1970) These graphs were transferred to coordinate paper having 10 graduations per inch, and scale factors of weight, food intake and age were determined. Coordinates of points on the curves were read at two graduation intervals along the age axes, and estimation of the ordinates of the curves was to 0.2 of a graduation. Using these estimates and the scale factors, tables of live weight in grams and feed rate in grams per day versus age after weaning were produced. Table A–17 (Appendix A) lists the data arrived at in this way for the High Males (HM). I realise the imprecision of data obtained in this rather primitive fashion, but frequently this is the only way to obtain interesting and useful data from the literature. Tables for the other populations HF, CM, and CF are in Appendix A–18 to A–20. Discussion of the Timon and Eisen experiment, which was a study on growth and feeding of the effects of selection for increased gain, of male and female mice from weaning to age 42 days,

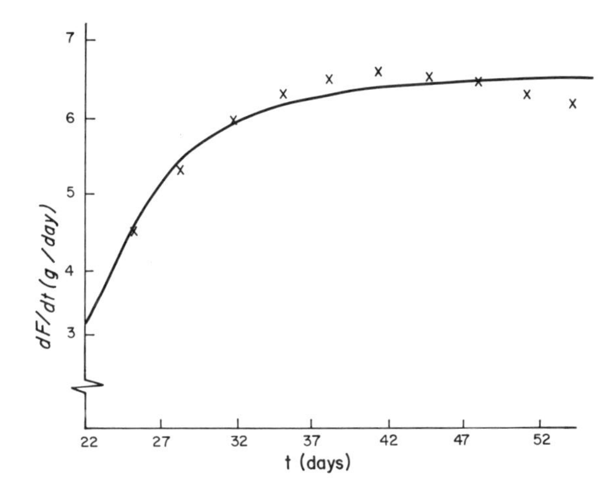


Fig. 4.8. An example of Class V ($t, dF/dt$) data read from curves in Fig. 4.7. Graph of feed rate versus age since weaning of selected HM male mice showing how the calculated curve tracks the data

is presented in Chap. 10 where changing growth curves through selected procedures is discussed in relation to changes in the growth parameters $A$, $(AB)$, $C$, and $t^*$.

The data in Appendix Table A-17 are of the same form as the example of Class II data discussed previously, except that the weights and feed rates were not estimated at the same ages. Using the methods outlined for the Class II data the results of regression are $A = 37.6$ g, $(AB) = 0.428$, $C = 6.56$ g/day, $t^* = 5.64$ days and residual standard deviation, rsd $= 0.61$ g. It is interesting to see that the values of $A$ and $t^*$ of these mice are comparable to the values for the mice discussed on p. 59.

Figures 4.8 and 4.9 show the fit of the calculated curves to the data. The undulations of the data points around the calculated curves are not interpreted here as real phenomena inherent in the growth of these mice for the reason pointed out by Brody. Suffice it to note that the calculated curves describe the main characteristics of the growth and ad libitum feeding of these mice.

Since the use of any least squares nonlinear regression technique requires initial estimates of the parameters $A$, $(AB)$, $C$ and $t^*$, it may not be remiss to say a few words about the problem of getting estimates from a table of data. The

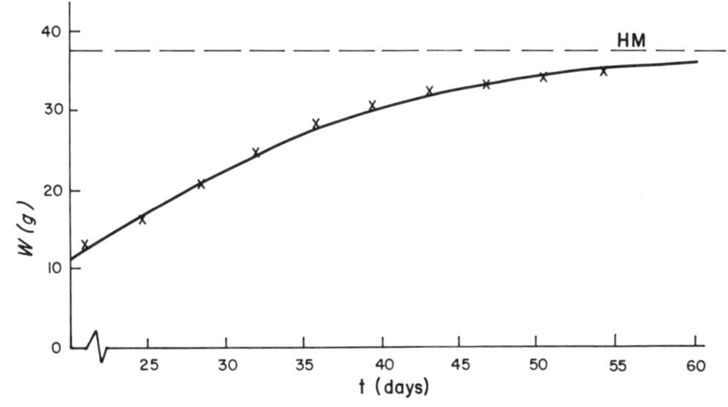


Fig. 4.9. An example of Class V ($t$, $W$) data read from the curves in Fig. 4.7. Graph showing the fit of the calculated curve to the data of growth since weaning of selected HM male mice

easiest method I have used depends on graphs. If the data are long term such as Appendix Table A–11 the plots of the $(i, f_i)$ and $(i, W_i)$ data (Figs. 4.1 and 4.2) clearly indicate approximate values of $C$ and $A$. The cumulative food $F_i$ is found from $\sum_{j=1}^{i} f_j$ and a graph of $(F_i, W_i)$ is made. From the property of the exponential function in Eq. (4.3) an estimate of $(AB)$ is $A \ln 2/F_{\frac{1}{2}}$ where $F_{\frac{1}{2}}$ is the point on the cumulative food axis of the graph where $W = (A + W_0)/2$ on the $W_i$ axis. The same technique on the $(i, f_i)$ graph will give an estimate of $t^* = T_{\frac{1}{2}}/\ln 2$ where $T_{\frac{1}{2}}$ is the time at which $f$ is about $C/2$. Brody's Table 16.1 of constants is another good source of estimates of $A$ and $t^*$ for different species and breeds, his values of $k$ are also good estimates for use in Eq. (4.4) when one is faced only with tables of $(t, W)$ data to express the growth of animals.

The initial average weight $W_0$ at $i=0$ can be considered either a parameter to be determined in the regression or a known number to be entered in Eqs. (4.3) or (4.4). If in a group of infant animals each animal is individually weighed, a subgroup can be selected in which the live weights are more sharply distributed than in the original group, then the starting weight $W_0$ can be considered known. In essence this is done by assigning weights $W_1 < W_2$ and discarding animals whose weights are less than $W_1$ and greater than $W_2$, then $W_0$ is the mean weight per animal of this subgroup with the weight range $W_2 - W_1$. If $W_0$ is not explicitly stated by the experimenter it should be considered a parameter to be determined by regression, but in this case $W_0$ is not to be considered as being among the growth parameters; it is only an initial condition for the experiment (Chap. 2.3).

The measurement of live weight of an animal at particular times presents little experimental difficulty but measurement of food consumed is not so easy. The feed rate $dF/dt$ at a particular time is not directly measurable. It can only be estimated by averaging daily intake for a number of days with the particular time as a midpoint of the interval. A direct method of getting cumulative food, $F$, data is by presenting the animals with excess food, weighing the remainder at specified times and correcting for wastage.

However the problems of equipment and manpower to handle large quantities of food and the possibilities of food spoilage and contamination may make this procedure unattractive.

A better method is to feed by periods of time and weigh the animals at the ends of the periods. Previous knowledge of the animal's feeding characteristics can be used to estimate the food required in each period so that it can be prepared in advance. In this way the problems of feeding animals can be made more manageable. As shown in the example treating Class I data, the errors due to measurement and to variability of environmental factors during each period are all distributed to the regression coefficients.

In this Section the problem of applying the new growth and feeding equations to data gathered from various sources has been discussed. Each source was usually an experiment carried out for purposes other than developing functions to describe the growth and feeding of the animals involved. The Class I data for male rats in Appendix Table A–11 are an exception. The design of this experiment and treatment of the data were guided by my studies of feeding and growth as presented in Chaps. 1 and 2.

The summary of the results of regression in Table 4.1 display the numerical stability of $(AB)$ and the variation of $A$, $C$, and $t^*$ over the mature weight range of about 30,000:1, and lays a foundation for generalisation of the equations of "ad libitum" feeding and growth of all homeotherms. I have not yet found long-term feeding and growth data on any poikilotherm to which the ideas thus far discussed may be extended in some rational manner.

The phrase ad libitum in the preceding paragraph was placed in inverted commas because it is generally interpreted as permitting the animal to feed freely on the food placed before it. This interpretation can be misleading. The food placed before an animal may inhibit intake (appetite) more or less, even though the diet may be dietetically "correct" for "normal" growth. Further, the technicians in charge of feeding animals can tend to offer the amount of food which in their judgement the animals will eat. Such factors can operate to vitiate an otherwise good ad libitum feeding experiment. Here attention is focused on the possibility that an experiment reported as ad libitum may inadvertently be a controlled feeding experiment and some of the results discussed in Chap. 4.2 may be doubtful for this reason. The possibility of identifying dubious ad libitum data is discussed in Chap. 5.

## 4.2 Summary and Discussion of the Growth Parameters Across Species and Within Breeds Fed Ad Libitum

Table 4.1 displays values of the growth parameters $A$, $(AB)$, $C$, and $t^*$, obtained by applying regression techniques discussed in Chap. 4.1 to tables of long term growth data collected in Appendix A, and shows the variabilities and trends of the growth parameters in passing from mice to cattle. The column headed "Table No." contains the table numbers of the original data in the Appendix. The parameters are listed according to decreasing mature weight $A$, starting with steers, for later comparisons with Brody's growth parameters.

For more information about the species, type of food, management etc. of some of the animals named in Table 4.1 the reader should consult the references cited.

The mature food intake, $C$, decreases monotonically with the mature weight, $A$. This is to be expected on grounds of increased basal metabolic rate plus all the energetic demands on the animal to live within its environmental constraints. A definite relation between $A$ and $C$ will be given in Chap. 6. In spite of the variability of $t^*$ for breeds within species there is a trend to higher values for heavier animals. However, the trend is broken by chickens which have values about equal to or larger than that of the pig. Within species $t^*$ is of the same order of magnitude but differences are present. Since $t^*$ expresses the dynamics of the changes of appetite as the animal grows, it is easily influences by various environmental and dietary restraints. The variability shown here may be due to such factors. No doubt there is a minimum value of $t^*$ greater than zero for each species reared under the best growing conditions on the most palatable and nutritious diet. A value of $t^*=0$ means an infant animal would eat at a rate equal to $C$: this seems impossible.

The near constancy of the growth efficiency factor, $(AB)$, over the greater than 30,000 to 1 range of $A$ is quite remarkable. These animals were not all on compa-

Table 4.2. Growth parameters $A$, $(AB)$, $C$ and $t^*$ for various species of animals fed nearly similar diets and known to have been fed ad libitum[a]

| Animal | $A$ kg | $(AB)$ | $C$ kg/wk | $t^*$ wks | Table |
|---|---|---|---|---|---|
| Mouse | | | | | |
| Control male | 0.033 | 0.41 | 0.044 | 1.1 | A-18 |
| Control female | 0.028 | 0.30 | 0.040 | 0.7 | A-20 |
| Male rat | 0.422 | 0.62 | 0.125 | 1.6 | A-15 |
| Male rat | 0.439 | 0.50 | 0.153 | 1.4 | A-8 |
| Male rat | 0.431 | 0.39 | 0.175 | 2.5 | A-11 |
| Male chicken | 3.31 | 0.31 | 0.956 | 7.8 | A-8 |
| Pig | 380.0 | 0.41 | 61.0 | 8.7 | A-5 |

[a] Diets are in range 20%–25% protein and specific metabolizable energy 3–4 kcal/g

rable diets nor comparably managed. Later chapters will show the effects on $(AB)$ of diet composition, especially percentage dietary protein, and ambient conditions. Table 4.2 is a selection from Table 4.1 of parameters of the animals which were fed diets with protein contents in the range 20%–25% and metabolizable energy levels in the range 3–4 kcal/g. This table emphasizes previous remarks about the variability of the growth parameters, especially the quasi-constancy of the feed efficiency factor, $(AB)$, across species grown under more comparable circumstances.

This property of $(AB)$ may be related to Rubner's law, i.e., the amount of energy required for doubling birth weight is the same per kilogram for all animals except man (Brody 1945, p. 47). A paraphrase of Rubner's law is: near birth the energetic growth efficiency of all animals (except man) is a constant. The growth parameter $(AB)$ quite closely follows this statement of Rubner's law. Differential Eq. (2.7) (Chap. 2) shows that near birth when $W_0/A$ is near 0, the efficiency of using food for growth, $dW/dF$, is approximated by $(AB)$, which is here seen to be almost constant, as required by Rubner's law. The growth efficiency, $dW/dF$, should not be confused with thermochemical efficiency of growth. The relationship between these measures of efficiency will be discussed in Chap. 11.

Rubner's law and the near constancy of $(AB)$ imply the ubiquity of the process of metabolism in animals. The growth efficiency factor, $(AB)$, appears constant in time and across species with the exception of man. It has also been said that $(AB)$ may be related to Fraps (1946) productive energy (PE) of a ration (private conversation with G.C. Emmans). The possibility of such a relationship is examined in Chap. 7.

The near constancy of $(AB)$ makes possible a useful equation for estimating the amount of food required to grow an animal from weight $W$ to weight $(A+W)/2$. In the law of diminishing return, $W=A[1-\exp(-BF)]$, parameter $B$ can be defined in terms of $F_{1/2}$, namely the food required to grow an animal from $W$ to $(A+W)/2$. The defining equation is

$$B=\ln 2/F_{1/2}.$$

Multiplying both sides of this equation by $A$ and solving for $F_{1/2}$ yields

$$F_{1/2}=A\ln 2/(AB).$$

If the readers will run their eyes over the values of $(AB)$ in Table 4.1 they will see $(AB)=0.43$ falls about midway in the range 0.235 for steers to 0.618 for Mayer's and Vitale's rats on a 25% protein diet. The estimation equation for $F_{1/2}$ then becomes, with $\ln 2/0.43$ approximately equal to 1.6,

$$F_{1/2}=1.6\,A,$$

i.e., it requires 1.6 $A$ units of food to grow an animal from weight $W$ to weight $(A+W)/2$. Thus it takes approximately 1,600 kg of TDN to get a 500 kg steer up to 750 kg, assuming $A=1{,}000$ kg. It would take about the same amount of food to get a 100 kg steer up to 550 kg, illustrating the decrease in growth efficiency as the animal matures. This estimating equation overestimates $F_{1/2}$ for those animals with $(AB)>0.43$ and underestimates for those with $(AB)<0.43$.

The growth efficiency factor also has a feature of theoretical interest in its relationship to $(aB)$ in the stochastic model of animal growth discussed in Chap. 3.

## 4.3 The Experimental Error $s(W)$

The study of the inherent errors, $\sigma(W)$, of live weight $W$ in connection with the stochastic model (Chap. 3) aroused my curiosity about the manner in which the standard deviation, $s(W)$, of the live weight $W$, at a particular value of $F$, of a population of animals varies with $F$. Because $(AB)$ is virtually constant over a 33,000-fold range of $A$, the variability of weights can be tentatively attributed to variability of mature weight $A$ among the individuals of the population as shown by $s(W)=(\partial W/\partial A)s(A)$, where $s(A)$ is the standard error of the mature weights. Hence $W=A[1-\exp(-BF)]$ indicates that $s(W)$ can be estimated as

$$s(W)=s(A)[1-\exp(-bF)], \tag{4.8}$$

where $b$ plays the same role as $B$ in the law of diminishing returns, though not necessarily equal to it, because the variabilities of $B$ and the environment are lumped into $s(A)$. Since the animals were weighed at $t=0$, $F=0$ the population will have a mean initial weight and $s(W_0)$, so that Eq. (4.8) should be modified to

$$s(W)=[s(A)-s(W_0)][1-\exp(-bF)]+s(W_0). \tag{4.9}$$

This equation suggests that $s(W)$ also follows the law of diminishing returns and it remains to examine experimental data where mean weights $\bar{W}$ are accompanied by $s(W)$ of the population as animals feed and grow. The literature provides a few cases of $[F, s(W)]$ data and a few cases in which the data are $[t, s(W)]$. These latter cases, involving $t$, suggest

$$s(W)=[s(A)-s(W_0)][1-\exp(-k'T)]+s(W_0), \tag{4.10}$$

where $T=t-t^*(1-D/C)[1-\exp(-t/t^*)]$ when $t^*$ and $D/C$ are known [see Eq. (4.4)] and $k'$ plays the same role as Brody's $k$ but not necessarily equal to it for the same reasons just given for $b \neq B$. It is assumed that $t^*$ and $D/C$ are already known from previous regression on $(t, F)$ data.

The experiment of Jull and Titus (1928), performed to test the validity of Spillman's (1924) hypothesis, has been discussed in Chap. 2.1 as part of the foundation

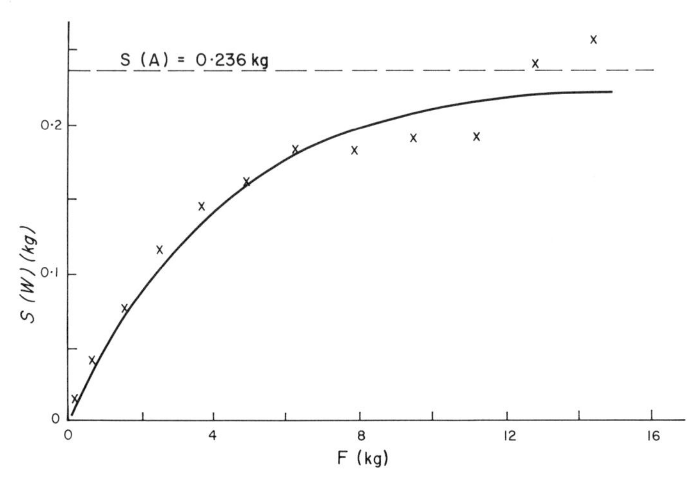


Fig. 4.10. The law of diminishing returns applied to the experimental error, $S(W)$, of life weight of lot 4 male chickens by Jull and Titus (1928)

Table 4.3. Results of regression on standard deviations of live weight measurements on four populations of 25 rats fed diets of differing energy densities

| RED % | $s\,(A)$ kg | $b$ kg$^{-1}$ | $s\,(w)$ kg | rsd $(s)$ kg |
|---|---|---|---|---|
| 70 | 0.0340 | 0.918 | 0.0048 | 0.0023 |
| 80 | 0.0390 | 1.19 | 0.0052 | 0.0042 |
| 90 | 0.0420 | 0.918 | 0.0056 | 0.0029 |
| 100 | 0.0425 | 1.18 | 0.0039 | 0.0034 |

of the work reported in this book. Appendix Table B–1 shows $s(W)$ versus age for the four lots of chickens which in combination with Table 2.2 of Chap. 2 gives $s(W)$ versus $F$ data. Figure 4.10 shows the plot of the $[F, s(W)]$ data for the lot 4 males. Except for 2 points at about $F=13$ and 14 kg, the points appear to rise asymptotically to a maximum value which can be presumed to estimate $s(A)$. Equation (4.9) was applied to these data, and the results of regression are $s(A)=0.236$ kg, $b=0.231$ kg$^{-1}$ and $s(W_0)=0.0022$ kg, with the residual standard deviation, rsd, $=0.017$ kg. The solid curve in Fig. 4.10 is calculated from the regression results. The deviations of the data points from the curve in the range of $F=6$ to 16 kg could be due primarily to the two exceptional data points mentioned previously and a better fit to the data would be had by calling the two points aberrant and deleting them prior to regression. However, I retained them because I had no basis for calling them aberrant. Since Jull and Titus were themselves dissatisfied with the experiment (it was their first attempt to verify the Spillman hypothesis), I decided not to work with the other three lots of their data. The results on lot 4 reported here were promising in showing the law of diminishing returns applicable to these $[F, s(W)]$ data.

I had $[F, s(W)]$ data for each of the four sets of 25 Sprague-Dawley male rats fed ad libitum LAB-BLOX (Allied Mills Inc.) laboratory rat "chow" in meal form and for the meal diluted to 90%, 80%, and 70% with a finely powdered form of

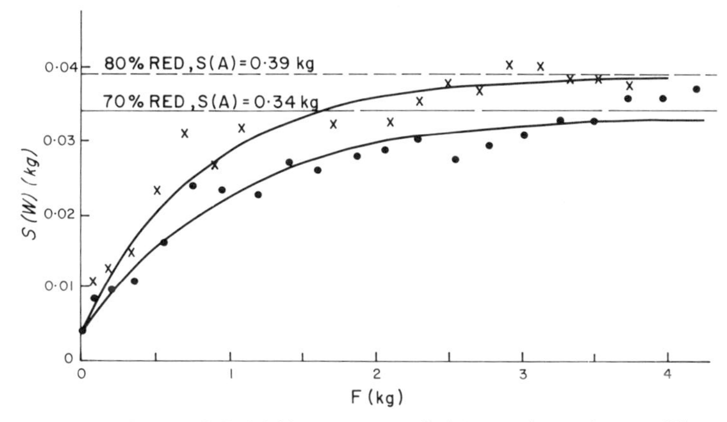


Fig. 4.11. The law of diminishing returns applied to experimental error of live weight, $S(W)$, of albino rats fed 70% and 80% RED diets

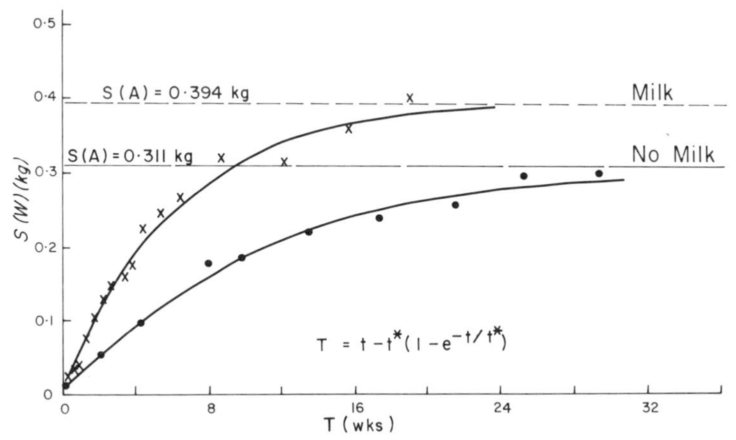


Fig. 4.12. The law of diminishing returns applied to experimental error of live weight, $S(W)$, versus $T$ of Rhode Island Red chicks with and without access to milk

silicic acid (see Chap. 9). The growth of the rats on 100% meal over a 20-week interval from weaning was discussed in Chap. 4.1. Tables B–2 and B–3 in the Appendix show $[F, s(W)]$ data for each of the other populations. Using Eq. (4.9) to fit the data the values of the coefficients $s(A)$ and $b$ were found as shown in Table 4.3 along with the rsd of fit. Figure 4.11 shows the data for rats fed the 70% and 80% relative energy density (RED) diets. The solid curves were calculated using values of $s(A)$ and $b$ from Table 4.3. I have no explanation for the wild swings of the data points around the 80% curve. These large deviations from computed value were absent from the data of the other groups of rats. The data and curve for the 70% RED diet are shown in Fig. 4.11 as an example of a better fit of Eq. (4.9).

There are $[t, s(W)]$ data from long term experiments on populations of chickens. Examples are shown in Appendix Tables B. Table B–4 shows 34 week data from two populations of Rhode Island Red males. The populations, which had milk or no milk available, contained 191 and 174 individuals, respectively. Both groups

Table 4.4. Regression coefficients $s(A)$ and $k$, and residual standard deviation (rsd) for standard deviations of live weight measurement of chickens

| Breed | $s(A)$, kg | $k'$, $wk^{-1}$ | rsd, kg |
|---|---|---|---|
| RIR, Male, no milk | 0.3946 | 0.1735 | 0.014 |
| RIR, Male, milk | 0.3112 | 0.0929 | 0.010 |
| RIR, Males, Grossman | 0.3555 | 0.2240 | 0.013 |
| RIR, Females, Grossman | 0.2832 | 0.1211 | 0.011 |
| WL, Males, Grossman | 0.2403 | 0.1386 | 0.007 |
| WL, Females, Grossman | 0.2476 | 0.0673 | 0.016 |

Table 4.5. Values of population parameters $b/B$ or $k'/k$ and $s(A)/A$ for long term growth experiments on populations of rats and chicks

| Animal | | $b/B$ or $k'/k$ | $s(A)/A$ |
|---|---|---|---|
| Rats, RED % | = 70 | 1.08 | 0.0912 |
| | = 80 | 1.27 | 0.102 |
| (1967) | = 90 | 0.977 | 0.100 |
| | = 100 | 1.18 | 0.101 |
| Chicks, RIR | males | 0.959 | 0.106 |
| | females | 0.841 | 0.116 |
| (1969) WL | males | 1.18 | 0.968 |
| | females | 0.852 | 0.130 |
| Chicks, males RIR, | milk | 2.40 | 0.070 |
| (1928) | no milk | 4.22 | 0.082 |
| | lot 4 | 2.38 | 0.0561 |

were fed the same basal food. The values of $t^*$ for transforming $t$ to $T$ for both cases were from the regressions of the data in Table A–48 of the Appendix. Equation (4.10) was used to find $s(A)$ and $k$ for both groups of chickens (Table 4.4). Figure 4.12 shows plots of the $[T, s(W)]$ data for the two groups in comparison with the calculated curves. The curves track the data very well. Here the availability of skim milk lowers the variability of $A$. Grossman (1969) also collected some $[t, s(W)]$ data in a 45 week experiment with Rhode Island Red and White Leghorn male and female chickens (Appendix Tables B–5 to B–8). These data were treated similarly to the Titus and Jull data after transforming age $t$ to $T$ using the values of $t^*$ from the regression of the $(t, W)$ Class IV data of these chickens in Appendix A. The values of $s(A)$ and $k$ for Grossman's birds are also shown in Table 4.4.

Tables 4.3 and 4.4 show quite clearly that the variability of weight can be described by the law of diminishing returns and it can be concluded that the growth of a population of animals of the same breed is not as unpredictable as some may believe. Since $k = BC$ we may try $k' = bC$ so that $k'/k \simeq b/B$. The values of these ratios for rats and chickens and of the coefficients of variation of mature weight, i.e., $s(A)/A$, are shown in Table 4.5 with the years when the data were obtained. These data show that in more recent times $k'/k \simeq b/B \simeq 1$ and the coefficient of variation of the mature weight $s(A)/A = 0.1$ with the result that the coefficient of variation of immature weight in a population of animals of the same breed estimates the coefficient of variation of mature weight. This last observation is strong evidence for the use of logarithm transformation of both the data and the dependent variable of the regression functions to minimise the sum of squares of the residuals ($\ln Y - \ln y$), where $Y$ is the $W$ or $F$ or $dF/dt$ data and $y$ is the dependent variable in the appropriate regression equation. This technique reports an rsd of fit which is analogous to a constant coefficient of variation over the range of the data, thereby giving a more homogeneous estimate of the variance. The similarity of ($\ln Y - \ln y$) to a coefficient of variation, $CV$, of $y$ is due to the approximate relation, that if $Y - y$ is small, say less than about 0.1, then ($\ln Y - \ln y$) is estimated by $(Y - y)/y$, which has the form of the CV of $y$. I now propose to use logarithm transformations of data and the dependent variable of the regression functions in all future studies of growth and feeding data.

## References

Bateman N, Slee N (1973) Private communication. Unpublished data used with their permission
Brody S (1945) Bioenergetics and growth. First published: Reinhold, New York (Reprinted: Hafner Press, New York
Card LE (1952) Poultry production, 9th edn. Lea and Febiger, Philadelphia
Draper NR, Smith H (1967) Applied regression analysis. Wiley, New York
Emmans GC (1974) Private communication. Unpublished data used with his permission
Fraps GS (1946) Composition and productive energy of poultry feeds and rations. Bull 678 Tex Agric Exp Stn
Grossman M (1969) A genetic and biometric study of growth in chickens. PhD Thesis, Purdue Univ
Hammond JC, Hendricks WA, Titus HW (1938) Effect of percentage of protein in the diet on growth and feed utilization of male chickens. J Agric Res 56:791–810
Headley VE, Miller ER, Ullrey DE, Hoefer JA (1961) Application of the equation of the curve of diminishing increments to swine nutrition. J Anim Sci 20:311–315
Hedammar A (1972) Private communication. Unpublished data used with his permission
Jull MA, Titus HW (1928) Growth of chickens in relation to feed consumption. J Agric Res 36:541–550
Marquart DW (1964) Least-squares estimation of nonlinear parameters. Dupont, Wilmington
Mayer J, Vitale JJ (1957) Thermochemical efficiency in rats. Am J Physiol 189:39–42
Monteiro L (1974) Private communication. Unpublished data used with his permission
Parks JR (1970) Growth curves and the physiology of growth. II. Effects of dietary energy. Am J Physiol 219:837–839
Rutledge J (1971) Private communication. Unpublished data used with his permission
Spillman WJ, Lang E (1924) The law of diminishing increment. World, Yonkers
Timon VM, Eisen EJ (1970) Comparisons of ad libitum and restricted feeding of mice selected and unselected for postweaning gain. I. Growth, food consumption and feed efficiency. Genetics 64:41–57

# Chapter 5 The Geometry of Ad Libitum Growth Curves

Chapter 4 has shown how the ad libitum feeding and growth equations in Chapter 2, empirically arrived at through a few cases of experimental long term growth, have been useful in the study of the long term growth of many species from mice to cattle managed under near laboratory conditions. This chapter is a transition from Chap. 4 to Chap. 6 in which data from experiments on animal growth under various restricted feeding regimes are studied to uncover some mathematical descriptions of the general response of animals to restricted feeding. Here I shall consolidate what we know of ad libitum feeding and growth and introduce some general and useful ideas expressed in the form of geometrical constructions.

## 5.1 Ad Libitum Growth as a Trajectory in a Three Dimensional Euclidean Space

A table of $(t, F, W)$ or $(t, dF/dt, W)$ number triplets is presented as a list of data from a completed experiment and, therefore, is apparently static in character. I think it takes little imagination to think of the table of data as having grown in time with each triplet representing a sample of a developing growth curve, but with the experimenter ignorant of the next triplet until he makes measurements of $W$ and $F$ or $dF/dt$ at the next instant of time in the schedule of the experiment. Turning the mind's eye away from the table of data one may visualize the developing growth curve steadily generated in a three dimensional space by the moving point $(t, F, W)$ analogous to the curve generated in a Euclidean space by a moving point $(x, y, z)$ where the axes of $x, y$, and $z$ are mutually perpendicular. Since the point $(x, y, z)$ is located in space in terms of a common unit of length (cm, in, etc.), we can convert the biological point $(t, F, W)$ to a geometrical point $(x, y, z)$ by choosing scale factors $a, b$, and $c$, where $a$ has units of length/unit of time, $b$ units of length/unit of weight of food consumed cumulatively and $c$ units of length/unit weight, so that $x = at$, $y = bF$, and $z = cW$ and plotting the geometrical point as $P$ in a Euclidean space with axes $X, Y, Z$ marked in units of length (Fig. 5.1). This figure also shows that $P$, beginning at $P_0$ (with coordinates $x_0$, $y_0$, and $z_0$) generates a space curve, in geometry generally called the trajectory $P$, in the direction of the arrow. It should be clear that for $P$ to generate a particular trajectory, $P$ must be under certain constraints stated by specific equations, such as $x = x$, $y = g(x)$, and $z = h(x)$. These equations constitute the equation of the particular trajectory generated by point $P$ in the $X, Y, Z$ space and are called the parametric equations of the trajectory with $x$ as the parameter. This is another meaning of the word "parameter" than that used in Chap. 1. Here the choice of $x$ as the parameter is arbitrary, i.e.

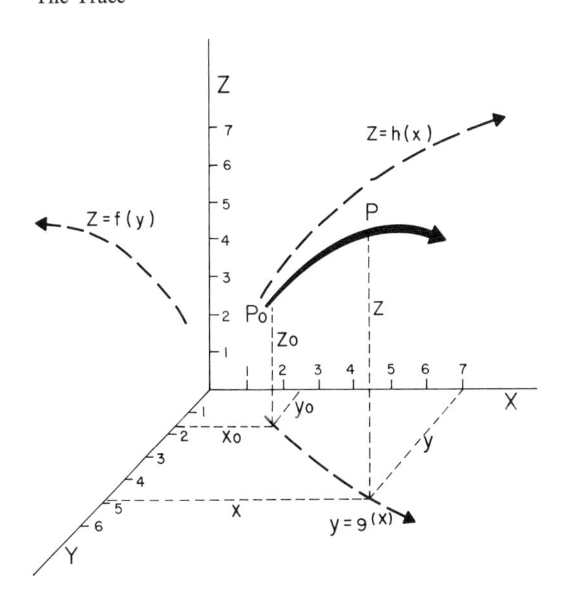


Fig. 5.1. As $x$ varies point $P$ generates the space curve $P_0P$, the projection of which, on the $ZX$, $YX$, and $ZY$ coordinate planes, are respectively the plane curves $z=h(x)$, $y=g(X)$, and $z=f(y)$

$y$ or $z$ or some other variable could have been chosen. A case in point is discussed in the following paragraphs.

Suppose point $P$ moves from $(x, y, z)$ to an adjacent point $(x+dx, y+dy, z+dz)$ along the trajectory; point $P$ has then moved a distance ds given by

$$ds=[(dx)^2+(dy)^2+(dz)^2]^{1/2},$$

which makes it possible to calculate the distance, $s$, $P$ has travelled along the trajectory from $P_0$ by the integral

$$s=\int_{x_0}^{x}\{1+[dg(x)/dx]^2+[dh(x)/dx]^2\}^{1/2}\,dx\,. \tag{5.1}$$

Equation (5.1) states $s$ as a function of $x$ which can be solved for $x$ as some function of $s$, say $f'(s)$. This means that the parametric equation $x=x$, $y=g(x)$, $z=h(x)$ can now be expressed as $x=f'(s)$, $y=g'(s)$, $z=h'(s)$ where $s$, the distance $P$ moves along the trajectory from $P_0$, is now the parameter.

This introduction has carried the reader, hopefully not too quickly, from the static realm of biology of growth as tables of data to the dynamic properties of growth curves in three dimensional geometry, with the objective of generalizing the usual concept of the growth curve expressed as a graph of $(t, W)$ data; which is shown here to be merely the shadow of the three dimensional trajectory of growth on the $XZ$ plane.

## 5.2 The Trace

Assuming $W_0$ negligible compared to $A$, and $D$ negligible compared to $C$, Eqs. (5.2), (5.3), and (5.4) constitute the parametric equation, with $t$ as the param-

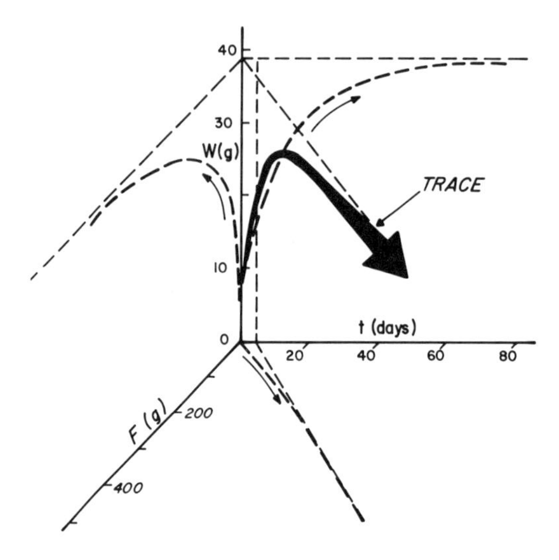


Fig. 5.2. The trace generated by Bateman's mice (Table 4.1) in the Euclidean space $(t, F, W)$

eter of the trajectory of animal growth under an ad libitum feeding regime.

$$W = A[1 - \exp\langle -k\{t - t^*[1 - \exp(-t/t^*)]\}\rangle] \tag{5.2}$$

$$F = C\{t - t^*[1 - \exp(-t/t^*)]\}, \tag{5.3}$$

$$t = t. \tag{5.4}$$

Figure 5.2 shows the trajectory in the $t, F, W$ space of feeding and growth since weaning of the mice of Bateman and Slee. The trajectory was calculated from the growth parameters of these mice in Table 4.1 with Brody's $k = (AB)C/A$. The progressive thickening of the trajectory serves two purposes: (1) the thickening may aid the reader to see the trajectory in depth, i.e., swinging up and out toward the viewer as the mice grow and feed from weaning towards adulthood, and (2) the thickening can represent the bundle of trajectories of the individual mice in the population Bateman and Slee used in their experiments. In Chap. 4 I discussed in detail, how the growth curve of animals diverges in time. The reader may think of Fig. 5.2 as a trajectory in a real three-dimensional box with solid faces but open in front, at the top and right side. A strong distant light thrown on the box exactly along the axis of $F$ the trajectory will throw a shadow on the $tW$ face of the model. This shadow is indicated by a dashed curve which represents the oft studied growth curve of $W$ versus age. The shadow has the mathematical name of projection. The distant light may now be dispensed with and the $W$ versus $t$ curve called the projection of the trajectory on the $tW$ plane; so also the dashed curves in the $FW$ and $tF$ planes are respectively the projections of the trajectory on these planes. The projection on the $FW$ plane is none other than the law of diminishing returns of live weight versus cumulative food consumed, which we met through Spillman in Chap. 2, and the projection on the $tF$ plane is the food consumed by the animal as it ages.

The word trajectory has wide meaning in the real world, but in this book I substitute the name "trace" for the phrases "growth space curve" and "growth trajec-

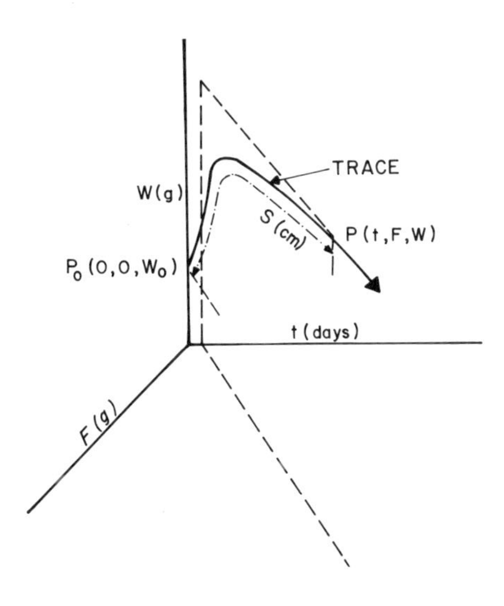


Fig. 5.3. Length, $s$, of the trace of Bateman's mice from $P_0$ at weaning to $P$ at time $t$, cumulative food consumed, $F$, and live-weight, $W$

tory" in discussing the biology of growth. Trace then is defined as the curve developing in the Euclidean space of time, cumulative food consumed and live weight; the projections on the various planes are ignored. Equations (5.2), (5.3), and (5.4) are the parametric equations of the trace (Fig. 5.3). Also indicated is the length, $s$, of the trace measured from weaning [Eq. (5.1)] in units of centimetres since the scale factors $a, b$, and $c$ for this figure are respectively 1 cm/20 d, 1 cm/100 g, and 1 cm/10 g. Using the parametric equations of the trace and these scale factors the length, $s$, by Eq. (5.1) is,

$$s = \int_0^t [1/400 + (dF/dt)^2/10{,}000 + (dW/dt)^2/100]^{\frac{1}{2}}\, dt\,,$$

measured in cm up to time $t$. This length is specific to the trace in Fig. 5.3. We can choose to express the length in any unit of length by assigning the scale factors arbitrarily. In this book I chose to express $s$ in centimetres by setting $a = 1$ cm/unit of time, $b = 1$ cm/unit of food consumed and $c = 1$ cm/unit of live weight so that Eq. (5.1) becomes

$$s = \int_0^t [1 + (dF/dt)^2 + (dW/dt)^2]^{\frac{1}{2}}\, dt\,, \quad \text{cm}\,, \tag{5.5}$$

where $dF/dt$ and $dW/dt$ are evaluated from Eqs. (5.3) and (5.2) respectively.

Although the integral in Eq. (5.5), expressed in $t$ only, appears quite formidable there are numerical methods of evaluating such integrals on desk calculators or, better, with electronic computers (Hamming 1962).

There is much in animal science literature about comparing animals of the same age (isochronomously), or at the same weight (isogravimetrically), or after the same cumulative food consumption (I have mentioned the possibilities of an alternative in Chap. 1). I now suggest that much of the confusion surrounding comparisons of this sort can be avoided by comparing animals of the same breed or by comparing different breeds of the same species at the same length (in centimetres) of their traces. For example percentage of carcase fat of animals is an important feature of interest to agriculturalists and has been compared among animals at the same age or same live weight or same carcase weight. My suggestion of comparing percentage of fat in the carcase or live weight of animals at the same length on their traces directly brings into the comparison how the live animals arrived at their particular body compositions at slaughter. Body ash or other measures can be treated similarly. It seems natural to me that the fraction of body fat, $f$, should be expressed as a function of $s$ and animals compared by comparing these functions and the growth and feeding parameters.

During its growth the animal is not only storing masses of different body tissues but also storing the chemical potential of these tissues, namely whole body energy. Stored body energy as a function of $s$ is a complete nutritional history of the animal as an energy storage system.

## 5.3 The Growth Phase Plane

Another set of parametric ad libitum feeding and growth equations with parameter $t$ is found by substituting Eq. (5.6) for Eq. (5.3).

$$dF/dt = C[1-\exp(-t/t^*)]. \tag{5.6}$$

The solid curve in Fig. 5.4, thickened progressively for the same reasons as in Fig. 5.2, is the growth trajectory of Hedammar's (1972) male Great Dane dogs in the $(t, dF/dt, W)$ Euclidean space. The curve swings up and to the right from the $(dF/dt, W)$ plane finally becoming asymptotic to the time line, $t$, perpendicular to the $(dF/dt, W)$ plane at the point $(C, A)$ in that plane. The projection of the trajectory on the $tW$ plane is the usual sigmoid growth curve and the projection on the $(t, dF/dt)$ plane is the law of diminishing returns the food intake follows [Eq. (5.6)]. The projection on the $(dF/dt, W)$ plane is a new expression of the ad libitum feeding and growth curve which will be shown to be highly useful in the study of how animals grow when fed according to any regime.

I have given the name "growth phase plane" (GPP) to the rectangular portion of the $(dF/dt, W)$ plane with sides $W=0$, $dF/dt=C$, $W=A$, and $dF/dt=0$. Figure 5.5 is the growth phase plane for the Great Dane dogs. Here it is seen that the response of the dogs to ad libitum feeding divides the GPP into two regions, a region generally infeasible, which can be entered only by extraordinary feeding of the dogs such as forced feeding, feeding goldthioglucose etc., and a region that can be entered experimentally by placing the animal on some defined controlled feeding regime. Time is implicit in the GPP and its passing is indicated by the direction of the arrows.

We know something about the controlled feeding region. For example if a mature dog, weighing 59 kg eating 1.2 kg/day, is placed on a constant 0.5 kg/day feed intake, i.e., moving point $S$ on the growth phase plane to point $R$ ($dF/dt=0.5$), the weight of the animal will fall. This is a case of partial starvation. We also know the weight will fall faster if point $S$ is moved to point $A$ where the animal is on a complete starvation regime ($dF/dt=0$). However, at this stage of the book we do not know what the response would be if the animals were grown on ad libitum feeding to point $P$ (say $dF/dt=0.90$) and then moved to point $P'$ where $dF/dt=0.6$.

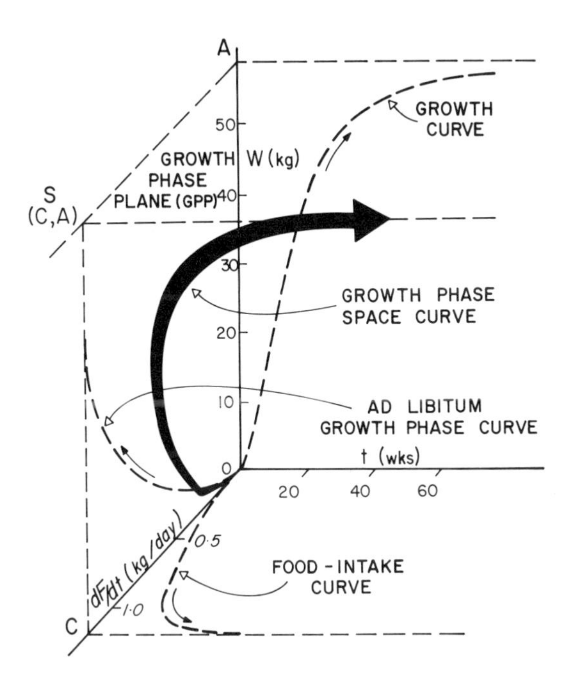


Fig. 5.4. The growth phase space curve of Hedammar's dogs (Table 4.1), the projections of which on the $(t, W)$, $(t, dF/dt)$ and $(dF/dt, W)$ co-ordinate planes are respectively the growth curve, the food intake curve and the ad libitum growth phase curve. The growth phase plane (GPP) is the rectangle OCSA

The diagonal, $OS$, of the GPP has the interesting feature that its slope is $A/C$, i.e., kg of mature weight per kg of food per week consumed at maturity, and it therefore has the units of time. Intuitively this is an important time constant related to the way a mature animal uses its food intake to maintain its weight. Examination of Table 4.1 where the growth parameters over many species are gathered, shows that $A/C$ increases as the mature weight increases but apparently in no particular systematic fashion.

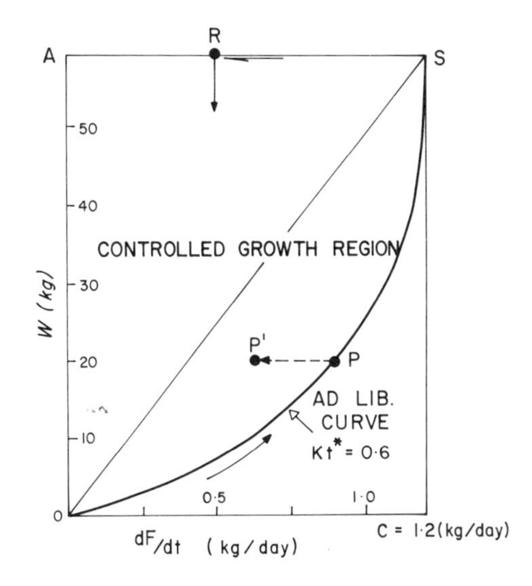


Fig. 5.5. The growth phase plane (GPP) showing the ad libitum growth phase curve OPS ($kt^* = 0.6$) calculated using Eq. (5.7). All of the points $(dF/dt, W)$ such as $P'$ and $R$ in region OPSA are points on possible controlled growth curves, for example it is well known that reduction of food intake of a mature animal from $C$ to $R$ will cause the weight to fall

The GPP will be shown in Chap. 6 to be a very useful device in organising data from many experiments on controlled feeding of animals and men. There, the diagonal will be shown to have special biological significance.

## 5.4 The Ad Libitum Feeding and Growth Discriminant ($\alpha$)

To use the growth phase plane one needs to know the equation of the response of live weight to ad libitum feeding (Fig. 5.5), namely the ad libitum growth phase curve $W$ as a function of $dF/dt$. For convenience let $q = dF/dt$, then the final equation for this function, namely Eq. (5.7), can be found in the following manner.

From Eq. (5.6), $1 - \exp(-t/t^*) = q/C$ and $-t^*\ln(1 - q/C) = t$, and substitution of these relations in Eq. (5.2) yields the desired $W(q)$,

$$W = A \langle 1 - \exp\{kt^*[\ln(1 - q/C) + q/C]\}\rangle. \tag{5.7}$$

On normalising live weight $W$ to degree of maturity $u = W/A$ and the food intake $q$ to degree of mature food intake $Q = q/C$, and after some algebraic reduction the above equation becomes a very interesting and informative dimensionless equation, namely

$$u = 1 - (1 - Q)^{\alpha}\exp(\alpha Q). \tag{5.8}$$

Here $\alpha$ is a dimensionless number equal to $kt^*$ or $(AB)Ct^*/A$. This is a rather surprising result, in as much as it states that the normalised responses of all animals to normalised ad libitum feed intakes are functions of $\alpha$ only and can be calculated from all the growth parameters $A$, $(AB)$, $C$, and $t^*$, or Brody's $k$ and $t^*$. Figure 5.6 shows the normalised GPP and how the growth phase curves, ad libitum feeding assumed, changed in shape for the values of $\alpha$ from 0.2 to 5.0. Alpha, $\alpha$, equals unity appears to be a critical value, in that the ad libitum growth phase curves lie totally below the diagonal $OS$, when $\alpha \leqq 1$, but lie partially below and partially above the diagonal for all $\alpha > 1$.

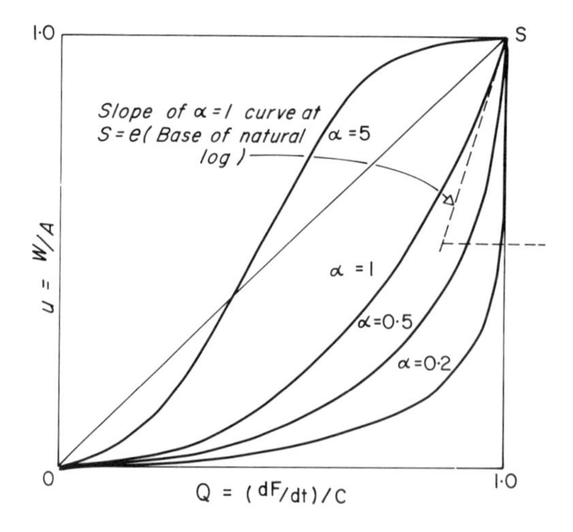


Fig. 5.6. The normalised GPP showing the possible biological significance of the diagonal $OS$ and the growth phase curves corresponding to $\alpha = 0.2$, 0.5, 1.0, and 5

Table 5.1. Values of $\alpha = (AB)Ct^*/A$ for the animals listed in Table 4.1

| Animal | α |
|---|---|
| Cattle | |
| Brody | |
| Jersey | 5.6 |
| Holstein | 13.0 |
| Monteiro | |
| Jersey | 0.50 |
| Friesians | 0.51 |
| Swine | 0.73 |
| Dogs | |
| Great Dane | |
| Ad libitum | 0.60 |
| 67% Ad libitum | 0.64 |
| Broiler | |
| Female (on lower quality diet) | 0.42 |
| Male chickens | |
| 25% P | 0.70 |
| 23 | 2.1 |
| 21 | 2.7 |
| 19 | 1.4 |
| 17 | 1.6 |
| 15 | 1.2 |
| 13 | 1.9 |
| 70% Ad libitum | |
| 25% P | 1.2 |
| 23 | 1.2 |
| 21 | 1.2 |
| 19 | 1.1 |
| 17 | 1.1 |
| 15 | 0.90 |
| 13 | 0.66 |

| Animal | α |
|---|---|
| Male rats | |
| 60% P | 0.29 |
| 25 | 0.30 |
| 10 | 0.13 |
| Male rats | |
| 100% RED | 0.40 |
| 90 | 0.43 |
| 80 | 0.39 |
| 70 | 0.53 |
| 36% P | 0.27 |
| 25 | 0.24 |
| 20 | 0.27 |
| 14 | 0.20 |
| Individual male rats (100% RED) | |
| E31 | 0.25 |
| D23 | 0.60 |
| A22 | 0.37 |
| A24 | 0.28 |
| D10 | 0.30 |
| Mice | |
| Male | 0.42 |
| Female | 0.44 |
| High male | 0.38 |
| Control male | 0.60 |
| High female | 0.43 |
| Control female | 0.33 |
| Male | 0.25 |

Table 5.1 shows the values of α calculated from the growth and feeding parameters $A$, the mature weight; $(AB)$, the growth efficiency factory; $C$, the mature food intake; $t^*$, for the species and breeds in Table 4.1. Table 5.2 shows values of α calculated from Brody's $k/100$ and $t^*$ ($\alpha = kt^*/100$) in his Table 16.1. In Table 5.1 the majority of the values of α are less than unity. The exceptions to this rule are some of the chickens with values of α above unity in the range 1.1 to 2.7 and Brody's Jersey and Holstein cattle with extreme values of α equal 5.6 and 13.0 respectively. In Table 5.2 the values of α less than unity predominate but range from 0.06 for the Ring doves to 1.8 for Leghorn × Brahma female chickens. For the cattle the values of α range from 0.37 to 0.51, which are comparable to those found from the data by Monteiro (Table 5.1). The values of α show some species dependence but the notable feature is that α is less than unity in most cases.

Table 5.3 displays the values of α calculated using the growth parameters in Table 4.2 for animals known to have been fed ad libitum, and here it is seen that α is less than unity in all cases. The extreme values of 5.6 and 13.0 for Brody's cattle

Table 5.2. Values of $\alpha = kt^*/100$ for animals listed by Brody (1945, Table 16.1)

| Animal | $\alpha$ |
|---|---|
| Beef cattle | |
| Castrated males | 0.37 |
| Dairy cows | |
| Holstein-Friesian | 0.38 |
| Ayrshire | 0.46 |
| Jersey | 0.48 |
| Register of merit Jerseys | 0.51 |
| Horse | |
| Percheron | 0.90 |
| Castrated males | 0.85 |
| Swine | |
| Duroc-Jersey females | 0.27 |
| Sheep | |
| Hampshire males | 0.64 |
| Suffolk females | 0.93 |
| Shropshire x Merino females | 0.88 |
| Rabbits | |
| Females | 0.17 |
| Males | 0.25 |
| Males and females | 1.2 |
| Flemish | 0.98 |
| Himalayan × Flemish | 0.98 |
| Polish × Flemish | 1.5 |
| Himalayan × Polish | 1.2 |
| Polish | 0.87 |
| Domestic fowl, females | |
| Rhode Island Red (RIR) | 0.38 |
| Plymouth Rock | 0.47 |
| Rhode Island White | 0.43 |
| White Leghorn | 0.39 |
| Ancona | 0.88 |
| RIR males | 0.38 |
| RIR males | 0.29 |
| RIR castrated males | 0.29 |
| RIR castrated females | 0.39 |
| Cornish males | 0.50 |
| Cornish females | 0.42 |
| Hamburg males | 0.32 |
| Hamburg females | 0.24 |
| Brahma males (B) | 0.68 |
| Brahma females | 0.72 |
| Leghorn males (L) | 0.80 |
| Leghorn females | 0.81 |
| L × B males | 1.4 |
| B × L males | 1.0 |
| L × B females | 1.8 |
| B × L females | 0.92 |
| Guinea pig | |
| $F_1$ Areq. × Race B, males | 0.41 |
| $F_2$ Areq. × Race B, males | 0.39 |
| $F_1$ Cav. Cut. × Race B, males | 0.55 |
| Race B males | 0.51 |
| Race B females | 0.64 |
| $F_1$ Cav. Cut. × Race B, females | 0.47 |
| $F_2$ Cav. Cut. × Race B, females | 0.42 |
| $F_2$ Cav. Cut. × Race B, females | 0.90 |
| Cav. Cut., males | 1.45 |
| Cav. Cut., females | 0.85 |
| Rat | |
| Norway, female | 0.12 |
| Norway, male | 0.11 |
| Albino, male, specially well fed | 0.62 |
| Albino, female (7–15 gen. series) | 0.54 |
| Albino, female (7–15 gen. series) | 0.69 |
| Albino, on whole milk and wheat, males | 0.88 |
| Albino, on whole milk and wheat, females | 1.4 |
| Albino (16–25 gen. series) males | 0.39 |
| Albino (16–25 gen. series) females | 0.48 |
| Stock rats, males | 0.24 |
| Stock rats, females | 1.3 |
| Control rats (inbreeding experiment) males | 0.51 |
| Control rats (inbreeding experiment) females | 0.49 |
| Males | 0.45 |
| Females | 0.39 |
| Stock rats, females | 0.45 |
| Stock rats, males | 0.39 |
| Stock rats, females | 0.62 |
| "Runt" female | 1.3 |
| Litter mate to runt female | 1.3 |
| White Mouse | |
| Males | 0.47 |
| Females | 0.75 |
| Albino mouse, male | 0.69 |
| Albino mouse, female | 0.56 |
| Pigeons | |
| Common, male and female | 0.11 |
| Ring dove, male and female | 0.06 |

Table 5.3. Values of $\alpha$ calculated using the growth parameters of the animals, listed in Table 4.2, which were known to have been fed ad libitum

| Animal | $\alpha=(AB)\,Ct^*/A$ |
|---|---|
| Mouse | |
| Control male | 0.60 |
| Control female | 0.30 |
| Male rat | 0.29 |
| Male rat | 0.24 |
| Male rat | 0.40 |
| Male chicken | 0.70 |
| Pig | 0.57 |
| | $\bar{\alpha}=0.443$ |

can be explained at least in part by the footnote on p. 51 of Brody (1945) indicating that the weights and feed intakes of his Jerseys and Holsteins were not taken simultaneously but came from independent sources. The values of $\alpha>1$ in Table 5.1 are associated with chicken experiments which Jull and Titus (1928) doubted were ad libitum although the animal keepers reported the experimental data as such. Jull and Titus remark that animal keepers often feed animals according to the amount of food which they consider the animals require. Since Brody's $k$ is stable under good management and constant diet composition, this kind of feeding "ad libitum" introduces delays, which increases $t^*$ thereby increasing $\alpha$, possibly to values greater than unity. Alpha greater than unity serves as a warning that a growth and feeding experiment reported as ad libitum is not in fact ad libitum and the experiment should be considered an inadvertent experiment in controlled feeding (see Chap. 6).

The near coincidence of the ranges of values of $\alpha$ in Table 5.1 and 5.2 shows that the growth parameters $A$, $(AB)$, $C$, and $t^*$, obtained from all the feeding and growth data from immaturity to near maturity, are consistent with Brody's $k$ and $t^*$ which are obtained from the last 20%–30% of the weight versus time data as the animal nears maturity. These results support the approach of considering growing animals as black boxes with input and output developed in previous chapters.

In view of the fact that $\alpha$, in Eq. (5.8), defines the kinds of ad libitum growth curves that occur in the normalised GPP, Fig. 5.6 is very instructive. When $\alpha$ is small (say 0.2) the animal reaches a high fraction $[Q=(dF/dt)/C]$ of its mature food intake while its degree of maturity $u$ is small, a generally observed phenomena, and as $\alpha$ increases toward unity this situation still holds, but to a lesser degree. For all $\alpha<1$ the curves become asymptotic to the vertical axis of the normalised GPP at $Q=1$. When $\alpha=1$ the curve, as it approaches point $S$, becomes tangent to a line of slope $e$, the base of natural logs. This can be shown by setting $\alpha=1$ in Eq. (5.8) and finding $du/dQ=Q\exp(Q)$ and allowing $Q$ to approach unity, $du/dQ$ then approaches $e$ (Fig. 5.6).

With a value of $\alpha>1$ the curve crosses the diagonal and becomes asymptotic to the horizontal line $u=1$ as $Q$ approaches unity as shown by $du/dQ=\alpha Q\exp(\alpha Q)(1-Q)^{\alpha-1}$ which approaches zero as $Q$ approaches unity.

When $\alpha$ is greater than unity the case for Eq. (5.8) being an ad libidum growth curve is weakened for two reasons, namely, (1) we have already remarked (Fig. 5.5) that when an animal is mature, reduction of its food intake below $C(Q<1)$ tends to cause loss of weight and, (2) intuitively it seems energetically unrealistic for an animal to approximate its mature weight while its food intake is something less than $C(Q<1)$. A stronger case for Eq. (5.8) not representing an ad libitum feeding and growth curve, with $\alpha>1$, is presented in Chap. 6.

For these reasons I have called $\alpha$ the ad libitum feeding and growth discriminant.

## 5.5 The *Z* Function

We have already seen the importance of the law of diminishing returns in the domain of cumulative food consumed $F$, namely $W=A\{1-\exp[-(AB)F/A]\}$. In previous chapters $F$ was considered the independent variable while $(AB)$ and $A$ were considered parameters. Here I shall consider $(AB)F/A$ as the independent variable and call it the $Z$ function of $F$. Because $[(AB)/A]$ has units of inverse kg of food consumed, it is the normalising factor for $F$ which makes $Z$ a dimensionless variable in the same sense that $u$, $T=t/t^*$ in Eq. (5.9) and $\alpha$ are dimensionless.

Substituting $F=C\{t-t^*[1-\exp(-t/t^*)]\}$ and $T=t/t^*$ in $(AB)/F/A$ the $Z$ function becomes

$$Z=\alpha[T-1+\exp(-T)]. \tag{5.9}$$

Here it is seen, as in the case for the degree of maturity, $u$, in Eq. (5.8), that the $Z$ function depends only on $\alpha$.

Substituting $Z$ for its equivalent in the law of diminishing returns gives the following parametric equation for the normalized space growth curve with normalized age $T$ as the parameter.

$$u=1-\exp\{-\alpha[T-1+\exp(-T)]\} \tag{5.10}$$

$$Z=\alpha[T-1+\exp(-T)] \tag{5.11}$$

$$T=T. \tag{5.12}$$

Because this space curve is fundamentally different from that which I called the trace in Chap. 5.2, I shall name the normalized space growth curve, defined by Eqs. (5.10), (5.11), and (5.12), the biotrace. And since the biotrace is in Euclidean space (Fig. 5.7), we can speak of its length in centimetres in the same way we speak of the length, $s$, of the trace for breeds and species of animals in Chap. 5.2. Letting $\Sigma$ be the length of the biotrace from point $(T_0, Z_0, u_0)$ to point $(T, Z, u)$ we can calculate $\Sigma$ by

$$\Sigma=\int_{T_0}^{T}[(du/dT)^2+(dZ/dT)^2+1]^{\frac{1}{2}}\,dT, \quad \text{cms}. \tag{5.13}$$

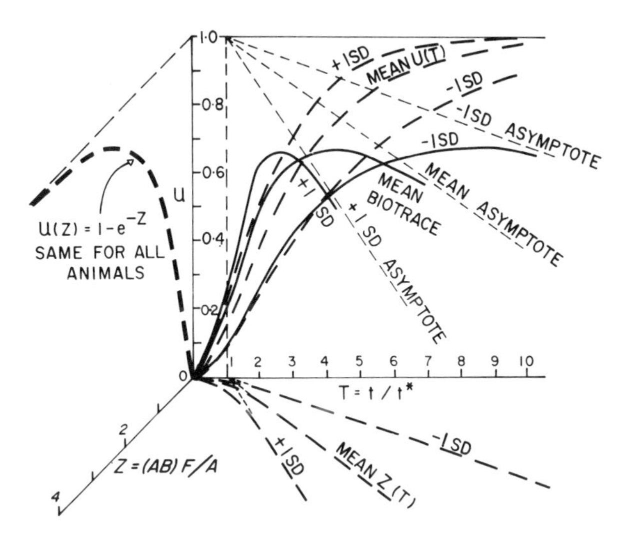


Fig. 5.7. The mean biotrace and the plus and minus one standard deviation biotraces of all animals with $0<\alpha \leqq 1$

## 5.6 Comparing Ad Libitum Feeding and Growth Across Species Excepting Man

The necessity for using the concept of the biotrace is seen in comparing the traces of large and small species such as pigs and mice. In the time it takes a pig to double its weaning weight, the mouse has almost reached its mature weight which is about 1/10,000 of the pig's mature weight. In a three-dimensional plot of the $(t, F, W)$ data for a pig, the data for the mouse on the same scale would be lost in the corner near the origin.

The equations of the biotrace show that if any two or more species, large or small, have the same value of $\alpha$ their biotraces coincide. In the case of mice and pigs the values of $\alpha$ are respectively 0.41 and 0.73 with coincidence of their biotraces appearing impossible. However the mean and standard deviation of the 32 values of $\alpha \leqq 1$ in Table 5.1 are $0.428 \pm 0.180$ with 31 degrees of freedom. This indicates that the biotraces of the mouse and pig are not all that different, suggesting the possibility that all animals except man have the same biotrace but with variability due most probably to difference in quality of food management (as implied in Tables 5.1 and 5.3). The animals which have the extreme values of $\alpha$ in Table 5.1 are the male rats fed a 10.5% protein diet ($\alpha = 0.13$) and the cockerels fed a 15% protein diet at 70% of ad libitum food intake ($\alpha = 0.90$), but the mice and Monteiro's cattle have values of $\alpha$ within one standard deviation of the mean values of $\alpha$, thereby casting doubt on the influence of widely differing species on $\alpha$. Figure 5.7 shows the mean biotrace for all animals together with the one standard deviation biotraces. About 63% of the biotraces of all animals fall between the biotraces of $\alpha = 0.428 + 0.180$ and $\alpha = 0.428 – 0.180$.

The implication of the near constancy for $\alpha$ for animals of today fed ad libitum is that, if there were many different types of metabolic chemical input-output systems typical of the animals aeons ago, evolution sorted out the system that all animals (except man) have today and in the process of adaption and natural selection sealed the systems, at least from mice to cattle. Alpha, $\alpha$, appears to have some special physiological significance in the animal world, at least among the homeotherms which I have studied.

These last remarks may have led the reader to think I believe $\alpha$ is a constant, that is, for animals on good foods and handled well at the management level $\alpha$ will approximate the mean value of 0.428 regardless of species. However $\alpha$ can be changed by changing the protein quality and energy level of the diet, thereby changing $(AB)$ (see Chap. 9), by increasing $t^*$ (see Chaps. 9 and 10), or by changing ambient temperature (see Chap. 11). However in true ad libitum growth experiments, $\alpha$ must lie in the range $>0$ and $\leqq 1$.

The mean curve in Fig. 5.7, the parametric equations of which are Eqs. (5.10), (5.11), and (5.12) with $\alpha = 0.428$ substituted in them, and the length $\Sigma$ (in centimetres) along the mean curve can be used by comparative physiologists to establish a scale for comparing physiological characters of different species at the same values of $\Sigma$ [Eq. (5.13)]. Theoretically the biotrace can be extrapolated to the origin 0 in Fig. 5.7 at which point $\Sigma = 0$ which may or may not coincide with conception of the animal. In this way $\Sigma$ can be thought of as "physiological age" even though it is measured in centimetres.

This is not the first suggestion of "age" measured in units other than time. Fermi defined an area in square centimetres, as expressing the "age" of the slowing down of high energy neutrons to thermal energies in nuclear reactors (The International Dictionary of Applied Mathematics, 1960).

## References

Brody S (1945) Bioenergetics and growth. First published: Reinhold, New York (Reprinted: Hafner Press, New York, 1974)
Hamming RW (1962) Numerical methods for scientists and engineers. McGraw-Hill, New York
Jull MA, Titus HW (1928) Growth of chickens in relation to feed consumption. J Agric Res 36:541–550

# Chapter 6 Growth Response to Controlled Feeding

The notion that the plasticity of growth curves in the time domain may be due in large part to diet and/or feeding management has been in the background of animal science for a long time (Chap. 3). Hopkins (1912, p. 441) showed the importance of accurate measurement of food consumed in interpretation of the effects of diet deficiencies on growth. Mitchell and Beadles (1930) discussed the prime importance of methods of control of conditions under which accurate measurements are made, with special attention to how food consumed by an animal is controlled. Blaxter (1968, p. 329) discusses the importance of food intake to growth. But only recently has some work been done on predefining how an animal was to be fed and then observing the growth response of the animal to this predefined feeding regime.

This chapter deals exclusively with some experimental growth responses to well-defined controlled feeding regimes, i.e., defined so that they could be precalculated, including partial and complete starvation. The experimentally controlled feeding regimes include holding food intakes constant for a long period of time (Fig. 6.1 a); holding food intake constant for a time and then stepping the intake up or down to some new level held constant for a time (Fig. 6.1 b); increasing food intake as a linear function of time (Fig. 6.1 c); and varying intake in time as a sine wave with a preselected period (Fig. 6.1 d).

*Chapter 6.1* will deal with the response of cattle to feeding regimes that include all of the simple functions depicted in Fig. 6.1. For convenience of presentation this section is subdivided into two parts with the first dealing only with constant feeding at several levels over a period of years, and the second dealing with an experiment in which the feeding regimes are combinations of all the simple functions of time depicted in Fig. 6.1.

*Chapter 6.2* discusses the responses of sheep to the predetermined feeding functions depicted in Fig. 6.1 b, namely the step function in which the food intakes are stepped up and down, at a preselected time, to some new levels of constant food intakes.

*Chapter 6.3* is a study of the response of chickens, grown from hatching over nearly a year, for which the preselected feeding functions formed the one parameter, namely $f_i$, family of food intake curves the function of which is

$$(dF/dt)_i = f_i C[1 - \exp(-t/t^*)], \tag{6.1}$$

in which $C$ and $t^*$ are the ad libitum food intake parameters known from previous ad libitum feeding experiments with this breed of chickens and $f_i$ is a set of preselected fractions from the range $0 < f \leqq 1$. With $f_i = 1$, the reader will recognise Eq. (6.1) as the ad libitum feeding equation developed in Chap. 2.

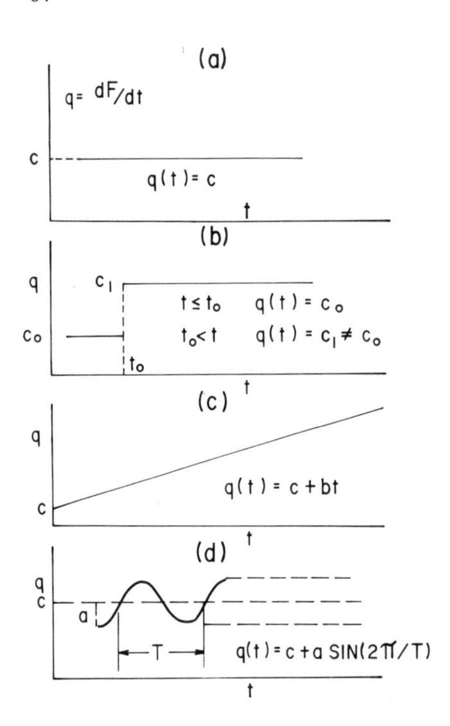


Fig. 6.1 (a–d). Four simple functions as possible controlled feeding functions; (a) constant, (b) step, (c) ramp, (d) sine wave

*Chapter 6.4* puts before the reader some results obtained from complete and partial starvation of several species of animals from the simple jellyfish to higher animals such as man. The results of these studies (Chap. 6.1 to 6.4) led me to a quite astonishing intuitive conclusion, namely that all the live weight responses were solutions of a simple first order differential equation with given initial conditions and given feeding functions.

*Chapter 6.5* is a detailed discussion of the relation of the diagonal of the GPP to the various regions of the GPP.

*Chapter 6.6* concludes this chapter with a discussion of a differential equation considered basic to controlled growth and feeding.

## 6.1 Controlled Feeding of Cattle

### 6.1.1 Responses to Constant Food Intake

Taylor and Young (1966, 1967, 1968) published three important papers on their experiments controlling the food intake of monozygotic and dizygotic twin Ayrshire cattle. The main objective was the investigation of genetic differences in growth response to controlled feeding, which the researchers discussed in 1966, followed in 1967 and 1968 by two papers on the transient response to feed rates fixed at constant levels. The reader's attention is here called particularly to the transient

Table 6.1. Equilibrium weights of Ayrshire cattle at constant levels of food intake. (Derived from data by Taylor and Young 1968, p. 395)

| $c$<br>Food intake<br>(lbs/wks) | $a$<br>Equilibrium wt<br>(lbs) |
|---|---|
| 35 | 355.7 |
| 56 | 500.7 |
| 77 | 751.0 |
| 98 | 969.3 |
| 119 | 1,162.9 |
| 140 | 1,359.4 |

Number of animals = 22
The animals were kept on the specified food intakes for periods of years

growth and equilibrium weights attained on constant levels of food intake discussed in the 1967 and 1968 papers.

Taylor and Young (1968) reported the results of holding Ayrshire cattle on various constant levels of food intake until equilibrium weights were attained. The equilibrium weights of these cattle, which had nearly reached constant weight, were estimated. Data on all the cattle were gathered in a table of mean weights versus levels of food intake at 12 week intervals from 270 to 342 weeks of age. The appearance of the data (there are minor variations about a mean in time) led me to summarise them as equilibrium weights $a$, versus constant intake $c$ in Table 6.1. Figure 6.2 shows a graph of equilibrium weight, $a$, versus constant food intake, $c$. The disposition of the data points about the straight line $a = T_0 c$ through the origin is unmistakable. The slope $T_0$ is in units of time if body weight and food consumed are expressed in the same units of mass. By linear regression, $T_0$ was estimated as $9.82 \pm 0.11$ weeks from the data in Table 6.1. Taylor (1968, p. 397, 398), using a different line of analysis, obtained $T_0 = 9.85 \pm 0.23$ weeks. These values for $T_0$ are not significantly better.

The linear relation between equilibrium weight and the level of constant feeding implies $T_0 = A/C$. Taylor confirms this (1968, p. 403) and defines $T_0$ as the efficiency of maintenance of equilibrium weight, which it is, when the units are considered, namely lbs live weight/lbs of food per week. However I have preferred to call it the Taylor time constant, because it was these experiments which first gave me an idea of the interpretation of the diagonal of the GPP discussed in Chap. 5, and because $T_0$ has the essential nature of a time constant characteristic of these cattle. Furthermore $T_0 = A/C$ suggests that all the animals I have studied and reported on in Table 4.1 of Chap. 4 have characteristic values of $T_0$, but it remains to be seen whether the straight line through the origin, in Fig. 6.2, having slope $= T_0$, is applicable to animals generally.

The 1967 paper by Taylor and Young discussed the transient growth of these cattle after they were put on the constant levels of food intake shown in Table 6.1.

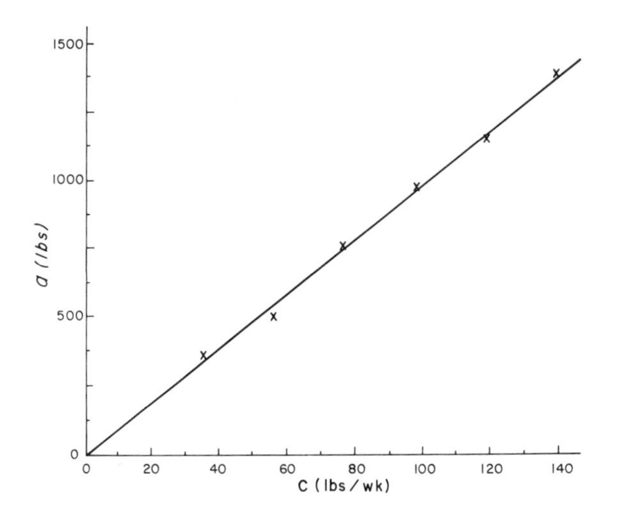


Fig. 6.2. Graph of equilibrium weight, $a$, versus constant food intake, $c$, of female Ayrshire cattle. *Crosses*($\times$) are data points from Table 6.1. (Data by Taylor and Young 1968). *Solid line* represents the linear relation $a = T_0 c$

Figure 6.3 shows how one of the cattle placed on 77 lbs $\mathrm{wk}^{-1}$ of food responded from its initial weight of about 300 lbs. Included with this response, the constant food level of 77 lbs $\mathrm{wk}^{-1}$ is also shown with two "wiggles" in it at about 1 year and 5 years respectively. These "wiggles" indicate the meticulous care that was taken to insure the average food intake remained constant over the period of the experiment. At the beginning of each "wiggle" the animal refused to eat all the food offered and continued to do this in increasing amounts until it began to consume more. When the animal had regained its appetite to 77 lbs of food per week it was offered more than 77 lbs $\mathrm{wk}^{-1}$ until it reached a precalculated peak after which the food offered was reduced back to 77 lbs $\mathrm{wk}^{-1}$ in such a manner that the average food intake during the "wiggle" was 77 lbs $\mathrm{wk}^{-1}$ of food intake. This manoeuvre was used on all the cattle in this experiment to ensure they were on the assigned constant food intakes (Table 6.1). As seen in Fig. 6.3 these "wiggles" of intake appeared to have no readily observed effects on the responses. Figure 6.3 was typical of the responses of the cattle on constant food intake. A very important feature of responses such as these is the signal to noise ratio of about 10.

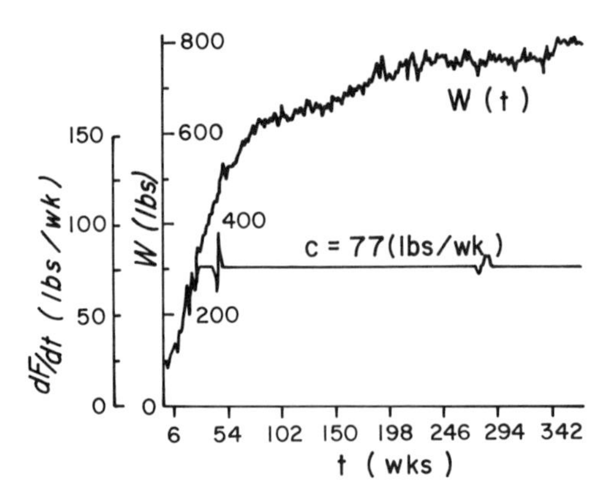


Fig. 6.3. Transient weight response of one animal given 77 lbs food/week from about 30 weeks of age. [Reproduced from Fig. 2 of Taylor and Young (1967) with permission of the publishers]

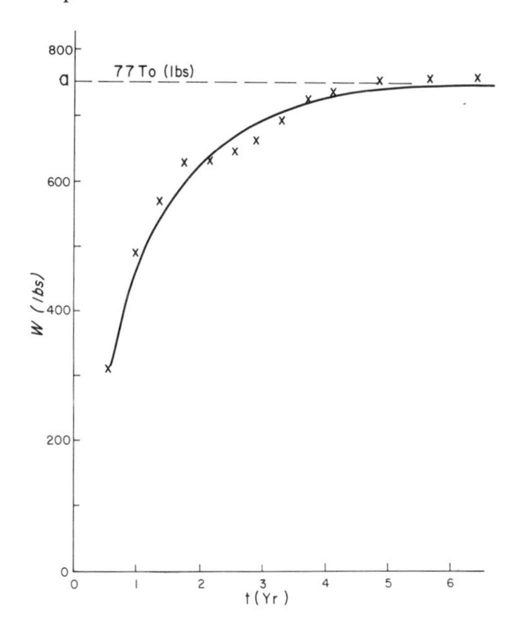


Fig. 6.4. Plot of points selected from the curve in Fig. 6.3. *Crosses* (×) are selected points and the *solid curve* is calculated from the equation of diminishing return [Eq. (6.2)]

The signal is the growth from 300 lbs live weight to about 800 lbs and the noise is the collection of rapid fluctuations in weight during this growth, having a peak to peak average of about 50 lbs. This gives some reason for trying the diminishing returns law in time as a possible deterministic function for describing the mean growth of this animal on 77 lbs $\mathrm{wk}^{-1}$ constant food intake.

The data plotted in Fig. 6.3 were treated as Class 5 data as described in Chap. 4. The data points selected from Fig. 6.3 are the × signs shown in Fig. 6.4 with the initial weight estimated at 306 lbs. These data were analysed using Eq. (6.2) as the regression equation, with $T_0$ and $b$ as the parameters to be determined,

$$W = (77\,T_0 - 306)[1 - \exp(-bt)] + 306, \tag{6.2}$$

with the results $T_0 = 9.77$ wks and $b = 0.88$ $\mathrm{years}^{-1}$. Dividing $b$ by 52 weeks gives $b = 0.017$ $\mathrm{wks}^{-1}$. This value of $T_0$ is within one half of the standard deviation of the value determined from fitting $a = T_0 c$ to all the data in Table 6.1, and $b = 0.017$ $\mathrm{wks}^{-1}$ is approximately Brody's $k = (AB)C/A = 0.020$ $\mathrm{wks}^{-1}$ and 0.022 $\mathrm{wks}^{-1}$ for Monteiro's Friesian and Jersey cattle respectively (Table 4.1, Chap. 4). This is only a suggestion that $b$ in Eq. (6.2) may be Brody's $k$ and requires much more supporting (or refuting) evidence. Some support will be found as this chapter unfolds.

This study of transient response and final equilibrium weights on constant feeding makes possible some additional interpretation of the growth phase plane (GPP) introduced in Chap. 5. Figure 6.5 is a three dimensional $(t, q, W)$ graph, similar to Fig. 5.4 but with the $t$ and $q = dF/dt$ axes interchanged for a better view of the transient due to holding food intake constant, thereby illustrating how the controlled growth of Taylor's cattle in the time domain relates to ad libitum growth in the GPP. The animal is brought along the ad libitum curve from birth weight to weight $W_0$ at which point it is put on constant food intake $dF/dt = q_0$. Defining the instant

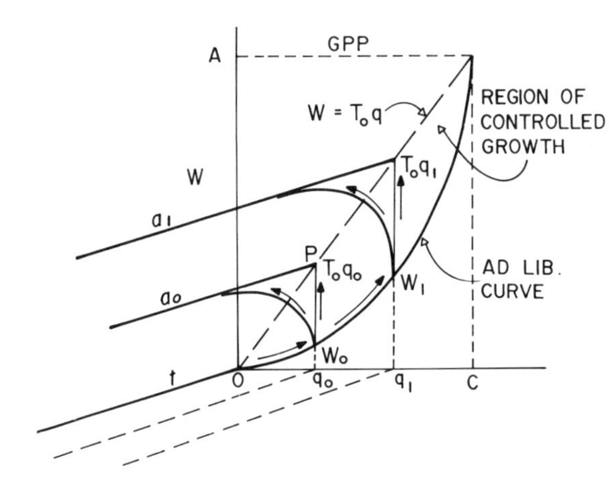


Fig. 6.5. A three-dimensional $(t, q, W)$ diagram illustrating Taylor's experiment. Under constant feed rates $q_0$ and $q_1$ the weights follow the equation of diminishing returns as functions of time and asymptotically approaching maximum weights $T_0q_0$ and $T_0q_1$ respectively

the animal is put on $q_0$ as the time origin $(t=0)$, it is seen that $W$ will follow a diminishing increment curve approaching asymptotically a time line parallel to the $t$-axis and passing through the point $P$ having coordinates $(q_0, T_0q_0)$. Projection of this curve on the GPP or $(q, W)$ plane gives a line from $W_0$ to point $P$; hence the progess of $W$ along the space curve in time can be represented as a moving point in the GPP and approaching $T_0q_0$ asymptotically from $W_0$. The same description applies to an animal brought from birth weight to weight $W_1$ and put on constant food intake $q_1$ etc.

Figure 6.5 gives a meaning to the Taylor diagonal of the GPP, the equation of which is $a = T_0q$. At every point on this diagonal the growth of the animal is zero $(dW/dt = 0)$ indicating that it cannot be a growth curve because every point on it is an equilibrium point. Here the importance of the diagonal is seen because it divides the GPP into two regions; namely the lower region, bounded by the diagonal and the ad libitum growth curve, in which controlled feeding and growth experiments are possible, and the upper region, the significance of which will be shown when in Chap. 6.4 I consider the experimental data on partial and complete starvation of men and beasts.

It seems clear that the intake of energy by an animal, when its $(q, W)$ point is within the controlled growth region of the GPP, is sufficiently above maintenance to permit the animal to grow more or less rapidly depending on how close the $(q, W)$ point is to the Taylor diagonal. More importantly it appears that the growth rate $dW/dt$ is greater than zero for any point within the controlled growth region or on the ad libitum growth curve. At this point these conclusions are tentative but they will find more support in the next part of this section and the sections that follow.

### 6.1.2 Sinusoidal and Ramp Function Feeding

Taylor and Young (1968) designed and executed a remarkable experiment involving combinations of all the simple functions depicted in Fig. 6.1 as food intake functions $q(t)$ for five groups of four Ayrshire twin cattle. Even though the concept

Table 6.2. Pre-assigned $q(t)$ functions for controlled feeding of Ayrshire cattle in the age range $4 \leqq t \leqq 108$ weeks. At age 4 weeks the mean weight of the cattle was 82.5 lbs

| Index No. $j$ | Identifying symbols[a] | $q(t)_j$, lbs/wk |
|---|---|---|
| 0 | I | $23.25, 4 \leqq t \leqq 12$ |
| 1 | HC | $31.0+1.0(t-12), 12 \leqq t \leqq 108$ |
| 2 | MC | $23.25+0.75(t-12)$ |
| 3 | LC | $15.5+0.5(t-12)$ |
| 4 | S | $23.25+0.75(t-12)+(7.75+0.25(t-12)) \sin[2\pi(t-12)/48]$ |
| 5 | NS | $23.25+0.75(t-12)-(7.75+0.25(t-12)) \sin[2\pi(t-12)/48]$ |

[a] I = initial constant food intake. HC = high constant acceleration of food intake. MC = medium constant acceleration. LC = low constant acceleration. S = sine wave. NS = negative sine wave, i.e., $\pi$ radians out of phase with S

of the GPP was not available at the time, Taylor somehow chose the constants of the food intake functions so that the live weight responses would be wholly within the controlled growth region of the GPP. The experiment was long term in that the cattle were subjected to controlled feeding from 4 to 108 weeks of age.

The pre-assigned $q(t)$ feeding functions are listed in Table 6.2. The plan was to bring the five groups from age 4 to age 12 weeks on a constant feeding regime of 23.25 lbs/wk designated as function *I* and index number 0. At 12 weeks each of the groups would be subject to one of the feeding functions designated as HC, MC, LC, S, and NS. The three functions HC, MC, and LC are ramps with feeding accelerations differing by 0.25 lbs/wk/wk, with the HC line having the highest acceleration of 1.0 lb/wk/wk. The other functions (S and NS) are sine waves with periods $T=48$ wks but with NS $\pi$ radians out of phase with S. The MC line is the axis of the sine waves which are periodically tangent to the HC and LC lines thus making their amplitudes linear functions of time. Step changes of 7.75 lbs/wk plus and minus in food intakes at 12 weeks were required to get animals to the HC and LC feeding regimes. Such abrupt changes would, no doubt, produce brief transient behaviour of the feeding and growth of the animals at 12 weeks, but the smallness of the changes reduces the seriousness of such transients.

The actual food intakes, shown in Fig. 6.6, show accurately the forms of the pre-assigned functions, except for some infrequent refusals of food during the 96 weeks of the controlled feeding period. The LC and MC groups ate accurately according to their preassigned regimes. The HC groups showed more frequent refusals especially at about 90 weeks when they began to deviate quite widely from the assigned feeding level. These refusals were to be expected because the assigned level, nearing the ad libitum mature food intake at about 60% of the mature weights, probably caused some metabolic upset. The same could be said of the refusals of the NS groups after about 90 weeks. However, the S group during the same period of time also showed refusal to eat when its prescribed level was considerably lower than that of the NS group. The intended periodic tangency of the S and NS curves to the HC and LC ramps was not obtained because of refusals near the times of tangency, but the deviations were not considered serious. At 12 weeks of age the LC group had to accept the 7.75 lbs/wk abrupt decrease of

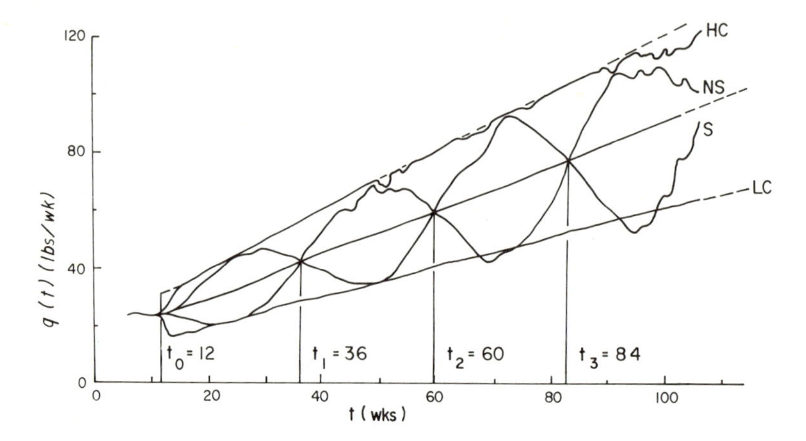


Fig. 6.6. Experimental food intakes showing deviations from the pre-assigned food intake functions in Table 6.2, but accurately depicting the forms of these controlled feeding functions

their food intakes but the HC group refused to have their food intakes abruptly increased by the same amount and it took them about 4 weeks to build their intakes up to the prescribed level.

The symmetry of the preassigned functions served two purposes, (1) ease of calculating and checking the schedules of the feeding regimes beforehand and, (2) establishing time markers of important events such as the times $t_0$ to $t_3$ when the S, MC, and NS curves cross over each other, namely, 12, 36, 60, and 84 weeks. The 108 week cross over time was obscured by the refusals towards the end of the experiment. As will be seen shortly, Taylor showed eminent foresight in this design, and Fig. 6.6 shows the large measure of success in achieving the designed controlled feeding regimes which resulted from the excellent collaboration between Taylor, Young and their many coworkers at the A.R.C. Animal Breeding Research Organisation, Edinburgh.

Figure 6.7 shows the astonishing responses of the mean body weight in each group $j$, $W_j(t)$, to the preassigned food intake $q_j(t)$, $j=1,2,\dots,5$ (Table 6.2). The HC, MC, and LC animals grew along practically straight lines, $W=W_0+c_jt$, $j=1,2,3$, and if there was any expectation of the relative locations of HC, MC, and LC to each other from the designed symmetry of $q_j(t)$, $j=1,2,3$, it was not fulfilled because the MC response does not bisect the angle between the HC and LC responses. The sinusodial character of the S and NS responses is readily apparent but again they lack the symmetry about the MC response which might be expected from the symmetry of the pre-assigned feeding functions $q_j(t)$, $j=2,4,5$. The most striking character displayed by the sinusoidal responses is the delay of the crossover times $t_1=45$, $t_2=65$, and $t_3=94$ wks respectively, beyond the corresponding crossover times of the food intake functions. These delays are respectively $54-36=9$, $65-60=5$, and $94-84=10$ weeks.

Overall, the responses show that Taylor and coworkers were successful in carrying out this experiment in the controlled growth region of the Growth Phase Plane. Excepting for a period of time around $T_2=65$ weeks when the NS group

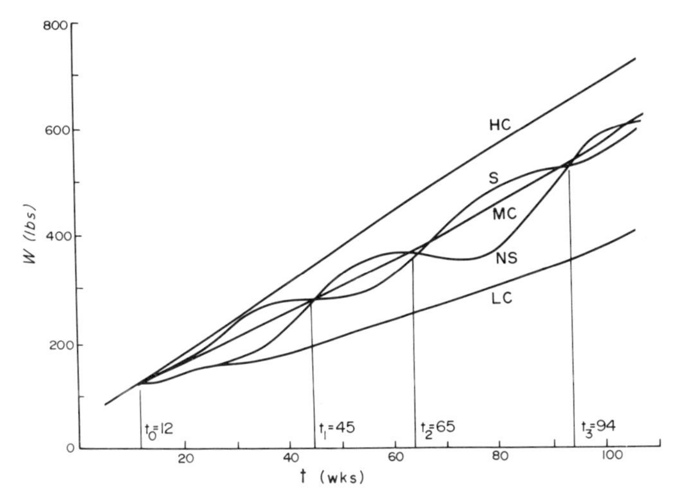


Fig. 6.7. Experimental weight responses of the cattle to the actual food intakes depicted in Fig. 6.6

lost some weight, all the groups continued growing in spite of the changes in the food intakes (Fig. 6.6).

Study of the experiments discussed in this section gave rise to some of my most interesting (and later most integrating) ideas on the phenomenology of animal growth. These experimental responses to constant, ramp, and sinusoidal feeding functions, especially the delay in the response crossover times, suggest that growth under controlled feeding might be largely described by the simple first order linear inhomogeneous differential Eq. (6.3),

$$dW/dt + gW = hq(t). \tag{6.3}$$

Here $g$ and $h$ might be some sort of grouped metabolic parameters, possibly constant but most likely varying with time. It is well known in physics that the solutions of Eq. (6.3), with various simple definitions of $q(t)$, have properties displayed by the experimental data from the Taylor and Young cattle under the constant and variable feeding regimes presented here.

However, detailed study and application of Eq. (6.3) in the biology of controlled growth, as I have applied the differential equations of the ad libitum feeding and growth, had to await my finding data from controlled growth experiments on other animals on more or less well defined feeding regimes. Chapters 6.2, 6.3, and 6.4 add information which permits a more detailed study and application of Eq. (6.3) in the final Chap. 6.6.

## 6.2 Controlled Feeding of Sheep

The following discussion is of an experiment on sheep reported by Clapperton and Blaxter (1965) and Blaxter (1968), which throws some light on the responses to step function changes of food intake (Fig. 6.1 b). The group of animals consisted of 16

Table 6.3. Results of constant feeding of sheep by Clapperton and Blaxter (1966). Reproduced by permission of the authors and the publishers, Plenum Press

| $c$ kg/wks | Remark on feeding level | Time $t$ (wks) | | |
|---|---|---|---|---|
| | | 0 | 21 | 42 |
| | | Weights (kg) | | |
| | | $W_0$ | $W$ | $W$ |
| 2.8 | Below main. | 43.8 | 37.2 | 33.2 |
| 4.9 | Maintenance | 50.0 | 50.8 | 50.6 |
| 7.0 | Above main. | 50.0 | 58.0 | 64.3 |
| 9.1 | Above main. | 54.5 | 69.1 | 78.7 |

wether sheep aged 2.5 years, half of which were Romney Marsh breed with the others being Hampshire.

The sheep had been maintained for some time on 4.9 kg of dried grass/wk. At time $t_0$ subgroups of four sheep were fed at constant rates of 2.8, 4.9, 7.0, and 9.1 kg of dried grass/wk (Table 6.3). From energy considerations Blaxter (1968) had expected the transient weight responses $W$ to these step changes of food intake to be described by the law of diminishing increments in the time domain. He calculated the new equilibrium weight $a$ and the exponential decay constant $b$ for each response. It is not the intent here to question Blaxter's theory or his values of $a$ and $b$. In the light of Taylor's work, discussed in the preceeding section, I accept Blaxter's conjecture that the law of diminishing increments applies. From Taylor's work this is expected for the two levels of feeding above 4.9 kg/wk, the prestep maintenance feeding level. The level of feeding below 4.9 kg/wk, namely 2.8 kg/wk, is a new experimental situation.

Tables 6.3 gives the weights of the subgroups of sheep at $t=0$, 21, and 42 weeks on the various levels of feeding. Minor variation of weight $W$, on the 4.9 kg/wk feeding level is evidence that 4.9 kg of dried grass/wk is the prestep maintenance feeding level. On the higher levels; $W$ increased with time, and on the lower level $W$ fell with time. The latter is expected since these animals, being put on submaintenance feeding level, were on a partial starvation ration relative to the maintenance ration of 4.9 kg/wk (Fig. 6.8).

Assuming the law of diminishing increments in time to hold in the three nonmaintenance cases and passing exactly through the initial weights ($W_0$), I graphically fitted the law to the weights at times 21 and 42 weeks after the step changes, for each feed level on 1-cycle semilog co-ordinate paper. It took three to four guesses of the equilibrium weights, $a$, for each feed level, $c$, to obtain a straight line passing through initial weights at $t=0$ and approximating the two measured live weights for $t=21$ and 42 weeks (Fig. 6.8). The slope of the line, $b$, is the exponential time decay constant of the law of diminishing return (Table 6.4). It is unfortunate that Blaxter did not present more data so that the law of diminishing increment could have been put in jeopardy. Later sections will show that Blaxter's conjecture

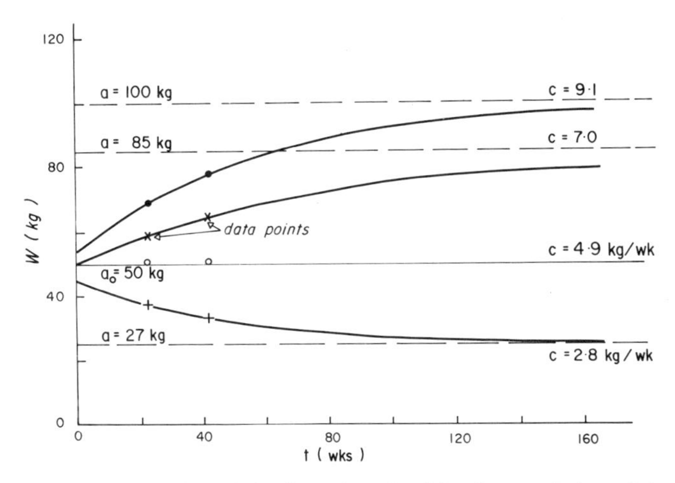


Fig. 6.8. Illustration of the dynamics of weight change of sheep fed on the step functions above and below maintenance

is well founded, though possibly not through his theoretical derivation of the law using a version (Kcal/day) of power balance.

Table 6.4 displays the values of $a$ and $b$ obtained for this experiment. On levels of feeding above maintenance, the equilibrium weight increased over the maintenance weight by 35 and 50 kg. The equilibrium weight on the submaintenance level was reduced 23 kg below maintenance. The values of $c$ and corresponding $a$, can now be used to derive an order of magnitude value of Taylor's time constant $T_0$ for sheep.

Figure 6.9 is the Taylor time constant plot of the $(c, a)$ data. The straight line was positioned approximately. This is crude but gives an approximate value of $T_0$; the scatter of the points about the line is not bad and the cumulated first moments of the points around the line is nearly zero. The slope of the line gives $T_0 = 10.8$ weeks. This is a tentative estimate of the Taylor time constant for sheep. If $A$ and $C$ were known from long term ad libitum feeding and growth data for the sheep used by Clapperton and Blaxter, this value of $T_0 = 10.8$ weeks could be checked against $A/C$.

Table 6.4. Values of parameters $a$ and $b$ for the transient growth of sheep after being put on various constant food intakes $c$

| $c$ kg/wks | $W_0$ kg | $a$ kg | $b$ 1/wks |
|---|---|---|---|
| 2.8 | 43.8 | 27 | 0.023 |
| 4.9 | 50.0 | Maintenance | — |
| 7.0 | 50.0 | 85 | 0.013 |
| 9.1 | 54.5 | 100 | 0.018 |
| | | Average $= 0.018$ wks$^{-1}$ | |

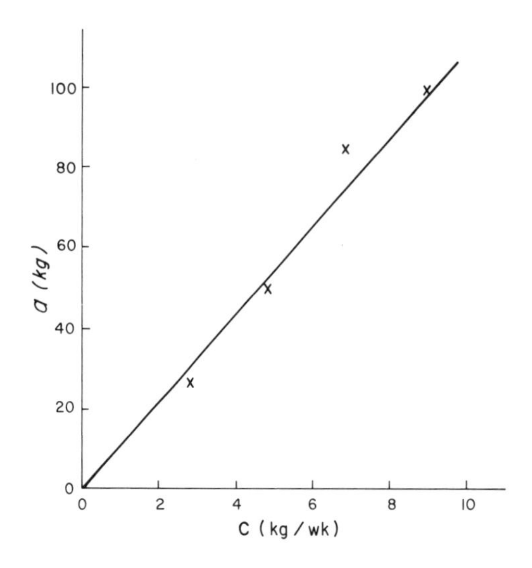


Fig. 6.9. Taylor time constant, $T_0$, graph of feeding levels, $c$, versus equilibrium weights, $a$, for sheep of Clappterton and Blaxter. *Crosses* (×) are data points from Table 6.4. Straight line drawn by eye

Figure 6.8 shows an estimate of the transient growth of the sheep from the time they were put on constant food intake, using the law of diminishing returns. The solid curves were calculated using $a$ and $b$ from Table 6.4. The intake of 4.9 kg of dried grass per week is the maintenance ration for body weight of 50 kg. The figure shows that when the food intake is stepped up or down from maintenance the body weight is exponentially adjusted to higher or lower equilibrium weights. Assuming the values of the exponential decay constant, $b$, in Table 6.4 are estimates of the same decay constant, the average value, namely $b=0.018$ wks$^{-1}$ can be taken as representative for the sheep used by Clapperton and Blaxter, until I have found long term ad libitum feeding and growth data on sheep from which $A, (AB), C, t^*$ can be determined and $b=0.018$ wks$^{-1}$ compared to $(AB)C/A$.

The drop in weight to a new equilibrium when the animal is partially starved is expected, but the possibilities that the law of diminishing returns applies in partial starvation, and that the lower equilibrium weight is the product of $T_0$ times the submaintenance constant food intake are new ideas. Figure 6.10 illustrates how the

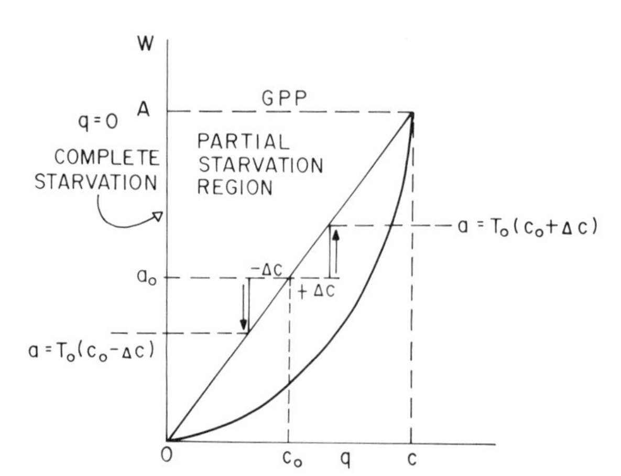


Fig. 6.10. Blaxter's experiments with sheep illustrated on the GPP

Clapperton and Blaxter experiments may be related to the ad libitum growth curve for sheep in the GPP. At feeding level $c_0$ the weight remains at $W_0$ as time passes, i.e., $c_0$ is the maintenance intake. At an above maintenance feeding level $(c_0+\Delta c)$ the weight increases asymptotically to $T_0(c_0+\Delta c)$ and on a submaintenance level the weight decreases asymptotically to $T_0(c_0-\Delta c)$. There are hints in Fig. 6.10 concerning the region in the GPP for possible experiments involving controlled growth and starvation. The diagonal appears to be a line through all the equilibrium points ($dW/dt=0$, approached from above and below) in the GPP, thus giving more meaning to the diagonal tentatively drawn in Fig. 5.5, Chap. 5, and Fig. 6.5.

The region between the ad libitum growth curve and the Taylor diagonal is a region for controlled feeding experiments of the Taylor and Blaxter types. The region above the diagonal may be available for partial starvation experiments increasing in severity as the food intake approaches zero. The submaintenance constant feeding case reported by Clapperton and Blaxter is an experiment in this region. The $W$ axis is a one dimensional region of complete starvation, food but not water is withheld. In the GPP the Taylor diagonal divides the experimental region into subregions of growth and inanition when the food intake is controlled. The preceding ideas are tentative at this point, but are interesting and possibly useful if generally substantiated. The possibility that the Taylor diagonal may not be a line as such, but may be a third region, is discussed in Sect. 6.5.

## 6.3 Controlled Feeding of Chickens

Titus et al. (1934) performed an experiment on seven pens each of male and female chickens in which one pen of each sex was ad libitum fed while the other six pens were fed at constant fractions $f_i$ of the levels of the mean ad libitum feeding rates of the ad libitum fed pens as functions of time [Eq. (6.1)]. This means the six controlled food intakes, including the ad libitum case having mature food intake, $c$, form a one parameter family of food intake curves with food intakes, $f_iC$, varied from pen to pen by varying $f_i$ but with $t^*$ kept constant. Figure 6.11 illustrates this method of controlled feeding. Curve A represents ad libitum feeding with $C$ as the mature intake, and curve B represents 50% of ad libitum with $C/2$ as the asymptotic maximum. These curves are from a one parameter family of diminishing returns curves only if $t^*$ has the same value for each. The experimental feed curves for the males and females of Titus et al. show the asymptotic food intake $c_i=f_iC$ was the only variable between pens with $t^*$ about equal to 14 wks for the males and 11 wks for the females (Fig. 6.11).

In as much as the experiment was continued for about a year, the chickens had ample time to asymptotically approach equilibrium weights, $a_i$, and final controlled food intakes $c_i=f_iC$ for both sexes.

The set of equilibrium weights, $a_i$, supported by the maximum food rates, $c_i$, for males and females are of interest here. Titus et al. analysed the data by the law of diminishing return of $W$ in the cumulative food consumed domain $F$ and found values of $a_i$ corresponding to the values of $c_i$ as shown in Table 6.5. The values of $c_i$ in kg/wk were found by reading the asymptotic ad libitum values from the graphs

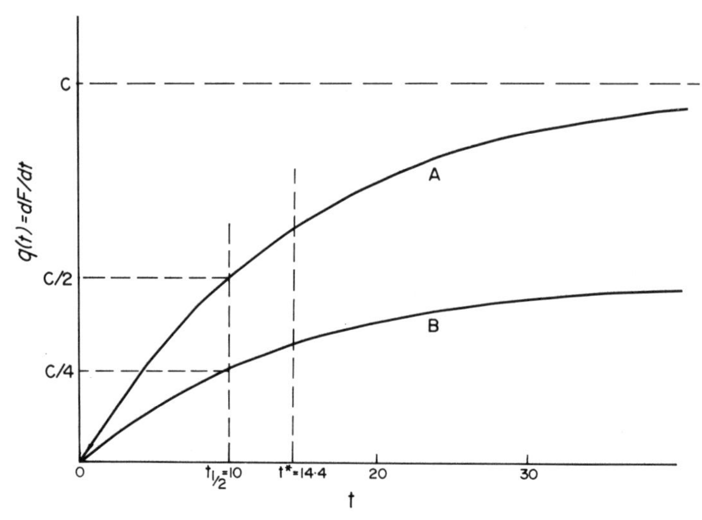


Fig. 6.11. Two feeding curves from a one parameter family of food intake curves of the type used by Titus et al. (1934). *Curve A* has an asymptotic maximum of $C$ and *curve B* has $C/2$ as its asymptotic maximum. The half times $t_{1/2}$ and therefore $t^* = t_{1/2}/\ln 2$ are the same for both curves, as they were for all the experimental pens

Table 6.5. Equilibrium weights, $a_i$, of male and female chickens versus the allowed mature food intake, $c_i$

| Males | | Females | |
|---|---|---|---|
| $c_i$ kg/wks | $a_i$ kg | $c_i$ kg/wk | $a_i$ kg |
| 0.91 (ad lib.) | 3.68 | 0.63 (ad lib.) | 2.17 |
| 0.62 | 2.96 | 0.53 | 2.36 |
| 0.54 | 2.62 | 0.46 | 2.16 |
| 0.45 | 2.24 | 0.38 | 1.97 |
| 0.36 | 1.65 | 0.31 | 1.64 |
| 0.27 | 0.985 | 0.23 | 1.23 |
| 0.18 | 0.638 | 0.15 | 0.645 |

in Figs. 7 and 8 of Titus et al. and multiplying by the percentage of ad libitum food intake of the other pens given in their Table 3, p. 822.

Figure 6.12 shows a plot of $a_i$ versus $c_i$ for both sexes. The scatter is probably evidence of the difficulties of managing an experiment of this size and duration. Regardless of the possibility of some statistical outliers, regression was used on all the data to get $T_0 = 4.25 \pm 0.19$ wks for males and females.

Titus et al. went further by fitting the law of diminishing increments to the data of $W$ versus $F$ for all the pens. They found that $(AB)$ was virtually constant over the degree of control imposed on the long term food intake (Table 6.6). I was surprised at this result, since I had no a priori reason to expect it. These were controlled feeding experiments and I would not have expected the $W$ versus $F$ data of each pen to be described by the law of diminishing returns with $(AB)$ constant.

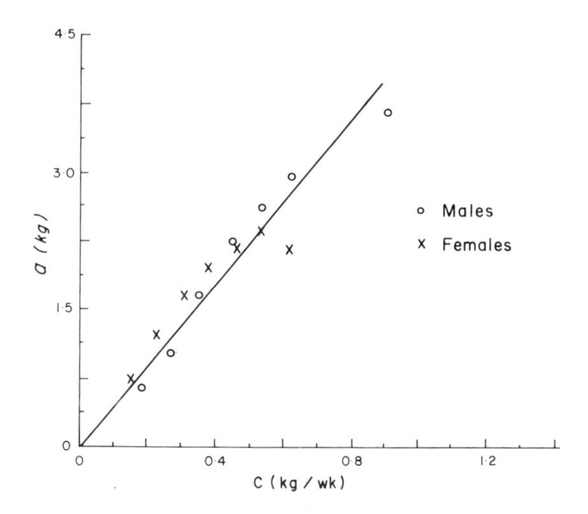


Fig. 6.12. Plot of equilibrium weights, $a_i$, of chickens versus the asymptotic maximum food intakes, $c_i$, in the experiments of Titus et al. (1934)

However the following reasoning may show why my expectations were not fulfilled.

Titus et al. had modified the appetency factor $C/t^*$ by reducing $C$, but kept $t^*$ constant, thereby reducing the cumulative food $F$ in proportion to $C$ only. Since $A$ was also reduced in proportion to $C/T_0$ being about constant for all pens, Fig. 6.12), the $Z$–function $[(AB)F/A]$ therefore was invariant among all the pens, so that the normalised weights, $u$, had to follow the law of diminishing returns, namely $u=1-\exp(-Z)$ (Chap. 5). By not changing $t^*$ between the ad libitum pens and all the others, Titus et al. had not altered the appetite controls of their chickens as completely as Taylor, Young, Clapperton, and Blaxter did in their experiments with cattle and sheep, where the appetites of the animals were completely overruled. Furthermore, the value of $\alpha$, the ad libitum feeding and growth discriminant, computed by $(AB)c_i t^*/a_i$ for each pen of both sexes, is high but not greater than unity, so that this experiment is not truly in the category of controlled growth as defined in this chapter.

Table 6.6. Values of the growth efficiency factor $(AB)$ from Table 3 of Titus et al. (1934, p. 822)

| Males | | Females | |
|---|---|---|---|
| % Ad lib. intake | $(AB)$ | % Ad lib. intake | $(AB)$ |
| 100.0 | 0.340 | 100.0 | 0.347 |
| 68.6 | 0.355 | 85.4 | 0.351 |
| 58.9 | 0.364 | 73.5 | 0.363 |
| 49.0 | 0.365 | 60.9 | 0.355 |
| 39.2 | 0.337 | 49.2 | 0.359 |
| 29.4 | 0.366 | 36.8 | 0.344 |
| 19.6 | 0.329 | 24.5 | 0.309 |

The overall average of $(AB)=0.349$

## 6.4 Complete and Partial Starvation

Animals in starved states have been studied for decades. The literature abounds with articles, texts, and reference works on the physiology, histology, biochemistry, etc. of starvation. However, little of this literature discusses the dynamics of weight change during complete and partial starvation. In the following, a few cases are presented with nonlinear regression analysis of data for animals from the jellyfish to man, using Eq. (6.4). The tables of data are in Appendix C.

$$W=(W_0-W_f)\exp(-bt)+W_f. \tag{6.4}$$

It will be seen shortly that this equation has been used in describing weight loss with time during starvation of lower and higher forms of life. Here $W$ is weight at time $t$; $W_0$ is initial weight when complete or partial starvation began and $W_f$ is the final asymptotic weight; $b$ is the exponential decay constant. The final weight, $W_f$, is zero for cases of complete starvation. Equation (6.4) states that $W$ decreases in such a manner that $W-W_f=(W_0-W_f)/2$ when $t=\ln 2/b$ usually designated $T_{1/2}$, i.e. the half time for decrease of weight $W$. The differential equation of (6.4), namely

$$dW/dt=-b(W-W_f), \tag{6.5}$$

shows that the rate of loss of weight is proportional to the weight remaining to be lost. Equations (6.4) and (6.5) are common to many natural phenomenon such as loss of charge from an electrical capacitor, radioactive decay (see Chap. 1), diffusion of a fluid from a porous container and the loss of heat from a thermal reservoir.

The equations proposed are empirical because there are no more fundamental biological quantities for calculation of $W_f$ and $b$ without appeal to the experiments discussed here. Some effort has been expended to make them more biologically plausible, such as the work by A.G. Mayer (1914) in his exhaustive study of the complete starvation ($W_f=0$) of the coelenterate *Cassiopea Xamachana* (jellyfish). He found that the chemical composition of the jellyfish was independent of the weight loss, which was known not to be true of higher animals. The chemical composition of the body mass of higher animals is drastically changed during starvation (Kleiber 1961, Table 3.3, p. 39). Mayer thought the simple Eq. (6.4) fitted his data with $W_f=0$, because the chemical composition of the body mass of *Cassiopea Xamachana* was constant throughout starvation (Fig. 6.13). Robertson (1923) discussing Mayer's results appeared convinced that the good fit of Eq. (6.4) signified the involvement of a monomolecular chemical basis of growth and loss of weight by organisms. See Chap. 1 for the critical assessment of Robertson's concepts. The implication of all these nonquantitative arguments is that equations such as Eq. (6.4) and (6.5) may be expected not to apply to weight variation of starving higher types of animals. I readily admit that these equations and those previously presented may not hold in biology with the apparent finality that engineering scientists have found in certain classes of physical phenomena but I cannot accept the ruling out of their use in the biology of feeding and growth solely because of the complexity of biological systems.

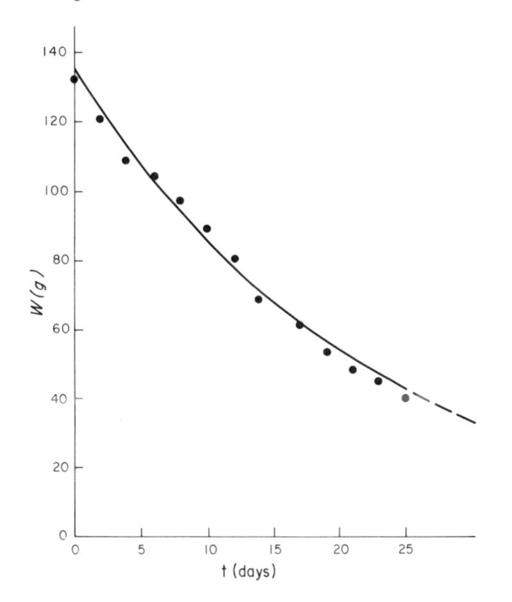


Fig. 6.13. Complete starvation of jellyfish *Cassiopea xamachana* in diffuse light. The ● are data points (Appendix Table C-1). *Solid curve* is calculated showing exponential decay to zero

Fourteen years before Mayer, Avrorov had published data on the starvation of an 8.8 kg dog for 67 days under well controlled conditions. However it was not until 1923 that Dr. Hecht, at the request of Morgulis (1923, p. 86) graphically fitted Eq. (6.4) ($W_f = 0$) to Avrorov's data. Using Fig. 3, p. 87 of Morgulis, I calculated $b = 0.024$ weeks. Except for the data points at 60, 65, and 67 days, the fit of Eq. (6.4) to the data is excellent. Lang presents the argument of complexity of the starving process against the use of Eq. (6.4) (Morgulis 1923, p. 89). I neither accept Mayer's argument of simplicity nor reject Lang's argument of the complexity of the dynamics of loss of body mass during partial and complete starvation: it is in such arguments that biomathematics can lose much of its power.

Tables C-1 to C-7 of the Appendix display data for several cases of complete and partial starvation of a wide range of species. All the cases, except that of the group of 32 men in the Minnesota experiments (Keys et al. 1950), are of individual animals. Keys presents the data for individuals as well as the group, but Tables C-6 and C-7 display only the group means which I used to study the dynamics of weight loss of some humans.

Table 6.7 summarises the results of regression on the data in Tables C-1 to C-7. All the parameters are expressed in units of kg and weeks with the use of appropriate conversion factors when the units in the tables in the Appendix differ. Only the mean values of parameters and the residual standard deviation [rsd($W$)] of each fit of Eq. (6.4) are shown. The results are arranged in descending order of starting weight $W_0$. The experimental food intakes as percentages of the pre-experimental food intakes are also shown. The results in Table 6.7 indicate the applicability of Eq. (6.4) to the dynamic weight changes of animals during starvation. The parameter $b$ may indicate something about how animals metabolize themselves during starvation: small animals appear to do this more rapidly than large ones because $b$ for them is about 10 times greater, indicating more rapid decline of $W$ when they are starved.

Table 6.7. Mean value of parameters $W_0$, $b$, and $W_f$, for various species, obtained by fitting Eq. (6.4) to data on complete and partial starvation given in Tables C–42 to C–48 of the Appendix

| Animal | $dF/dt$ %[a] | $W_0$ kg | $b$ $wk^{-1}$ | $W_f$ kg | rsd ($W$) kg | Appendix Table |
|---|---|---|---|---|---|---|
| *Cassiopea xamachana* | | | | | | |
| in dark | 0 | 0.0856 | 0.546 | 0 | 0.0018 | C–1 |
| Same in diffuse light | 0 | 0.1308 | 0.326 | 0 | 0.0026 | C–2 |
| Dog | 0 | 8.8 | 0.024 | 0 | | Morgulis (1923) |
| Dog (Oscar) | 0 | 25.8 | 0.060 | 0 | 0.39 | C–3 |
| Sheep | 58 | 43.8 | 0.023 | 27.1 | | Section 6.2 |
| Human – Succi | | | | | | |
| 30-day fast | 0 | 61.4 | 0.052 | 0 | 0.72 | C–4 |
| 40-day fast | 0 | 54.9 | 0.051 | 0 | 0.72 | C–5 |
| Human | | | | | | |
| 32 men | 44.9 | 69.8 | 0.074 | 48.5 | 0.27 | C–7 |

[a] Percent of pre-experimental intake

Mayer's (1914) experiments on *Cassiopea Xamachana* covered the effects of a wide range of environmental factor plus some experiments on jellyfish with excised stomachs. He found that Eq. (6.4) satisfactorily described the general exponential decrease of body weight with time without food. However, in all the cases there were systematic oscillations of the data around the mean curve as shown in Fig. 6.13.

Kleiber (1961, pp. 25–28) discussed the case of starvation of dog Oscar in detail as an example of linear regression using the log transformation of Eq. (6.4) with $W_f = 0$ as a "working hypothesis." He found a satisfactory fit to the data as shown in Fig. 6.14. The calculated curve tracks the major variation of weight in 110 days. Here again are seen oscillatory systematic deviations from the mean curve. Kleiber hints at the applicability of Eq. (6.4) to the loss of weight of a professional human starver named Succi during a 30 day fast, leaving the numerical analysis to the reader as an exercise in curve fitting. Using the average of Succi's daily losses at Florence and Naples (Benedict 1907, p. 304, Table 185) I obtained the data in Table C-2. The data for Succi's 40 day fast came from Table 2, p. 912, Sect. 4 of the Handbook of Physiology (see Morgulis 1923, p. 90, Table 1). It is interesting to note in Table 6.7 that the different starting weights of Succi had no effect on the value of $b$. Mayer (1914) also showed $b$ independent of the starting weights of his starving jellyfish.

By far the best experiment I have found on semistarvation is that discussed by Keys et al. (1950) in which a group of 32 men were held on 44.9% of normal daily calorie intake for 24 weeks (Tables C-6 and C-7). Figure 6.15 gives satisfactory evidence that the average weight of the group fell exponentially towards a non zero equilibrium weight.

All the cases of starvation discussed here are unique experiments unrelated to the feeding and growth of the animals to get the animals to the initial weights used in the experiments. Yet these experiments point to the conclusions that animals in complete starvation lose weight exponentially towards zero equilibrium weight and semistarved animals exponentially lose weight towards some non zero equilibrium

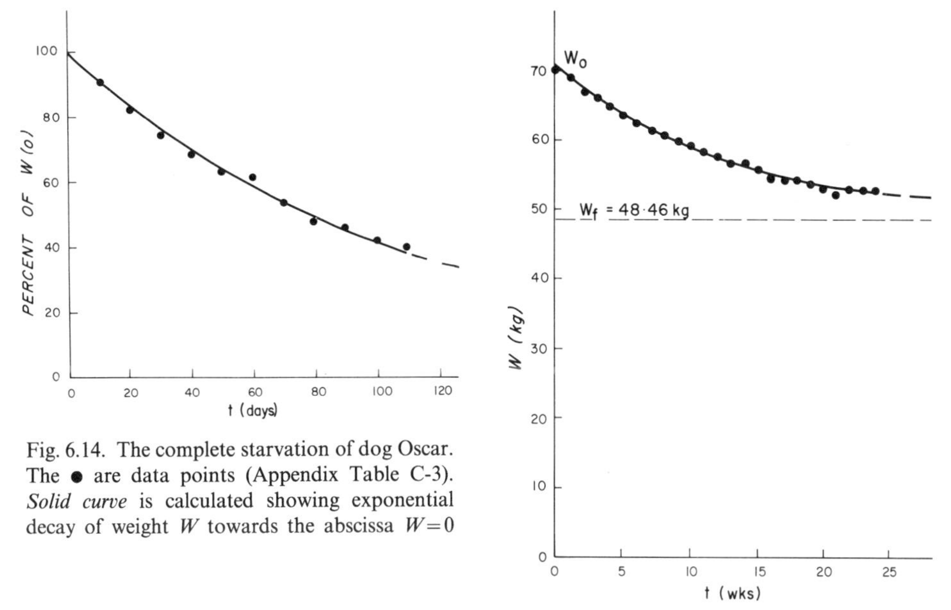


Fig. 6.14. The complete starvation of dog Oscar. The ● are data points (Appendix Table C-3). *Solid curve* is calculated showing exponential decay of weight $W$ towards the abscissa $W=0$

Fig. 6.15. Semi-starvation of 32 men in the Minnesota experiment on an energy intake of 1,569.7 kcal/day. The ● are data points (Appendix Table C-7). *Solid curve* is calculated showing exponential decrease of weight $W$ to a nonzero equilibrium weight, $W_f$

weights, thereby implying the possibility of the existence of the Taylor diagonal in the GPP in the process of starvation as well as in the processes of controlled growth reported in the preceding sections of this chapter.

## 6.5 The Growth Phase Plane and the Taylor Diagonal

The experimental evidence presented here, although not nearly complete, led me to the intuitive belief that the Taylor diagonal had some real significance in the GPP because of its apparent separation of the regions of controlled growth and partial starvation, with the $W$ axis $(q=0)$ being the line of complete starvation (Fig. 6.16). When the phase point $(q, W)$ is in the controlled growth region it will tend to drift upward towards the diagonal with a diminishing rate $(dW/dt)$ approaching zero, and when the phase point is in the partial starvation region it tends to drift to the diagonal with a diminishing rate $(-dW/dt)$ approaching zero. Further, when the phase point is on the ad libitum growth curve, and the animal is continually fed ad libitum, the phase point will generate a curve which is the lower bound of the controlled growth region and finally drift to the extreme point $S(C, A)$ on the diagonal where the growth rate $(dW/dt)$ approaches zero and the animal is considered mature. The reasons for these drifts toward the Taylor diagonal are of an energetic nature, i.e. when the phase point $(q, W)$ is in the controlled growth region there is sufficient energy above energetic maintenance for the animal to grow until at the diagonal the energy for growth is exhausted and $dW/dt$ goes

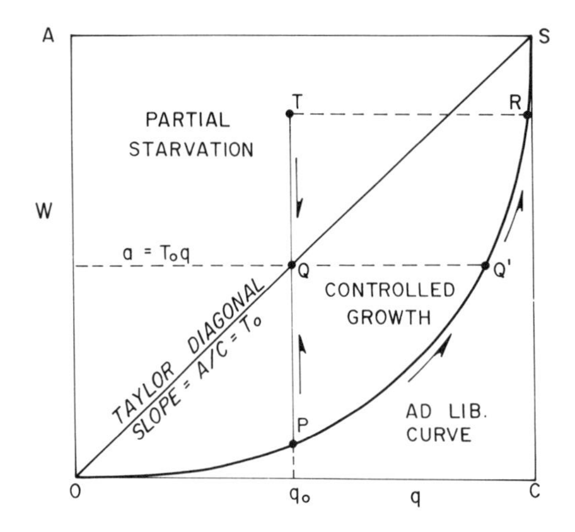


Fig. 6.16. The GPP is divided into regions of control growth and partial starvation by the Taylor diagonal, $OS$, and the ad libitum growth curve, $OPS$. With constant food intake $q_0$ animals at $P$ grow towards $Q$ and at $T$ lose weight towards $Q$. *Arrows* show direction of response in time. The time axis is perpendicular to the plane of this figure

to zero with the final weight given by

$$W_f = T_0 q, \tag{6.6}$$

and when the phase point is in the partial starvation region the energy intake is insufficient to maintain $W$ and it will fall until it reaches the diagonal where $-dW/dt$ goes to zero with final weight also given by Eq. (6.6).

The GPP conceptually integrates growth under ad libitum and controlled feeding with weight loss under partial and complete starvation. The Taylor diagonal can be thought of as a line at the bottom of a trough into which the phase points tend to settle eventually, unless the food intake is manipulated to place them on either side of the diagonal. Even so, the phase point will always tend toward the Taylor diagonal. By feeding freely the phase point $P$ on the ad libitum growth phase curve skirts, from near point 0 (immaturity), around the edge of the trough until it finally settles at the equilibrium point $S$ (maturity) which is a known point on the Taylor diagonal. It will probably be found that the Taylor diagonal, instead of being a sharp line as approached from both sides, may be a region of varying breadth especially marked by death in the approach from the partial and complete starvation side.

Equation (6.6) is a surprising experimental conclusion (Taylor and Young 1967, 1968) regarding the Taylor diagonal, since it has long been held that to maintain body weight, animals should be fed in proportion to their "metabolic" body size $W^{0.75}$, i.e. $q = hW^{0.75}$ where $h$ is some constant factor (Kleiber 1961). In order to check this discrepancy from one of the laws of animal science Taylor and Young performed linear regressions of log $q$ versus log $W$ of their data (Table 6.1) and found that the coefficient of log $W$ was not significantly different from unity but significantly different from 0.75. Since the beginning of my studies of animal feeding and growth I have strongly suspected the validity of extrapolating the relation between basal metabolic rate and $A^{0.75}$ (which had been shown quite valid for mature animals of weight $A$) to immature growing animals – my suspicions are now much deepened.

Recently Thonney et al. (1976) have re-evaluated the concept of metabolic size being proportional to $W^{0.75}$. A conclusion from their study of the fasting heat production, of males and females of species from rats to cattle, versus body weight, is that the heat production is adequately predicted in most cases by a linear relation to body weight which has a nonzero intercept.

I should point out that I have found in my studies of feeding and growth of animals, a fasting animal (that is one the phase point of which is (0, $W$) has no theoretical significance. A detailed study of the relation of my theory to the power balance to which living animals must adhere is in Chap. 11.

## 6.6 The Controlled Feeding and Growth Differential Equation

I have already remarked how the solutions of Eq. (6.3), repeated here for convenience,

$$dW/dt + gW = hq(t) \tag{6.3}$$

have been found to simulate the response to controlled feeding regimes $q(t)$. For example the exponential fall of weight [Eq. (6.4)] when the phase point is in the partial starvation region, and the food intake is the constant $q_0$, is a solution of Eq. (6.3). This can be shown by substituting Eq. (6.4) in (6.3) and letting time approach infinity thereby getting

$$W_f = (h/g)q_0, \tag{6.7}$$

where $q_0$ is the constant submaintenance food intake for the initial weight $W_0$. I have already suggested that when the phase point is in the controlled growth region and $q_0$ is a constant food intake the law of diminishing returns in the time domain, i.e., Eq. (6.2) applies

$$W = (T_0 q_0 - W_0)[1 - \exp(-bt)] + W_0, \tag{6.2}$$

where $b$ may be an estimate of Brody's $k$. This also is a solution of Eq. (6.3) with the final weight $W_f$, when $t$ approaches infinity, given by

$$W_f = T_0 q_0. \tag{6.8}$$

Equations (6.7) and (6.8) suggest that the differential Eq. (6.3) may hold in both the starvation and controlled growth regions with $h = gT_0$. However, parameters $h$ and $g$ may not be the same in both regions, but when some point on the Taylor diagonal is approached from either side, where $-(dW/dt) = 0$ or $dW/dt = 0$, then $W_f = T_0 q_0$ and $T_0 = h/g$, implying that if the values of parameters $g$ and $h$ are different in both of the regions their ratios $(h/g)$ must approach equality at least in some region bordering the Taylor diagonal if not on the diagonal per se.

To my knowledge no experiment has been reported in which the dynamics of the approach to the Taylor diagonal from both sides has been carefully measured for the same breed of animals, so that the equalities of $g$ and $h$ and the ratio $h/g$ between the regions could be tested. Such an experiment is suggested by the experimental plan shown in the GPP shown in Fig. 6.16. But, first ad libitum studies must have been done to establish the value of the Taylor time constant $T_0$ for the animals to be subjected to experiment. Here the animals are taken to point $P$ along the ad libitum feeding and growth curve, where the food intake is fixed at $q=q_0$. The live weight data are then taken at equal time intervals until the phase point is near $Q$, where the animals are rehabilitated to ad libitum food intake at $Q'$ and allowed to grow to $R$ where they are put back on food intake $q_0$ at point $T$ in the partial starvation region. Live weight data are then taken at equal time intervals as the weight falls from $T$ to $Q$. Data from experiments such as this repeated at various points along the Taylor diagonal could then be used to test the hypotheses that the law of diminishing returns in the time domain, with various constant food intakes, applies in both the controlled growth and partial starvation regions of the GPP, and that $g, h$, and $h/g=T_0$ are the same in both regions. Until such experiments are performed I shall take $T_0=A/C$ as the slope of a line in the GPP.

## 6.7 The Differential Equation of Controlled Feeding and Growth

The units of Eq. (6.3) are the same for all its terms $(dW/dt)$, $gW$, and $hq(t)$, namely kg of live weight per unit of time (say week), if and only if $q$ has the units of $\text{wk}^{-1}$ and $h$ the units of live weight per unit of food consumed, i.e. h is an efficiency factor. With this understanding Eq. (6.3) can be read as saying the following.

The term $hq(t)$ is the live weight equivalent of the food intake $q(t)$ which is distributed between growth $dW/dt$ and no-growth $gW$. Here no-growth must not be confused with the term maintenance, because traditionally maintenance has been associated with energetics. Here $gW$ means only that it is a portion of the live weight equivalent of the food intake $q(t)$ which, for some unknown metabolic reasons, does not contribute to the growth of the animal. Equation (6.3) is a balance equation which applies to animals the phase points $(q, W)$ of which lie either in the controlled growth region of the GPP or in the starvation region, i.e., it is applicable for all the points of the GPP except those on the ad libitum growth curve bounding the GPP.

Intuitively I take $g$ as being Brody's $k$, which has been shown in Chap. 5 to be $(AB)C/A$. But this quantity from what has been said in the preceeding sections about $T_0=A/C$, then becomes $k=(AB)/T_0$. On taking $h$ to be the efficiency factor $(AB)$ Eq. (6.3) becomes

$$dW/dt+[(AB)/T_0]W=(AB)q(t), \tag{6.9}$$

which can be solved for $W(t)$ as the response to any given controlling feeding function $q(t)$, given the initial conditions that at $t=0$, $W(0)=W_0$, and $q(0)=q_0$. The solution has already been given for $q(t)=q_0$ for all $t>0$, namely Eq. (6.2). But what are the solutions when $q(t)$ is a ramp feeding function, or a ramp with a superimposed sine wave having a linearly increasing amplitude? Answers to these questions can be found and should be useful in the study of the responses Taylor and Young got in their experiments with Ayrshire twin cattle using the experimental feeding functions given in Table 6.2 (Chap. 6.1.2).

The live weight response as a general solution of Eq. (6.9) is given by the following integral

$$W(t) = (AB)\exp[-(AB)t/T_0]\int_0^t q(t)\exp[(AB)t/T_0]\,dt$$
$$+ W_0\exp[-(AB)t/T_0]\,. \qquad (6.10)$$

This equation, though it appears formidable, is reducible to sums of simple integral functions depending on how the feeding functions, $q(t)$, are defined. Table 6.2 shows the various definitions of $q(t)$ on the age intervals of 4–12 weeks and 12–108 weeks. Equation (6.10) can then be separately evaluated over the ranges of time, $t$, from 0–8 weeks and 0–96 weeks respectively with function $I$ in Table 6.2 applying in the first range and the other functions applying in the second. The initial weight, $W_0$, for the first range is given as 82.5 lbs. The initial weight, $W_0$, for the second range must be calculated as the weight attained after holding $q(t)$ constant at 23.25 lbs of food per week over the time range 0–8 weeks. In other words, $W_0$, for all the other functions defined on the second range is $W(8)$ at the end of the first range and calculated by

$$W_0 = W(8) = (23.25\,T_0 - 82.5)\{1 - \exp[-(AB)8T_0]\} + 82.5,$$

using Eq. (6.2), already shown to be a solution of Eq. (6.10) when $q(t)$ is a constant feeding rate. For convenience of writing the various integrals of the $q(t)$ functions, let us put aside for later reference the factor $(AB)\exp[-(AB)t/T_0]$ and the second term $W_0\exp[-(AB)t/T_0]$ on the right hand side of Eq. (6.10). Also it is convenient to represent the numbers 23.25, 15.5, 1.0, 0.75 etc. in Table 6.2 by the symbols $p$, $r$, $s$, $u$ etc. With the above ranges of integration, their initial weights, $W_0$, and the preceding steps agreed upon, Table 6.8 lists the simple integrals that are required to build up a solution $W(t)$ of Eq. (6.10) corresponding to any $q(t)$ in Table 6.2. Table 6.9 provides the evaluation of each of the integrals in Table 6.8 so that after a solution for $W(t)$ is assembled for a particular feeding function $q(t)$ in Table 6.2, for which values of $p, r, s$, and/or $u$ are known, a value of $W(t)$, at a particular $t$,

Table 6.8. The list of four simple integrals occurring in the equations for $W(t)$ associated with the food intake function $q_j(t)$ in Table 6.2

| No. | Integral |
|---|---|
| 1. | $p\int_0^t \exp[(AB)t/T_0]\,dt$ |
| 2. | $r\int_0^t t\exp[(AB)t/T_0]\,dt$ |
| 3.[a] | $s\int_0^t \exp[(AB)t/T_0]\sin(2\pi t/T)\,dt$ |
| 4.[a] | $u\int_0^t t\exp[(AB)t/T_0]\sin(2\pi t/T)\,dt$ |

[a] $T$ = period of sine wave

Table 6.9. Evaluations of the definitive integrals in Table 6.8 versus number

| No. | Functional form of the evaluation |
|---|---|
| 1. | $[pT_0/(AB)]\{\exp[(AB)t/T_0]-1\}$ |
| 2. | $[rT_0/(AB)]\{\exp[(AB)t/T_0]\,[t-T_0/(AB)]+T_0/(AB)\}$ |
| 3. | $\{s/[(AB)^2/T_0^2+4\pi^2/T^2]^{1/2}\}\{\exp[(AB)t/T_0]\sin(2\pi t/T-\phi)+\sin\phi\}$, where $\phi=\arctan[2\pi T_0/(AB)T]$ |
| 4. | $\{u/[(AB)^2/T_0^2+4\pi^2/T^2]^{1/2}\}$ $\cdot\langle t\exp[(AB)t/T_0]\sin(2\pi t/T-\Upsilon)-\{\exp[(AB)t/T_0]\sin(2\pi t/T-\psi)+\sin\psi\}/[(AB)^2/T_0^2+4\pi^2/T^2]^{1/2}\rangle$, where again $\phi=\arctan[2\pi T_0/(AB)T]$ and $\psi=2\phi$. |

Note: In 2. there is a time lag $=T_0/(AB)$. In 3. and 4. $\phi$ is a phase lag equivalent to a time lag of $\phi T/2\pi$ and the phase lag $\psi$ is equivalent to the time lag $\phi T/\pi$. $T$ is the period of the sine waves

can be calculated after the assembly is multiplied by $(AB)\exp[-(AB)t/T_0]$, i.e. the factor of the integral in Eq. (6.10) which was set aside at the beginning, and the proper $W_0\exp[-(AB)t/T_0]$ added. However the computations cannot be made, as the reader has no doubt noticed, until values of $(AB)$ and $T_0$ are known for Taylor's Ayrshire cattle. In as much as no ad libitum feeding experiments on these cattle have been reported, I chose $T_0=10$ wks (Fig. 6.2) and $(AB)=0.19$ (Table 4.1, Chap. 4, Monteiro's results on Friesians) as the values of these parameters with the hope that they are close enough for calculating the responses of Taylor's Ayrshires as solutions of differential Eq. (6.9) with the initial weight of 82.5 lbs and the given values of $q(t)$ in Table 6.2.

Using the recipes for calculating the various responses from the combinations of the proper evaluations of the integrals in Table 6.9 and the proper values of $p, r, s$, and $u$ from Table 6.2, I computed $W(t)$ for each of the $q(t)$ on the given ranges of $t$ at 2 week intervals. The results of these calculations are shown in Fig. 6.17.

Comparison of Fig. 6.17 with Fig. 6.7 shows the strong overall resemblances of the computed responses to Taylor's experimental responses, thereby giving some support to the intuition that differential Eq. (6.9) does in fact apply in the controlled growth region of the GPP. This is especially apparent in the comparison of the crossover times $T_1$, $T_2$, and $T_3$ and how they are delayed behind the crossover times in the specified feeding functions MC, S, and NS. These lag times are dependent on the phase lags $\phi$ and $\Psi=2\phi$ shown in Table 6.9 for the integral evaluations 3 and 4. The values of $\phi$ and $\Psi$ are respectively calculated as 1.43 and 2.86 radians corresponding to lag times 10.9 and 21.8 weeks. Since the evaluation number 4 (Table 6.9) shows that the term involving $\phi$ is weighted by time $t$ and the term involving $\Psi$ has no corresponding weight it can be concluded that the delay time of 21.8 weeks will dominate the 10.9 weeks delay near $t=0$ and therefore tend to elongate the first loop of the response. But by the time $t=40$ weeks the delay of 10.9 weeks will be dominant and the crossover time of the response should approximate the feeding crossover time of 36 weeks (Fig. 6.6) plus 10.9 weeks equal to 46.9 weeks which is near that observed in Fig. 6.7, namely 45 weeks. However

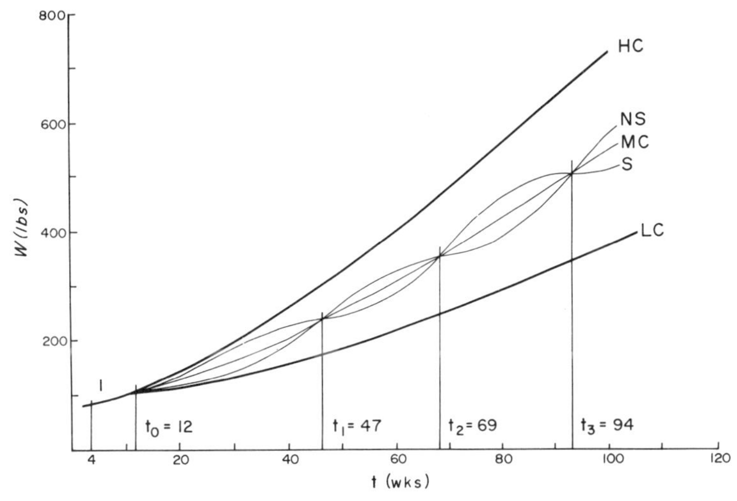


Fig. 6.17. Computed weight responses as solutions of the controlled feeding differential equation, Eq. (6.9), with the theoretical intake functions, $q_j(t)$, given in Table 6.2 for comparison with the actual responses shown in Fig. 6.7

there is a large difference between the computed and observed crossover times $T_2$, namely about 5 weeks at a time $t$ of about 60 weeks. In Fig. 6.7 it is seen that the observed crossover time is quite indeterminate. If the solutions of Eq. (6.9) are adhered to, then around 60 weeks the animals experienced some kind of metabolic and/or management changes affecting temporarily the values of $(AB)$ and/or $T_0$. I say temporarily because by about 90 weeks into the experiment the computed crossover time at $T_3$ is the feeding crossover time 84 plus 10.9 weeks equal to 94.9 weeks which compares favorably with the observed crossover time of about 94 weeks. That there were some kind of metabolic changes can best be seen in the observed response to the NS feeding regime (Fig. 6.7). Here in the time period between 60 and 90 weeks the animals show a loss of weight which they appear to regain sufficiently to give approximately the expected crossover time of 95 weeks. Furthermore the apparent straightness of the observed responses to the HC, MC, and LC feeding regimes, when these responses are computed to be monotonically curving upwards, suggest that, again if Eq. (6.9) is strictly adhered to, $dW/dt$ for the observed responses being three different constants must indicate variation of some sort in $(AB)$ and/or $T_0$ as the animals age. Letting $dW/dt$ be some constant, say $a$, for the HC response, and choosing the HC feeding regime $q(t) = 31.0 + 1.0\,t$ (Table 6.2), Eq. (6.9) would give the computed response as

$$W(t) = 31.0\,T_0 + T_0 t - T_0 a/(AB),$$

which indicates that from 12 weeks of age onward the values of $(AB)$ and $T_0$ must be changing during the experiment, because the computed HC response in Fig. 6.17 is far from linear. Despite these observed local differences between the observed and computed responses the overall similarities encouraged me to change the growth parameters from $A, (AB), C$, and $t^*$ to $A, (AB), T_0$, and $t^*$ as being biologically more fundamental and to incorporate them in the differential Eq. (6.9). I was also encouraged to couple Eq. (6.9) with the differential equations for ad libitum feeding and growth (Chap. 2) to formulate an empirical theory of the feeding and growth of animals, which I shall discuss in detail in the next chapter.

## References

Benedict FG (1907) Publ Carnegie Inst 77:304

Blaxter KL (1968) The effect of dietary energy supply on growth. In: Lodge GA, Lamming GE (eds) Growth and development of mammals. Plenum Press, New York

Clapperton JL, Blaxter KL (1965) Absence of long term adaption in the energy metabolism of sheep on constant feed. Proc Nutr Soc 24:33

Hopkins FG (1912) Feeding experiments illustrating the importance of accessory factors in normal dietaries. J Physiol (London) 44:425–460

Keys A, Brozek J, Henschel A, Mickelsen O, Taylor HS (1950) The biology of human starvation, vol II. Univ Minnesota, Minneapolis, p 1133

Kleiber M (1961) The fire of life. Wiley, New York

Mayer AG (1914) The law of governing the loss of weight in starving Cassiopea. Pap Totugas Lab 6:55–82

Mitchell HH, Beadles JR (1930) The paired feeding method in nutrition experiments and its application to the problem of cystine deficiencies in food proteins. J Nutr Soc 2:225–243

Morgulis S (1923) Fasting and undernutrition. Dutton, New York

Robertson TB (1923) The chemical basis of growth and senescence. Lippincott, Philadelphia

Taylor St CS, Young GB (1966) Variation in growth and efficiency in twin cattle with live weight and food intake controlled. J Agric Sci 66:67–85

Taylor St CS, Young GB (1967) Variation in growth and efficiency of twin cattle on constant feed levels. Anim Prod 9:295–311

Taylor St CS, Young GB (1968) Equilibrium weight in relation to food intake and genotype in twin cattle. Anim Prod 10:393–412

Thonney ML, Touchberry RW, Goodrich RD, Meiske JC (1976) Intraspecies relationship between fasting heat production and body weight; a reevaluation of $W^{0.75}$. J Anim Sci 43:692–703

Titus HW, Jull MA, Hendricks WA (1934) Growth of chickens as a function of the feed consumed. J Agric Res 48:817–835

# Chapter 7 The Theory

After examination of past efforts to develop growth equations for accurate description of live weight versus age data, a combination of three continuous functions, namely cumulative ad libitum food consumed versus age, live weight versus cumulative food consumed, and live weight versus age, was suggested in Chaps. 1–3 as an alternative description of the ad libitum feeding and growth of animals.

These functions were shown to be not independent of each other, since, given any two of them the third could be found by substituting either of the given ones in the other, thereby suggesting that the growth curve is three dimensional rather than the two dimensional curve it is usually considered to be. Furthermore the three functions are parsimonious descriptions of growth and feeding data in that they involve only four characteristic parameters, namely $A$, $(AB)$, C, and $t^*$, with the age of the animal as the independent variable. Here $A$ is the mature weight; $(AB)$ the growth efficiency factor; $C$ the mature food intake; and $t^*$ is from Brody. The differential equations of these functions were briefly discussed in Chap. 2 to make clear the distinction between the four parameters characteristic of the animal and its environment, and the initial conditions, namely $D$, the initial food intake of the animal and $W_0$, its initial live weight at the time food intake and live weight data collection began.

Chapter 4 reported the values of the parameters obtained by using the growth and feeding functions as nonlinear regression equations on long term $(t, F, W)$ experimental data collected on domestic and laboratory species and varieties of animals ranging in size from mice to steers. Chapter 5 introduced the notion of plotting $(t, F, W)$ data of an animal in three dimensional Euclidean space and naming the resulting curve, the trace, in order to distinguish it from the traditional two dimensional sigmoid growth curve of live weight versus age. When the growth and feeding parameters of any animal have been determined, the dimensionless variables $u = W/A$, $Z = (AB)CF/A$, and $T = t/t^*$ can be plotted to give a three dimensional space curve which was named the biotrace of the animal. In this way the biotraces of the animals from mice to steers could be drawn for comparison in the same $(T, Z, u)$ space.

Chapter 5 also introduced the notion of plotting $(t, q^*, W)$ data in a three dimensional Euclidean space to obtain a curve the projection of which on the $(q^*, W)$ plane is a curve which I called the ad libitum growth phase curve. The rectangle, in the $W$ versus $q$ plane, with height $A$ and width $C$, was named the Growth Phase Plane (GPP). The notable features of the GPP were the diagonal from point (0,0) to point $(C, A)$ with slope $A/C = T_0$ and the ad libitum growth phase curve $W = W(q^*)$. Study of this function revealed a dimensionless relation between the parameters $(AB)$, $T_0$, and $t^*$ designated as the ad libitum feeding and growth discriminant, $\alpha$, which is forbidden to be greater than unity.

Chapter 6 extended the study to the growth of various animals under well specified controlled feeding regimes. Here it was found that stable live weights of an animal, $a<A$, required food intakes of $q=a/T_0$, where $T_0$ (the slope of the diagonal of the GPP) is a time constant characteristic of the animal and its food and environment. Since the mature food intake $C$ maintains the mature live weight $A$, the time constant is then given by $T_0=A/C$. Here, the physical significance of $T_0$ may be illustrated by an electrical analogue of $a=qT_0$, namely: $V=IR$. If $a \leqq A$ is analogous to the electrical potential drop, $V$, across a resistor and if $q \leqq C$ is analogous to the electric current, $I$, through the resistor, then $T_0$ is analogous to the resistance, $R$, of the resistor. Biophysically $T_0$ appears to have the property of a resistance to gain or loss of live weight at or near an equilibrium live weight, $dW/dt=0$.

Most importantly, it was conjectured that the responses of the various animals (including man) to various controlled feeding regimes were solutions of the linear inhomogeneous differential equation $dW/dt+gW=hq(t)$, given the feeding regime function $q(t)$ and the initial condition $W=W_0$ at $t=0$. The controlled feeding functions used by the various experimenters were zero (complete starvation of the animal), a constant (partial starvation), linearly increasing functions, and sinusoidal functions. Study of the live weight responses suggested that the coefficient $g$ on the left-hand side of the differential equation might very well be approximated by Brody's $k=(AB)C/A=(AB)/T_0$ and $h$ on the right hand side by $(AB)$. It was then decided that the growth parameters should be $A$, $(AB)$, $T_0$, and $t^*$. Here was the key to a more general empirical theory of the feeding and growth of animals which, unlike the theories of Robertson, von Bertallanfy, Laird etc. (Chap. 1), is based on feeding and growth data alone and does not involve any apriori biological notions.

## 7.1 The Theory as a Set of Differential Equations

The live weight response of an animal under good management is a solution of the following differential equations depending only on how it is fed and what its live weight and food intake are at the time experimental observations begin, i.e. depending only on the i.c.

If the animal is permitted to eat to its appetite, then the ad libitum food intake, $q^*(t)$, and its live weight, $W(t)$, are respectively solutions of

$$dq^*/dt+(1/t^*)q^*=A/T_0t^*, \tag{7.1}$$

with the initial condition of $q^*(0)=D$ at $t=0$, and

$$dW/dt+[(AB)q^*(t)/A]\,W=(AB)q^*(t)\,, \tag{7.2}$$

with the initial condition $W(0)=W_0$ at $t=0$. Here $q^*(t)=dF^*/dt$ where $F^*$ is the ad libitum cumulative food consumed and is equal to $\int_0^t q^*(t)dt$. It has been shown that the ad libitum feeding discriminant, $\alpha$, must be less than or equal to unity. Since $\alpha=(AB)Ct^*/A=kt^*$, it is apparent that $\alpha \leqq 1$ places a constraint on the relative values of $(AB)$, $t^*$ and $T_0$, if Eqs. (7.1) and (7.2) are indeed applicable to experimental data which are obtained under true ad libitum feeding conditions. It was stated in Chapter 5 that a possible biophysical meaning of $\alpha$ was not then apparent, however a meaning will be proposed later in this section.

If the animal is fed according to a well defined preselected regime, $q(t)$, which satisfies the inequality $0 \leq q(t) < q^*(t)$, its live weight, $W(t)$, is a solution of

$$dW/dt + [(AB)/T_0]W = (AB)q(t), \tag{7.3}$$

with initial condition $W(0) = W_0$ at $t=0$. Controlled feeding suppresses the expression of the animals natural appetite and its natural ability to reach its mature live weight. The growth parameters $t^*$ and $A$ are overridden, but the parameters $(AB)$ and $T_0$ remain independent of the mode of feeding. If the animal is at a later time rehabilitated to ad libitum feeding, all four parameters will again be operative and Eqs. (7.1) and (7.2) will apply but with new i.c.

In as much as both Eqs. (7.2) and (7.3) relate to the growth of the animal, i.e. each term has the units of kg of live weight per unit of time, their right-hand sides are the live weight equivalents of the ad libitum food intake, $q^*(t)$, and controlled food intake, $q(t)$, respectively. These live weight equivalents are distributed in both cases between growth, $dW/dt$, and no growth, i.e., $[(AB)q^*(t)/A]W$, when the animal is fed ad libitum and $[(AB)/T_0]W$, when the animal is fed on the controlled regime, $q(t)$. Calling these latter terms, on the left hand sides of Eqs. (7.2) and (7.3), no-growth terms, means only that these portions of the live weight equivalents of the food intakes are used by the animal for biological activities other than increasing its live weight at the rate of $dW/dt$. These other biological ativities are usually thought of as maintenance of $W(t)$, but in the context of this theory of growth it is best to retain the traditional bio-energetic meaning of maintenance and use the term no-growth in the sense just discussed.

Equations (7.1), (7.2), and (7.3) are specific examples of the general first order linear inhomogeneous differential equation

$$dv/dt + rv = s, \tag{7.4}$$

with initial conditions $v=0$ at $t=0$, first studied by Newton, 1642 (Book II of his great work *The Principia*), when he invented the differential and integral calculus as aids in his experimental studies of the flow of heat from hot to cold bodies and of motion in the resisting media. The question, whether Newton or Leibniz, 1646, was first to invent the calculus, was hotly debated by British and German intellectuals for quite a few years, but that Newton was the primogenitor of the modern theoretical and experimental scientist has never been in dispute.

Newton considered the coefficient $r$ with units of inverse time, as the coefficient of conductance of the medium to motion, $v$, through it. The inverse of the conductance is the resistance to the motion in units of time. Today this conductance is referred to as Newtonian viscosity of resisting fluids or thermal conductance in study of heat flow. He also considered $s$, on the right-hand side of Eq. (7.4), as the agent causing the motion.

Comparing Eqs. (7.1), (7.2), and (7.3) with Newton's interpretation of (7.4), $t^*$ may be interpreted as a natural biological resistance an animal may have to increasing its appetite, i.e. increasing $dq^*/dt$, to the maximum attainable by it, namely the mature appetite $A/T_0$; and $A/(AB)q^*(t)$ may be interpreted as the variable resistance the animal naturally has against the live weight equivalent of its ad libitum food intake going to no-growth. And $T_0/(AB)$ may be interpreted as the natural resistance the animal, under controlled feeding, has against the live weight equivalent of its food intake going to no-growth. Also following Newton's interpretation of $s$ in Eq. (7.4) as a causative agent, the term $A/T_0 t^*$ on the right-side of the Eq. (7.1) may be considered a natural biological drive of the animal to increase its appetite against the resistance $t^*$. Also the terms $(AB)q^*(t)$ and $(AB)q(t)$ on the right-hand sides of Eqs. (7.2) and (7.3) may be considered natural biological drives of the animal, under the conditions of ad libitum and controlled feeding respectively, to increase its live weight against the variable resistance $A/(AB)q^*(t)$ in the case of ad libitum feeding and against the constant resistance $T_0/(AB)$ in the case of controlled feeding. It seemed natural for me to give the biological drives (or urges) $A/T_0 t^*$, $(AB)q^*(t)$ and $(AB)q(t)$ respectively, the names appetency, the live weight equivalent of the ad libitum food intake, and the live weight equivalent of the controlled food intake.

The preceeding discussion indicates a possible biophysical meaning of the ad libitum feeding and growth discriminant $\alpha$, which must be less than or equal to unity if the feeding and growth of an animal is in fact ad libitum. Alpha, given by $kt^*$ or its equal $(AB)t^*/T_0$, places a constraint on the appetite resistance $t^*$ relative to the growth resistance $T_0/(AB)$ as shown in

$$\alpha = t^*/[T_0/(AB)] \leqq 1. \tag{7.5}$$

In as much as $T_0/(AB) = 1/k$ appears less susceptible to change than $t^*$, inequality Eq. (7.5) states that, if the resistance to increase of appetite $t^*$ is greater than the critical value $T_0/(AB)$ of the resistance to food going to no-growth under controlled feeding, the animal must be on a controlled feeding regime even if it appears to be feeding ad libitum. Inequality Eq. (7.5) forbids $t^*$ to be greater than $T_0/(AB)$ in a true ad libitum feeding and growth situation.

It is hoped that the preceeding discussions and illustrations of the meanings of $\alpha$ and the terms of the DE's, (7.1), (7.2), and (7.3) have led the reader to an understanding of the theory of feeding and growth of an animal or population of animals of the same breed. However, a theory is not complete without statement of some experimentally testable predictions. The predictions of this theory are two in number.

Firstly, given the way the animal is fed and the initial conditions at the start of an experiment, the live weight response, $W(t)$, is a solution of DE's (7.1) and (7.2) with $\alpha \leqq 1$ if the chosen feeding regime is ad libitum, or the response is a solution of DE (7.3) if the chosen feeding regime is other than ad libitum.

Secondly, the values of the parameters $(AB)$ and $T_0$, under any controlled feeding regime, are equal to the values applicable to ad libitum feeding and growth of the same animal under the same dietary and management conditions.

Combination of these predictions states that the parameters $(AB)$, $T_0$, are independent of the feeding regimes. It is presumed that there are no dietary or management changes in the experimental situation. Chapters 9 and 11 discuss the effects of dietary and other environmental variations on the growth parameters $A$, $(AB)$, $t^*$, and $T_0$.

It is most important that the reader understands clearly that although the growth and feeding parameters are constant during an experiment, they are phenotypic characters of the animal's growth and feeding. This is true because they subsume all the phenotypic properties of the feeding and growth data representing the dynamic course of growth of the animal in the three dimensional Euclidean space, the co-ordinate axes of which are time, $t$, cumulative food consumed, $F$, and live weight, $W$. With this statement in mind Chapter 10 examines the possibility that the phenotypic variance of the growth parameters within a population contain sufficient genetic variance to make possible changes in the course of growth by selecting on the growth and feeding parameters as quantitative genetic characters.

This empirical theory of the feeding and growth of animals has emerged from the study of experimental data taken by many observers on many animals under widely differing circumstances with objectives mostly unrelated to how any of the experimental animals might grow in the long run, whether controlled or freely fed. There is then a need to plan and perform an experiment which will subject the theory to such stringent conditions that the experimental results can be used to declare the theory useless for further research, useful if modified in some rational manner, or useful as presently formulated.

## 7.2 The Plan and Execution of Experiment BG 54

In September, 1974, G.C. Emmans of the East of Scotland School of Agriculture, and B.J. Wilson of the Agricultural Research Council's Poultry Research Centre in Edinburgh, approached me to explore the possibility of devising an experiment which would subject the theory to severe test. The facilities at the P.R.C. experimental station at Roslyn, Midlothian, would become available for a long term large scale experiment starting in November. The space could accommodate six treatments with two replicates per treatment of 30 animals per replicate for four different types of animals for a period of 30 weeks.

Since Emmans and Wilson were experienced with chickens it was decided that the four types of animals would be males and females of the commercial breeds of the Ross Ltd. large fryers and small Apollo layers. After considerable discussion of the use of the GPP (Fig. 7.1) in designing feeding and growth experiments, they understood the need for as good preliminary estimates of the growth and feeding parameters of each type of these chickens as our pooled experience permitted. To our knowledge these breeds of chickens, or any like them, had never been subjected to long term ad libitum or controlled feeding experiments. Detailed knowledge of the feeding and growth of these particular breeds could be economically, as well as scientifically, useful.

The estimates of $A$, $(AB)$, $T_0$, and $t^*$ finally agreed upon are listed in Table 7.1. It must be emphasised that these are not predicted values of the growth parameters; they serve only in the design of the experiment.

In order to avoid confusion of the estimates with the growth parameters yet to be determined in the experiment, the symbols, $a$, $(ab)$, $t_0$, and $t'$ respectively will symbolize the preliminary estimates of $A$, $(AB)$, $T_0$, and $t^*$ (Table 7.1). The last two

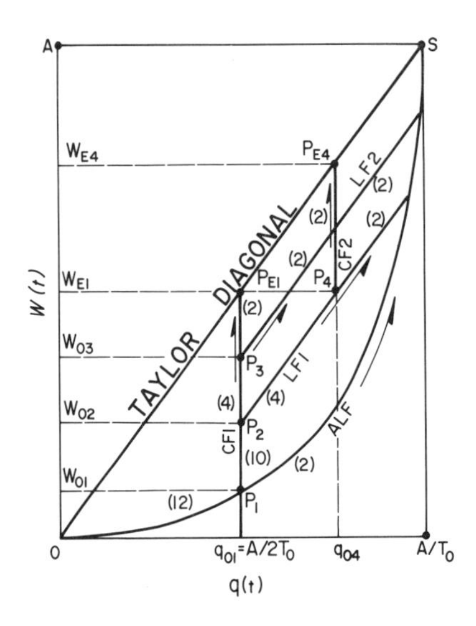


Fig. 7.1. The growth phase plane (GPP) used to plan Experiment BG 54 on both sexes of two genotypes of chickens

Table 7.1. Estimated values of the growth parameters of the Ross Ltd. fryer and layer, male and female chickens

| | Fryers | | Layers | |
|---|---|---|---|---|
| | Males | Females | Males | Females |
| $a$, kg | 4.9 | 3.9 | 2.1 | 1.8 |
| $(ab)$ | 0.54 | 0.54 | 0.54 | 0.54 |
| $t_0$, wks | 4.1 | 3.9 | 3.5 | 3.6 |
| $t'$, wks | 3.4 | 4.1 | 4.9 | 5.9 |
| $W_0$, kg | 0.045 | 0.045 | 0.035 | 0.035 |
| $\alpha$[a] | 0.45 | 0.54 | 0.76 | 0.89 |

[a] $\alpha=(ab)t'/t_0$

rows are the mean weights of the day old chicks taken as the initial weights, $W_0$, of each breed, and the values of $\alpha$ estimated by $(ab)t'/t_0$; which are all less than unity, with that of the Apollo female being highest, at 0.89.

Figure 7.1 facilitates understanding of the use of the GPP in the plan, organisation and conduct of the experiment by illustrating the kinds of treatment to which each of the four types of chickens would be subjected. Starting with 12 pens of day old chicks fed ad libitum (Trt. ALF), 10 pens would be split off, when the mean food intake per bird reached $q_{01}=a/2\,t_0$ and their mean live weight per bird became $W_{01}$ at age $t_{01}$ with the remaining two pens continued on ALF. The 10 pens would be held at constant food intake $q_{01}$ (Trt. CF 1). If the chicks were held on this treatment long enough their mean live weights would ultimately reach the equilibrium weight, $W_{E1}=t_0 q_{01}$ *at point* $P_{E1}$, on the Taylor diagonal.

When, at age $t_{02}$, the mean live weights of the 10 pens on CF 1 had reached $W_{02}=(W_{E1}-W_{01})/3+W_{01}$, four pens would be split off and fed so that their growth phase curve would be the straight line

$$W_2(t)=W_{02}+t_0[q_2(t)-q_{01}], \tag{7.6}$$

parallel to the Taylor diagonal, the estimated slope of which is $t_0$ (Trt. LF 1). At age $t_{02}$, live weight $W_{02}$ and food intake $q_{01}$, two additional pens would be split off the 10 pens on CF 1 and fed according to treatment five, RF, to be discussed later.

This leaves four pens to be carried on CF 1 until at age $t_{03}$ the mean live weight $W_{03}=2(W_{E1}-W_{01})/3+W_{01}$ would be reached, at which time two pens would be split off and fed so their growth phase curve would be the straight line

$$W_3(t)=W_{03}+t_0[q_3(t)-q_{01}], \tag{7.7}$$

also parallel to the Taylor diagonal LF 2. Two pens remain to be continued on treatment CF 1 to the end of the experiment.

Referring back to the four pens on treatment LF 1, two of these pens would be split off at age $t_{04}$, when the mean live weight reached $W_{04}=W_{E1}$ and the food intake became $q_{04}$. These pens would be kept on this food intake as treatment CF 2 to the end of the experiment. It would be expected that the mean weights would approach equilibrium $W_{E4}=t_0q_{04}$.

Controlled feeding treatment, RF, referred to earlier, followed from a significant suggestion by Emmans that at age $t_{02}$, live weight $W_{02}$ and food intake $q_{01}$, two of the pens split off from the ten pens on CF 1 be fed a weekly amount of food selected at random from treatments CF 1, LF 1, LF 2, and CF 2 present during that particular week. Figure 7.2 shows the initially available choices are the weekly foods, $f(t)$, consumed from treatments CF 1 and LF 1 with those for LF 2 and CF 2 becoming available at later times. This treatment could be a severe test of the theory because it has never before been applied to any animal and it may closely approximate a real world situation in which the food supply might well be subject to chance fluctuations. Once the rules of chance selection of weekly food were established,

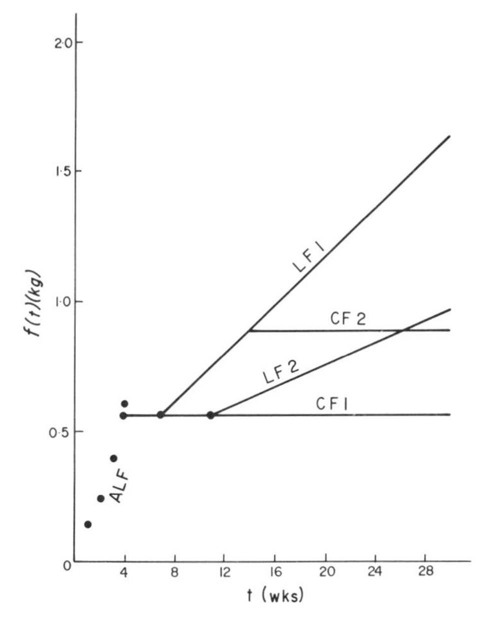


Fig. 7.2. The linear, LF 1 and LF 2, and constant, CF 1 and CF 2, controlled feed intakes as functions of age for the Ross fryer males as an example

the same rules of selecting the weekly food randomly would be applied to each of the four types of birds, using the controlled food intakes appropriate to each type.

Since the experiment depends on precalculating the weekly foods for the controlled feeding treatments CF 1, LF 1, LF 2, CF 2, and RF, the equations for calculating $q_i(t)$, $i=1,2,3,4,5$, where $i$ is an index number assigned to each controlled feeding treatment respectively, must be known in advance. From Fig. 7.1 and Table 7.1, the food intake for CF 1 of any of the four types of birds is

$$q_1(t)=q_{01}=a/2t_0, \text{ for ages } t \geqq t_{01},$$

and for CF 2 is

$$q_4(t)=q_{04}, \text{ for ages } t \geqq t_{04}, \text{ and live weight } W_{04}=W_{E1}, \text{ at age } t_{04}.$$

Here $t_{01}$, $q_{04}$, $W_{04}$, and $t_{04}$ are presently unknown. However the point ($t_{01}$, $q_{01}$, $W_{01}$) is on the ALF treatment so that the coordinates ($t_{01}$, $q_{01}$) and ($t_{01}$, $W_{01}$) should satisfy the solutions of DE's (7.1) and (7.2) with initial conditions $q^*(0)=0$ and $W(0)=W_0$, i.e. 0.045 or 0.035 kg depending on the type of chicken, and the estimated values of the growth parameters from Table 7.1 appropriate to the type of chicken under consideration. The solutions of DE's (7.1) and (7.2) have already been encountered and used in Chapters 2 and 4. They are repeated here for convenience and clarity as

$$q^*(t)=(a/t_0)[1-\exp(-t/t')], \tag{7.8}$$

and

$$W(t)=(a-W_0)[1-\exp\langle-[(ab)/t_0]\{t-t'[1-\exp(-t/t')]\}\rangle]+W_0\,. \tag{7.9}$$

Since $q^*(t_{01})=a/2t_0$, $t_{01}$ is found from (7.8) as

$$t_{01}=t'\ln 2, \tag{7.10}$$

and $W_{01}=W(t_{01})$ is given by

$$W_{01}=(a-W_0)\langle 1-\exp\{-[(ab)/t_0][t'(\ln 2-{}^1/_2)]\}\rangle+W_0\,. \tag{7.11}$$

Here $t_{01}$ and $W_{01}$ are the i.c. for solving DE (7.3) for $W_1(t)$ of treatment CF 1 with the constant food intake $q_1(t)=a/2t_0$.

The solution of DE (7.3) with constant food intake has already been seen in Chap. 6. With the estimated parameters in Table 7.1 and $q_1(t)=a/2t_0$, this solution is the law of diminishing returns, namely

$$W_1(t)=(a/2-W_{01})\langle 1-\exp\{-[(ab)/t_0](t-t_{01})\}\rangle+W_{01}\,. \tag{7.12}$$

When the age is $t_{02}$, $W_1(t_{02})=W_{02}=(a/2-W_{01})/3+W_{01}$, which when substituted in Eq. (7.12) gives the value of $t_{02}$ as

$$t_{02}=t_{01}+[t_0/(ab)]\ln(3/2). \tag{7.13}$$

These values of $t_{02}$ and $W_{02}$ are the initial values for treatment LF 1, the equation of which, in the GPP, is Eq. (7.6). Since $W_2(t)$ is a solution of Eq. (7.3) with the appropriate estimated parameters $(ab)$ and $t_0$, the form of the controlled feeding function $q_2(t)$ can be found by substituting $W_2(t)$ in DE (7.3). After some algebraic manipulation $q_2(t)$ is found to be a solution of the following DE with initial condition $q_2(t)=a/2\,t_0$ at $t=t_{02}$ from Eq. (7.13)

$$dq_2/dt=[(ab)/t_0^2](a/2-W_{02}). \tag{7.14}$$

The solution, for precalculating the feeding regime $q_2(t)$ for LF 1, is therefore the straight line

$$q_2(t)=a/2\,t_0+[(ab)/t_0^2](a/2-W_{02})(t-t_{02}). \tag{7.15}$$

Beginning again with Eq. (7.12) and knowing that $W_1(t)$, at $t=t_{03}$, is $W_{03}=2(a/2-W_{01})/3+W_{01}$, and following the same line of reasoning that was taken to reach Eq. (7.15) but using Eq. (7.7), it is found that $t_{03}$ is given by

$$t_{03}=t_{01}+[t_0/(ab)]\ln 3, \tag{7.16}$$

and the equation for precalculating $q_3(t)$ for LF 2 is

$$q_3(t)=a/2\,t_0+[(ab)/t_0^2](a/2-W_{03})(t-t_{03}). \tag{7.17}$$

It remains now to determine $q_{04}$ and $t_{04}$ for treatment CF 2 when $W_{04}=W_2(t_{04})=W_{E1}=a/2$ on treatment LF 1. The equation for $W_2(t)$ is found by substituting Eq. (7.15) in Eq. (7.6) to give

$$W_2(t)=W_{02}+[(ab)/t_0](a/2-W_{02})(t-t_{02}). \tag{7.18}$$

When the age of the chickens on LF 1 reaches $t_{04}$, $W_2(t_{04})=W_{E1}=a/2$ and substituting these in Eq. (7.18) and solving gives

$$t_{04}=t_{02}+t_0/(ab), \tag{7.19}$$

and $q_{04}=q_2(t_{04})$ becomes

$$q_4(t)=(a-W_{02})/t_0, \text{ namely} \tag{7.20}$$

the constant food intake for treatment CF 2.

The preceeding discussion developed the general equations to be used in calculating the weekly controlled feeding regimes, $q_i(t)$, and the initial ages, $t_{0i}$, live weights, $W_{0i}$, and food intakes, $q_{0i}$, $(i=1,2,3,4)$, from the theory of the GPP and the estimated values of the growth and feeding parameters listed in Table 7.1. Table 7.2 is a listing of the equations for precalculating the food intakes of the controlled feeding regimes for the fryer male as an example, and Fig. 7.2 illustrates these regimes as applied to these animals during the experiment.

Figure 7.3 illustrates the random feeding regime, RF, of the fryer males, as a step function of weekly feeds chosen randomly from the controlled feeding regimes shown in Fig. 7.2 as previously discussed. The same scheme scaled appropriately was followed for the other types of chickens.

In any feeding and growth experiment it is necessary to have means for estimating the total amount of food that will be required. This information can be found by integrating various food intake equations $q(t)$ for each type of bird over the experimental period. It has already been stated that the experimental facilities would be available for 30 weeks. It was decided to keep the males for 30 weeks and the females for 21 weeks only, in order to avoid complications due to egg laying. The composition of the food is shown in Table 7.3 and is considered nutritious for these chickens.

Table 7.2. Equations for precalculating the controlled food intake and the expected ages and weights at start. Equations for Ross Fryer males as examples

| Trt | Initial conditions and food intake functions |
|---|---|
| CF1 | $t_{01}=t'\ln 2=2.357$ wks |
| | $W_{01}=(a-W_0)\langle 1-\exp\{-[(ab)/t_0]t'(\ln 2-0.5)\}\rangle+W_0=0.447$ kg |
| | $q_1(t)=a/2t_0=0.598$ kg/wk |
| LF1 | $t_{02}=t_{01}+[t_0/(ab)]\ln 3/2=3.605$ wks |
| | $W_{02}=(a/2-W_{01})/3+W_{01}=1.115$ kg |
| | $q_2(t)=(a/2t_0)+[(ab)/t_0^2](a/2-W_{02})(t-t_{02})=0.598+0.0429(t-3.605)$ kg/wk |
| LF2 | $t_{03}=t_{01}+[t_0/(ab)]\ln 3=10.698$ wks |
| | $W_{03}=2(a/2-W_{01})/3+W_{01}=1.782$ kg |
| | $q_3(t)=a/2t_0+[(ab)/t_0^2](a/2-W_{03})(t-t_{03})=0.598+0.0215(t-10.698)$ kg/wk |
| CF2 | $t_{04}=t_{02}+t_0/(ab)=11.198$ wks |
| | $W_{04}=a/2=2.45$ kg |
| | $q_4(t)=q_2(t_{04})=a/2t_0+[(ab)/t_0^2](a/2-W_{02})(t_{04}-t_{02})=0.924$ kg/wk |
| RF | $t_{05}$ and $W_{05}$ same as $t_{02}$ and $W_{02}$ |
| | $q_5(t)$, see Fig. 7.3 for fryer male |

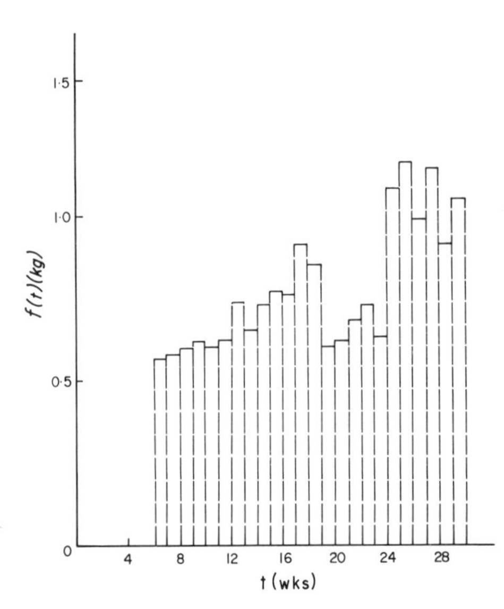


Fig. 7.3. An example of the random weekly feed intakes for treatment RF. Data is for the Ross fryer males (Table A-29)

In as much as the chickens on experiment could not be expected to perform exactly as estimated, the following schedule for applying the precalculated controlled feeding regimes was adopted. The actual age, $t_{01}$, and mean live weight, $W_{01}$, were to be noted as experimental data when the mean food intake of the 12 pens on treatment ALF was as close as possible to $q^*(t_{01})=q_{01}=a/2\,t_0$. The actual age, $t_{02}$, was to be noted as a datum when the mean live weight of the 10 pens on treatment CF 1 reached the estimated live weight, $W_{02}$ (Table 7.2); here four pens were to be

Table 7.3. Composition of the food in units of gram/grams, having 21% crude protein and 2.64 kcal/g metabolizable energy ($ME$)

| | | | |
|---|---|---|---|
| Barley | 0.1000 | Grass | 0.0500 |
| Maize | 0.3000 | Limestone | 0.0100 |
| Wheat | 0.2450 | Salt | 0.0025 |
| Herring | 0.0500 | Dical. Phos. | 0.0175 |
| Soya | 0.2200 | Supplements | 0.0050 |
| | | Total | 1.0000 g |

put on the precalculated feeding regime $q_2(t)$ for treatment LF 1 (Table 7.2). At this point in the program two pens were to be put on treatment RF (Fig. 7.3). The actual age, $t_{03}$, of the four pens remaining on CF 1 was to be recorded when their mean live weight reached the estimated weight, $W_{03}$ (Table 7.2); here two pens were to be put on the precalculated regime $q_3(t)$ for treatment LF 2 (Table 7.2). The two pens remaining on CF 1 were to be continued to the end of the experiment. Of the four pens already on LF 1 the actual age, $t_{04}$, and mean food intake, $q_{04}$, were to be recorded as data when the mean live weight reached the pre-estimated weight, $W_{04}$ (Table 7.2); here two pens were to be held at constant food intake, $q_{04}$, to the end of the experiment (Trt. CF 2). The two pens remaining on treatment LF 1 were to be carried on to the end of the experiment.

The data to be reported were mean weekly live weights, $W(t)$, and mean weekly food consumed, $f(t)$, for all 12 pens on the ALF treatment for each type of bird, from day old to age $t_{01}$, and on the two pens continued on ALF to the end of the experiment. Beginning at $t_{01}$, the data to be reported were the mean weekly live weights, $W(t)$, of those pens on the various controlled feeding regimes. Eight large graphs, four live weight versus age and four weekly food consumed versus age, were laid out for plotting the weekly data as it came from the experimental facility. As the experiment progressed, following the development of the curves in this manner offered the possibility of detecting gross errors immediately and rechecking could be facilitated.

In the cases of controlled feeding, three conditions were agreed on in order to as far as possible equalize the distribution of the limited amounts of food among the birds in each pen. Firstly, the amount of feeding trough space was to be double that normally used in ad libitum feeding; secondly, longer daylight than normal was to be provided; and thirdly, the weekly food was to be spread evenly in the troughs three times during the week providing amounts for two, two and three consecutive days respectively.

In as much as the estimates of the growth parameters of each type of bird listed in Table 7.1 yielded GPP's which were only approximations to the actual GPP's to be obtained experimentally, it could not be expected that the experimental slopes, $T_0$, of the actual Taylor diagonals could be equal to the estimated slopes, $t_0$ (Table 7.1). Therefore the theoretical equations for describing the experimental responses to treatments LF 1 and LF 2 are solutions of DE (7.3), with the driving function $q(t)$ given by the precalculated linearly increasing feeding regimes $q_2(t)$ or $q_3(t)$ respectively, with the initial conditions $W_{02}$ at $t_{02}$, or $W_{03}$ at $t_{03}$ respectively.

Table 7.4. Solutions of differential Eqs. (7.1), (7.2), and (7.3) with $\alpha$ considered in the ALF treatment

| Trt. | Theoretical functions |
|---|---|
| ALF | $f(t)=(A/T_0)\{1-t^*[\exp(1/t^*)-1]\exp(-t/t^*)\}$<br>$F(t)=(A/T_0)\{t-t^*[1-\exp(-t/t^*)]\}$<br>$W(t)=(A-W_0)\{1-\exp[-(AB)F/A]\}+W_0$<br>$\alpha=(AB)t^*/T_0<1$.<br>$W(t)=(A-W_0)\{1-\exp[-F(t)T_0/At^*]\}+W_0$;<br>$\alpha=1, \quad (AB)=T_0/t^*$ |
| CF1 | $W(t)=(T_0q_{01}-W_{01})\{1-\exp[-(AB)(t-t_{02})/T_0]\}+W_{01}$ |
| LF1 | $W(t)=(T_0q_{01}-T_0^2m_1/(AB)-W_{02})\{1-\exp[-(AB)(t-t_{02})/T_0]\}+W_{02}+T_0m_1(t-t_{02})$;<br>$m_1=[(AB)/t_0^2](a/2-W_{02})$ |
| LF2 | $W(t)=(T_0q_{01}-T_0^2m_2/(AB)-W_{03})\{1-\exp[-(AB)(t-t_{03})/T_0]\}+W_{03}+T_0m_2(t-t_{03})$;<br>$m_2=[(ab)/t_0^2](a/2-W_{03})$ |
| CF2 | $W(t)=(T_0q_{04}-W_{04})\{1-\exp[-(AB)(t-t_{04})/T_0]\}+W_{04}$ |
| RF | $W(t)=[T_0q_5(t)-W(t-1)]\{1-\exp[-(AB)/T_0]\}+W(t-1)$ |

A method of testing the first theoretical prediction is discussed in detail in the following paragraphs. Table 7.4 contains the solutions of DE's (7.1) and (7.2), for ad libitum feeding and DE (7.3) when $q(t)$ is a constant, a linearly increasing function or a random function of age, $t$. These functions are to be used with nonlinear regression techniques discussed in Chap. 4 to determine the actual values of the growth parameters $A$, $(AB)$, $T_0$, and $t^*$ from the $[t, f(t), W(t)]$ experimental data of the ALF treatments of the four types of chickens and the actual values of $(AB)$ and $T_0$ from the $[t, W(t)]$ experimental data of the five controlled feeding regimes.

In the following remarks the symbols $f(\hat{t})$, called "$f-$hat", and $\hat{W}(t)$, called "$W-$hat," are used to distinguish the computed value of the dependent variables from the experimental values $f(t)$, or $W(t)$, at age $t$. Since the experimental variables $f(t)$ and $W(t)$ have greater variabilities as the animals age (Chap. 4.4), the logarithm transformations, $\ln[f(t)]$ and $\ln[W(t)]$, of the data, were to be used in the regressions with the logarithm transformations of the computed values, namely $\ln[f(\hat{t})]$ and $\ln[\hat{W}(t)]$. The characteristic growth parameters $A$, $(AB)$, $T_0$, and $t^*$, or $(AB)$ and $T_0$ depending on the feeding regime, are determined from the minimization of the weighted sums of squares of the residuals, i.e., $\Sigma M_t\{\ln[f(t)]-\ln[f(\hat{t})]\}^2$, plus $\Sigma M_t\{\ln[W(t)]-\ln[\hat{W}(t)]\}^2$ using nonlinear regression. The statistical weight, $M_t$, is the appropriate number of pens used to calculate the means of each $f(t)$ or $W(t)$ divided by the sum of these numbers during the period of the experimental treatment. The numbers in parentheses, adjacent to each treatment in Fig. 7.1, are the number of pens used to get the mean weekly foods and live weights for that treatment.

The residual standard error, rsd, of fit will be referred to as the $CV$ because it is analogous to the coefficient of variation of fit of the theoretical functions in

Table 7.4 to the experimental data when rsd is less than 0.1. The analogy springs from rsd $\simeq \Delta \ln y = \Delta y / y$. However, the *CV* alone is not a sufficient criterion for judging whether or not the theoretical functions are best descriptors of the experimental data.

In regression, another criterion of the suitability of the fit of a theoretical function to the experimental data is the distribution of the runs of the signs of the residuals. If the distribution of the runs of signs is random, then it is expected that the residuals are a sample from a normal distribution with mean zero and variance $\sigma^2$. However, if the distribution of the runs of the signs of the residuals is not random, it is expected that there is a correlation between the residuals and the independent variable which will inflate the *CV* (Box and Jenkins 1970, p. 37). The existence of such a correlation could open the possibility of correcting the theoretical function and revising its differential equation in a rational way. Such a revision when reapplied to the actual data should reduce the *CV* and increase the probability that the distribution of the runs of the signs of the residuals is random. The mathematical discrete probability theory of the distribution of the runs of the signs of residuals is discussed in detail in Hogg and Craig (1970, p. 368). This theory will be used to estimate the probability, $P(R)$, that the distribution of the runs of signs of the residuals in a particular regression is random.

Another criterion of the desirability of a function as a descriptor of experimental data is the number, $n$, of the characteristic parameters the function contains. The smaller the value of $n$ the more parsimonious the function is as a descriptor.

The desirable features of a theoretical function are therefore three in number, namely (1) a small *CV*; (2) a large $P(R)$; and (3) a small $n$. Intuitively this suggests that the function for which the product of the values of *CV*, $[1-P(R)]$ and $n$ is least, is the most desirable descriptor of the data. If the product, $CV[1-P(R)]\,n$, is a minimum for the best functional descriptor, the inverse of this product will be a maximum when comparing competing functions. Here I shall adopt the inverse, $\{CV[1-P(R)]n\}^{-1}$, as the criterion for comparing competing functions and name it the Figure of Merit or FM. A value of FM less than unity will be considered sufficient to cast doubt on the first theoretical prediction. There is no competing theory, encompassing response to both ad libitum and controlled feeding treatments, against which the present theory may be tested using the values of FM as a criterion. However, the ad libitum live weight versus age data can be fitted using other popular sigmoid functions, such as the logistics, the generalised Richards function (1959), etc., and the values of FM compared with the values obtained to test the theory in this experiment. Discussion of this type of limited testing of the theory will be presented in Chap. 7.5.

The second theoretical prediction is that the values of $(AB)$ and $T_0$ obtained from fitting the predicted solutions of DE (7.3) (Table 7.4) to the actual data of each treatment of any one of the four genotypes are equal to the values of $(AB)$ and $T_0$ obtained from the ALF treatment of that genotype. This statement makes two assumptions in the context of the proposed experiment. Firstly it is assumed that the values of $(AB)$ and $T_0$ for the ALF treatment of a type of animal are the true values. Secondly it is assumed that the values for the controlled feeding regimes are a random sample drawn from a population normally distributed with the ALF value as the mean and an unknown variance estimated by the variance of the mean of the sample. Acceptance of these assumptions means a two tailed $t$-test, with the appropriate degrees of freedom, can be used to reject the assumed null hypothesis.

This section has been devoted to preplanning the entire experiment, i.e., how the controlled feeding regimes were to be precalculated, the collection of data, when and how the controlled feeding regimes were to begin and be carried out, and finally how the data will be used to test the theoretical predictions. The results of the experiment and the analysis of them by way of evaluating FM, and the student $t$-test to examine the strengths of the theoretical predictions are presented in the next section with a general discussion of the theory versus the results to be presented in Chap. 7.4.

## 7.3 The Results of Experiment and Their Analysis

The experimental data for all the animals and treatments are in Tables A-24 to A-47 of Appendix A. The data in each table were analysed by nonlinear regression using the theoretical function in Table 7.4 appropriate to the treatment and the estimated values of the parameters appropriate to the genotype and sex in Table 7.1 and the initial conditions, appropriate to the genotype, sex, and treatment, listed in each table in the Appendix. In the regressions the starting values of the parameters $A$, $(AB)$, $T_0$, and $t^*$ were the estimated values in Table 7.1.

Little difficulty was experienced in these regressions excepting those for the ALF treatments for both genotypes and sexes. When the theoretical functions $f(t)$ and $W(t)$ for the ALF treatments were used, the best values of $(AB)$, $T_0$, and $t^*$ were such that $\alpha=(AB)t^*/T_0>1$ for each case. Such values of $\alpha$ are forbidden. To avoid this difficulty $\alpha$ was fixed at unity and the appropriate theoretical function for $W(t)$ in Table 7.4 was used with the $f(t)$ function, to determine $A$, $T_0$, and $t^*$ from which $(AB)$ could be calculated as the ratio $T_0/t^*$.

Using the results of regression of the fryer male as an example, Figs. 7.4 to 7.6 show how well the $(t, W)$ data points are tracked by the solid curves calculated from the best values of $A$, $T_0$, and $t^*$ [with $(AB)$ estimated by $T_0/t^*$] for the ALF treatment and the best values of $(AB)$ and $T_0$ for the responses to the controlled feeding treatments. The results shown in these figures are typical for both genotypes and sexes with the exception of the experimental response of the Apollo layer male, during treatment LF 2, depicted in Fig. 7.7. The rapid rise of the response in the last few weeks is inexplicable unless there was some unaccountable change in the LF 2 feeding regime. In this case the regression could find no best values of $(AB)$ and $T_0$ when fitting the theoretical function for LF 2 (Table 7.4) with the appropriate initial conditions $t_{03}$, $q_{03}$, and $W_{03}$ (Table A-39 in the Appendix) and the value of $m_2$ calculated with the appropriate estimated values of $(ab)$ and $t_0$ in Table 7.1. The implication is that the first theoretical prediction is false for this case and the second prediction does not apply. The reason for this particular failure is unknown.

In the preceding discussions the phrase "best values" of the parameters, in a regression, are those values which minimize the sums of squares of the residuals, namely $(\ln y - \ln \hat{y})$. Table 7.5 lists the best values of $A$, $(AB)$, $T_0$, and $t^*$ (with $\alpha=1$)

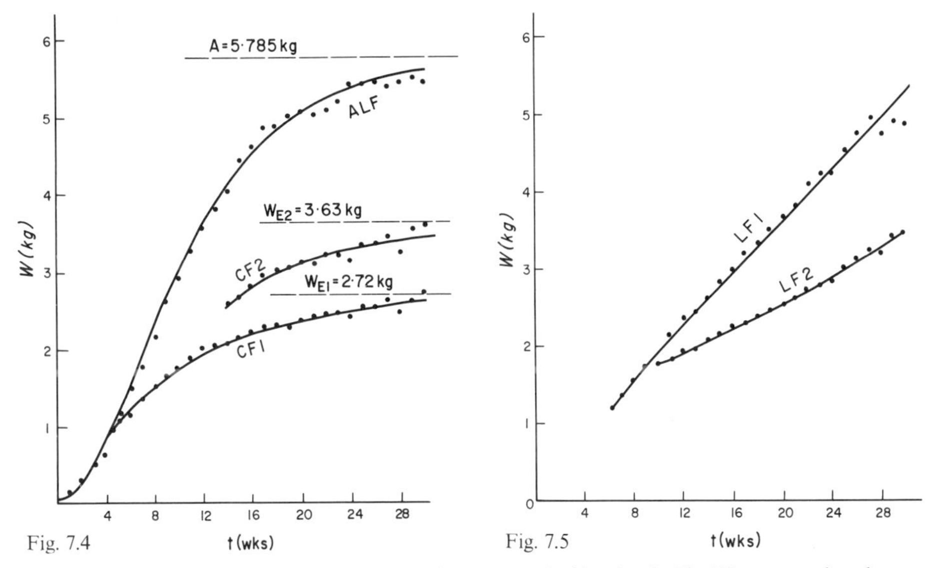


Fig. 7.4 Fig. 7.5

Fig. 7.4. Responses of the Ross fryer males to the constant feed intakes in Fig. 7.2 compared to the responses to ad libitum feeding

Fig. 7.5. Responses of the Ross fryer males to the linear feeding treatments (Fig. 7.2)

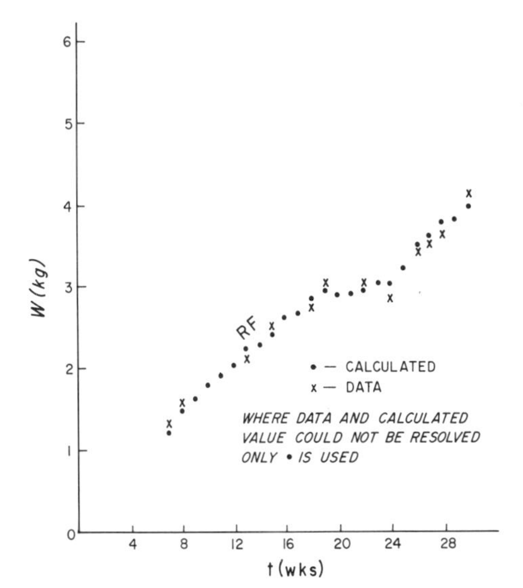


Fig. 7.6. Response of the Ross fryer males to the random weekly feeding shown in Fig. 7.3

obtained for the ALF treatments of both genotypes and sexes. Comparison of the values of the growth parameters with those in Table 7.1 shows that the estimated values are not too different from those found by experiment. This was borne out by the usefulness of the estimated values as starting values in the nonlinear regression procedure. The most notable features in comparing these two tables are, 1) that the estimated values of $(AB)$, namely $(ab) = 0.54$, for all the different animals, although close to the values in Table 7.5 for the fryers, appear to be different

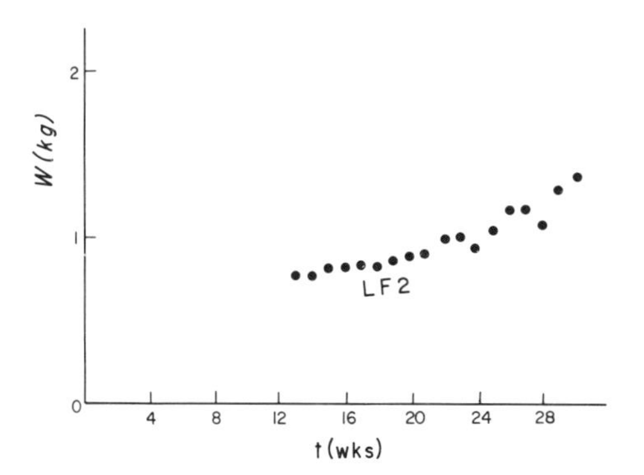


Fig. 7.7. An illustration of how the Apollo layer males failed in response to linear feeding treatment LF 2

Table 7.5. Best values of the growth parameters $A$, $(AB)$, $T_0$ and $t^*$ for the ALF treatments of both genotypes and sexes of Ross Ltd. chickens

| | Fryers | | Apollo layers | |
|---|---|---|---|---|
| | Males | Females | Males | Females |
| $A$ kg | 5.785 | 4.110 | 2.380 | 1.405 |
| $(AB)$ | 0.567 | 0.582 | 0.435 | 0.445 |
| $T_0$ wks | 3.745 | 3.910 | 2.747 | 2.691 |
| $t^*$ wks | 6.605 | 6.890 | 6.315 | 6.084 |

$\alpha = (AB)t^*/T_0 = 1$

in value for the layers, and 2) that the values of the appetite factor $t^*$ in Table 7.5 are found to be practically the same across genotypes and sexes whereas in Table 7.1 the estimated value of $t^*$, namely $t'$, varies from a low of 3.4 weeks for the fryer males to a high of 5.9 weeks for the layer females (a value near the experimental values of 6.048 for these animals). The values of $(AB)$ and $T_0$ in Table 7.5 will be modified when the results for $(AB)$ and $T_0$ across all the treatments are examined in relation to the second theoretical prediction later in this section. If the second prediction is acceptable we shall have more information about $(AB)$ and $T_0$ over the controlled growth region of the GPP.

The Figure of Merit, FM, was proposed in Chap. 7.2 for judging the comparative suitability of the theoretical functions (Table 7.4) as descriptors of the experimental data from the various treatments. Strictly speaking, the values of FM across the various treatments and animals are not to be compared among themselves, but their relative values may indicate the strength of the first theoretical prediction overall.

Table 7.6 illustrates the computation of FM across all treatments for the fryer male and Table 7.7 reports the FM's for the other animals. These sets of values of FM indicate that the first prediction is weakest for the fit of the theoretical feeding, $f(t)$, and growth, $W(t)$, functions (Table 7.4) to the ad libitum feeding and growth data. In a few of the controlled feeding cases the FM's are smaller than may be desired, for example FM = 9.743 for the CF 2 treatment of the Apollo layer male. All the other values indicate that the first prediction for the controlled feeding regime

Table 7.6. Calculation of the Figure of Merit (*FM*) illustrated using the values of *n*, *N*, *R* and *CV* obtained from using the functions in Table 7.4 on the Ross Ltd. fryer male

| Trt. | Para. $n$ | Data points $N$ | Runs $R$ | $P(R)$ | $CV$ | Fig. of Merit $FM$[a] |
|---|---|---|---|---|---|---|
| ALF | | | | | | |
| $f(t)$ | 2 | 30 | 12 | 0.195 | 0.163 | 3.811 |
| $W(t)$ | 2 | 30 | 6 | $0.252 \times 10^{-3}$ | $0.493 \times 10^{-1}$ | 10.145 |
| Total | 4 | 60 | 17 | $0.344 \times 10^{-3}$ | 0.145 | 1.725 |
| CF1 | 2 | 26 | 9 | $0.681 \times 10^{-1}$ | $0.214 \times 10^{-1}$ | 25.072 |
| LF1 | 2 | 24 | 8 | $0.593 \times 10^{-1}$ | $0.341 \times 10^{-1}$ | 15.587 |
| LF2 | 2 | 19 | 5 | $0.152 \times 10^{-1}$ | $0.144 \times 10^{-1}$ | 35.258 |
| CF2 | 2 | 17 | 9 | 0.617 | $0.277 \times 10^{-1}$ | 47.129 |
| RF | 2 | 24 | 13 | 0.842 | $0.280 \times 10^{-1}$ | 113.020 |

[a] $FM = \{n[1-P(R)]CV\}^{-1}$

Table 7.7. A listing of the values of *FM* for the other genotypes and sexes

| Trt. | Fryer Female | Apollo layers | |
|---|---|---|---|
| | | Male | Female |
| ALF | | | |
| $f(t)$ | 3.467 | 2.591 | 2.500 |
| $W(t)$ | 5.354 | 4.902 | 2.577 |
| Total | 0.502 | 0.792 | 1.270 |
| CF1 | 70.681 | 14.299 | 52.011 |
| LF1 | 120.147 | 12.542 | 14.055 |
| LF2 | 72.779 | — | 52.582 |
| CF2 | 497.576 | 9.743 | 40.092 |
| RF | 53.986 | 43.335 | 26.922 |

is strong. It was especially gratifying to find very high values of FM for the RF treatment of all four types of chickens. This adds additional weight to the first prediction. The reader will recall that the RF treatment was considered closer to the real world than the other treatments.

The weakness in the first prediction in all the ALF treatments is due to the poor fit of the theoretical $f(t)$ function to the data leading to low values of FM. It appears that there may be an additional deterministic element in the $f(t)$ data not accounted for in the $f(t)$ function, which would inflate the $CV$ and reduce FM. This problem will be discussed in detail in Chap. 7.4.3.

With regard to the second prediction, Table 7.8 sets forth the values of $(AB)$ and $T_0$ for all the treatments and all the types of chickens. As the second prediction is stated, the proposed statistical test is only applicable columnwise to the $(AB)$ and

Table 7.8. A listing of the values of $(AB)$ and $T_0$ for both genotypes and sexes for all the treatments

| Trt. | Fryers | | | | Apollo layers | | | |
|---|---|---|---|---|---|---|---|---|
| | Males | | Females | | Males | | Females | |
| | $(AB)$ | $T_0$ | $(AB)$ | $T_0$ | $(AB)$ | $T_0$ | $(AB)$ | $T_0$ |
| ALF | 0.567 | 3.745 | 0.582 | 3.267 | 0.435 | 2.953 | 0.445 | 2.256 |
| CF1 | 0.518 | 4.736 | 0.615 | 3.910 | 0.403 | 2.747 | 0.698 | 2.691 |
| LF1 | 0.878 | 3.615 | 0.966 | 3.457 | 1.000 | 2.454 | 1.000 | 2.517 |
| LF2 | 0.385 | 4.733 | 0.874 | 3.998 | Failed | | 0.125 | 1.000 |
| LF2 | 0.471 | 4.067 | 0.470 | 4.433 | 0.427 | 2.739 | 1.000 | 2.529 |
| RF | 0.587 | 4.410 | 0.697 | 3.940 | 0.337 | 3.277 | 0.841 | 2.600 |

Table 7.9. Final values of the growth and feeding parameters for the Ross Ltd. fryer and Apollo layer chickens

| | Fryers | | Apollo layers | |
|---|---|---|---|---|
| | Males | Females | Males | Females |
| $A$, kg | 5.785 | 4.110 | 2.380 | 1.405 |
| $(AB)$ | 0.622 ± 0.246 | | | |
| $T_0$, wks | 4.025 ± 0.476 | | 2.524 ± 0.572 | |
| $t^*$, wks | 6.019 ± 0.817 | | | |

$T_0$ values. Using the $t$-test in this fashion, the null hypothesis that the mean value of $T_0$ for all the treatments of the female Ross Fryers was equal to the value obtained for the AFL treatment, had to be rejected because the difference could have happened by chance alone in less than 5 trials in 1,000. The second prediction was acceptable in the other cases under the test.

I now realize that, due to this experiment not being properly replicated, the data in Table 7.8 cannot be the basis of a strong test to reject the null hypotheses of the second prediction when in fact they should be rejected. Experiment BG 54 does not challenge the second prediction as strongly as we thought during the design phase.

Table 7.9 is a presentation of the final values of the growth and feeding parameters of the Ross Ltd. Fryer males and females and the Apollo layer males and females obtained from the experiment. Here I have concluded that the first prediction passed the critical test and the second prediction failed the test; therefore the values in the table are biased by my apriori knowledge of the sexual dimorphism of $A$ in a genotype, the stability of $T_0$ within a genotype, the constancy of $(AB)$ in most circumstances, and the narrow range of $t^*$ in Table 7.5 across genotypes and sexes. With these assumptions I pooled the values of $(AB)$ over all the treatments, genotypes and sexes, the values of $T_0$ over all the treatments and the sexes within each genotype, and the values of $t^*$ for the ALF treatments across the genotypes and sexes listed in Table 7.5 to get the values of the growth and feeding parameters of these genotypes of chickens.

Although Table 7.9 does not fulfill the high hopes we had for BG 54, it is very useful in that the parameter values are far better estimates than the guessed esti-

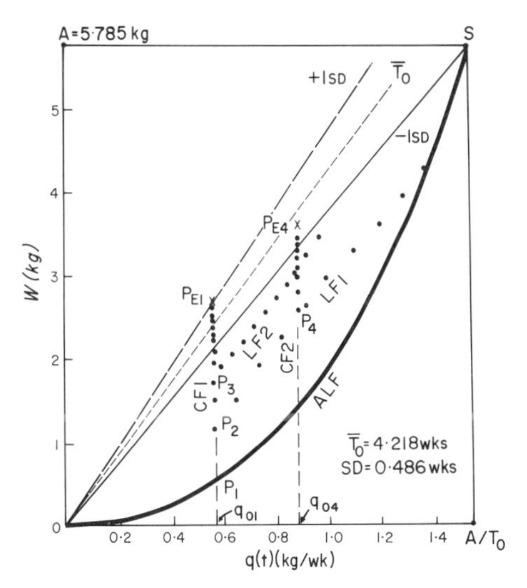


Fig. 7.8. The experimental GPP obtained from the Ross Fryer males. Compare with Fig. 7.1

mates listed in Table 7.1, and are therefore more useful for the design of any experiments using these chickens in their growing phases.

Figure 7.8 is the GPP using the computed weights and food intakes of the fryer male as an example for comparison with the assumed GPP in Fig. 7.1. The comparison is favourable to the theory as presented in Chap. 7.1.

## 7.4 Theory Versus the Results of the Experiment

A few months after the completion of the experiment, Wilson, Emmans, and Fawcett gave a preliminary review of the results at the Poultry Breeders Round Table in Birmingham (November 1975). Even though the results were far from complete they concluded that the theory appeared robust enough to be useful in the general study of animal feeding and growth. They pointed out the weakness of the theoretical weekly food consumed function, $f(t)$, as a descriptor of the actual values of $f(t)$. A more thorough discussion of the low descriptive power of the $f(t)$ function and a possible correction of it will be presented later in this section.

The usual notions, entertained by many animal scientists, of the possible importance of gut fill and lag times of response to changes of feeding regime are not substantiated by the data in Chap. 6 and the data and theory in this chapter. In general the theory states that the animal, plus the way it is fed, reacts as a linear system characterised by the feeding and growth parameters $A$, $(AB)$, $T_0$, and $t^*$.

Here the meanings of the mature weight $A$ and the mature food intake, $A/T_0 = C$, are sometimes obscured because of the confusion surrounding the use of the word "maturity" due to its association with the age when animals can beget young. In the context of the work discussed in this book, mature age will mean the age when an animal, male or female, has reached (or almost reached) its steady state live weight and food intake, but before the age the animal is considered senile.

The biophysical meanings of the parameters $(AB)$, $T_0$, and $t^*$ are obscured by the inability of the researchers to observe these properties of feeding and growth directly. However, in the context of the differential Eqs. (7.1), (7.2), and (7.3) these parameters have biophysical meanings as measures respectively of the efficiency animals have in converting food ingested to body weight, of an internal resistance to loss of body weight, and of an internal resistance to increase in the appetite.

It is readily seen that these parameters are lumped representations of the totality of the complexities of the internal activities of all the organs (down to the cellular level) which are regulated by the endocrine system under the feedback control of the hypothalamus through its production of peptide molecules. The whole body functions so controlled include deep body temperature, appetite, thirst, sleeping, and waking, the levels of blood sugar, salt, and water in the blood system and so on. Here reference can be made to any good book on general physiology or endocrinology or reference to the works of the 1978 Nobelists, R.S. Yalow, A.V. Schally, and R.C.L. Guillemin. It seems almost impossible that such complexity could be subsumed in at least four parameters that are constant and independent of age, body weight, the quantity of food consumed, and the rate of food consumption. By its very nature the theory is a simplification of complex issues, however in this simplification lies its usefulness in problems associated with nutrition (Chap. 9) and genetics (Chap. 10).

### 7.4.1 Use of the Theory to Examine Mature Weight and Appetite Variation

The theory, supported by the experiment designed to challenge it, can be used to examine two problems which, to my knowledge, have not been dealt with experimentally. The first deals with the unexpected variations in live weight during the time the animal is considered mature and the second introduces production of some product, for example wool or eggs. Production has not been ignored in developing the theory, it was only set aside until now.

It is assumed in the first problem that the variation of the mature body weight, $A$, is caused by the variation of the animal's appetite. Suppose the variation of mature body weight is denoted by $A+a(t)$, where $A$ represents the long term mean mature live weight and $a(t)$ represents the more or less rapid variation of the live weight about the mean. Also suppose that the mature food intake $C$ varies as $C+c(t)$. Now as $q^*(t)$ in Eq. (7.1) approaches $C$, $W$ in Eq. (7.2) approaches $A$ in such a manner that Eq. (7.2) approaches the controlled differential Eq. (7.3), which becomes in the limit

$$da(t)/dt+[(AB)/T_0]a(t)=(AB)c(t). \tag{7.21}$$

Here $c(t)$ is the variation of the appetite of the animal about the mean mature food intake. The variations of appetite, $c(t)$, considered due to a combination of internal physiological and environmental effects, may be expressed as a stationary time series, as in the following hypothetical example (Fig. 7.9).

This set of points, defining $c(t)$ as a discreet time series over the time span of $N$ equal periods of time, can be represented as a sum of sine waves, such as

$$c(t)=\sum_{i=1}^{N} b_i \sin(2\pi t/T_i+\phi_i),$$

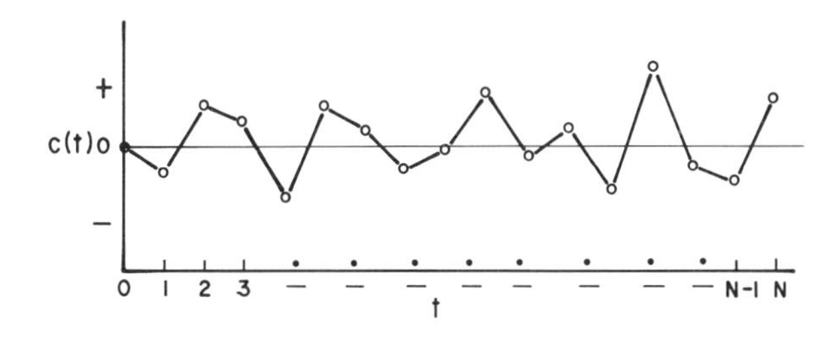


Fig. 7.9. Illustration of a possible variation, $c(t)$, of the appetite of a mature animal

(Box and Jenkins 1970, p. 36). The response, $a(t)$, to $c(t)$ so defined, will be the sum of the integrals of each of the components of $c(t)$, such as

$$a_i(t) = (AB)b_i \exp(-kt) \int_0^t \exp(kt) \sin(2\pi t/T_i + \phi_i)\, dt\,, \tag{7.22}$$

for example. Here $k$ is used instead of $(AB)/T_0$ for convenience, and the initial value of $a(t)$, at $t=0$, is taken as zero. Two other simplications of Eq. (7.22), namely dropping the subscript, $i$, and making $\phi_i$ equal zero, will have no effect on the question we wish to ask here. We have already encountered the evaluation of Eq. (7.22), with the above simplifications, in Chapter 6 (item 3 in Table 6.9 with the phase shift $\phi$ replaced by the time lag, $\Delta t$, equal to $\phi T/2\pi$ and since $\phi = \arctan [(2\pi/T)k]$, $\sin \phi = (2\pi/T)/[k^2 + (2\pi/T)^2]^{1/2}$), thus transforming Eq. (7.22) to

$$a(t) = \frac{(AB)b}{(k^2 + 4\pi^2/T^2)^{1/2}} \left\{ \sin[(2\pi/T)(t - \Delta t)] + \frac{(2\pi/T)\exp(-kt)}{(k^2 + 4\pi^2/T^2)^{1/2}} \right\}. \tag{7.23}$$

The question to which an answer is sought is: Given values of $b$, $(AB)$, $k$, and time, $t$, how does $a(t)$ and the time delay, $\Delta t$, of the response behind the appetite variation, change as the period, $T$, of the appetite sinewave, increases from low values (high angular frequency, $2\pi/T$) to high values (low angular frequency)? In arriving at the equation for the particular component of $a(t)$ given below we found the time lag to be given by

$$\Delta t = (T/2\pi)\arctan(2\pi/Tk).$$

Also a good measure of the effects of the variation of the period, $T$, on the component of $a(t)$, under consideration, is the amplititude of the sine wave response, namely

$$\text{amp} = (AB)b/(k^2 + 4\pi^2/T^2)^{1/2}.$$

Here it can be seen that as $T$ approaches zero, both $\Delta t$ and amp also approach zero, with the conclusion that high frequency variations of appetite will have little or no effect on the response $a(t)$. On the other hand when the period, $T$, becomes large, it can be shown that $\Delta t$ increases to the limit $T_0/(AB)$ and the value of amp increases to the limit $bT_0$, with the conclusion that low frequency variations of appetite tend to have greatest effect on the response.

A feasible experiment in this area is the imposition of a sine wave with a properly selected amplitude and period on the feeding function of a mature animal with known growth parameters. The above formulae could be used to decide upon values of $b$ and $T$, to give a delay time $\Delta t$ and a value of $a(t)$ both of which would be detectable over the chance variations of the response $a(t)$ during the experiment.

It would be prudent for the experimenter to begin his controlled $c(t)$ as a negative sinewave, in order to avoid refusal of food if the sinewave began as a positive increasing function at $t=0$.

An experiment of this kind would afford an opportunity to isolate more clearly the chance variations of $a(t)$ due to the steady state character of the maintenance of the mature weight. These chance variations could be followed, for example, by monitoring at equal periods of time the analysis for variations of the components, i.e., blood glucose, lipids and the various hormones, known or suspected to be involved in both the deterministic and chance fluctuations of $a(t)$.

### 7.4.2 Use of the Theory in the Problems of Production

The second feasible experiment, in which the production of some product is involved, also deals with mature animals with known growth parameters. Suppose the animals have a food intake, $C$ kg/wk, supporting a mature live weight, $A$, and the production of a product at the level of $p$ kg/wk. Now let the animal (or group of animals) be kept for $t'$ weeks on a constant food intake $fC$ where $0<f<1$. This puts the animal on a partial starvation ration and its growth phase curve will be a vertical straight line lying in the region of the GPP above the Taylor diagonal. The question to be answered experimentally is: "What is the food intake equivalent of the production?" Denote the food intake equivalent of the production by $a$, then the controlled feeding DE (7.3) can be written as

$$dW/dt+(AB)W/T_0=(AB)(fC-ap), \tag{7.24}$$

with initial conditions $W(0)=A$, at $t=0$. For $t=t'$ the solution is,

$$W(t')=T_0[fC_i-ap(t')]+\{A-T_0[fC-ap(t')]\exp[-(AB)t'/T_0]\}. \tag{7.25}$$

Here $p(t')$ is the mean production of the animal over the $t'$ weeks it is held on the partial starvation ration $fC$. In as much as it was already known the growth parameters are $A$, $T_0$, $(AB)$, and $C$, and the experimental data are $[t', f, W(t'), p(t')]$ the value of a is found by solving Eq. (7.25) to give

$$a=\frac{A-W(t')-(A-T_0fC)\langle 1-\exp\{-[(AB)/T_0]\,t'\}\rangle}{T_0p(t')\langle 1-\exp\{-[(AB)/T_0]\,t'\}\rangle}.$$

A better value of $a$ can of course be found by replicating the animals (or groups) at the same value of $f$ or at several different values of $f$.

However, it is most likely that the growth parameters of the experimental mature animals are not known from previous long term experiments. In this case, the data listed above replicated for several values of $f$ could be used with Eq. (7.25), as a nonlinear regression equation to determine $A$, $(AB)$, $T_0$, and $C$ as well as the desired food intake equivalent of the production, $a$. This experiment is certainly a feasible and economically interesting one.

In the past experiments of this kind have been done on cows and female chickens with the production being respectively kilograms of milk and eggs per week. In these experiments the expected results are formulated as an energy balance, i.e. the metabolizable energy consumed ($ME$) is partitioned between the

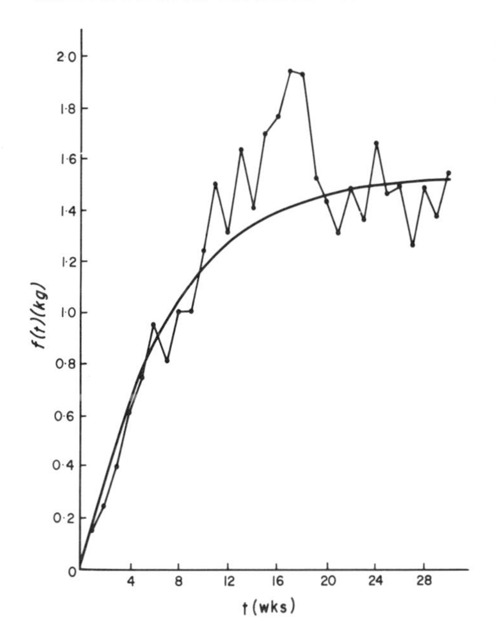


Fig. 7.10. Variations of the actual weekly food intake from theoretical for the Ross fryer males on treatment ALF

maintenance energy of the mean live weight over time period, $t'$; the energy equivalent of the weight lost; and the energy equivalent of the production. The model for this statement is expressed as the following linear equation,

$$(ME) = b[W(t')]^{0.73} + c[W(t') - A] + ap(t'),$$

which can be solved for values of $a, b$, and $c$ through the use of multivariate linear regression techniques given sufficient data (Brody 1945, p. 880).

The reader should note the fundamental difference between the theoretical approach to the problem of determining the food (or energy) equivalent of production, $a$, suggested above and the linear partition of the food with the extra unknown constants $b$ and $c$. The linear equation justifies itself by being arithmetically correct only through the ad hoc definitions of the factors $b$ and $c$.

### 7.4.3 The Poor Fit of the Theoretical Food Intake Function

Turning now to the poor fit of the theoretical ad libitum weekly food consumed to the experimental data. Figure 7.10 shows why the FM are so low for the four types of chickens (Tables 7.6 and 7.7). The FM for the fryer male is typical of the variation of the ad libitum weekly food consumed about the theoretical curve. Brody (1945) had suggested ignoring this kind of variation of the data from the expected, because it might be due more to the averaging procedure used to get the data per animal than to any biological response of the animal. Brody might well be correct and it is not necessary to worry about the low values of FM. However I have seen too many sets of averaged intake data, graphs of which are very similar to Fig. 7.10 (Fig. 7.11 for example) as well as the sets which do conform to DE (7.1) to feel comfortable with Brody's suggestion. In the following an attempt will be

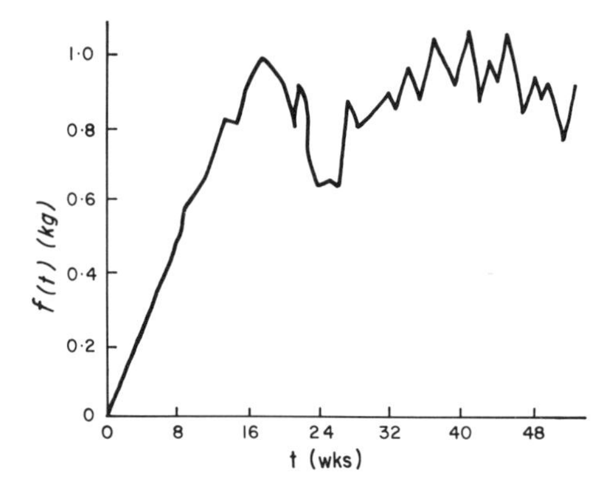


Fig. 7.11. Weekly food consumed by male chickens. Pen 84 was fed ad libitum. *Solid broken line* is observed data. (Data from Titus et al. 1934). Here is shown feeding behaviour similar to that in Fig. 7.10 41 years later

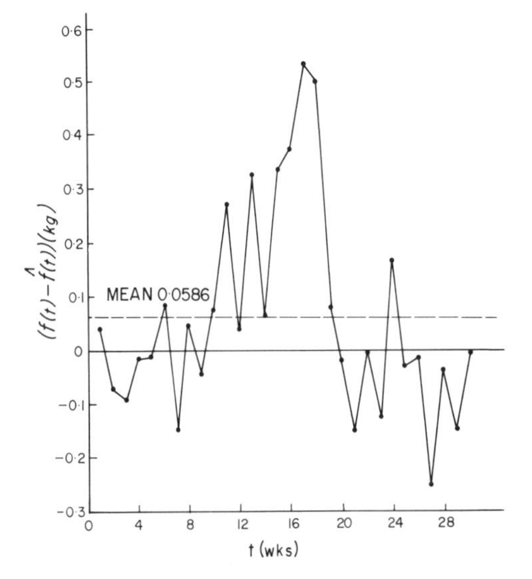


Fig. 7.12. The residuals as a function of time showing the possible need for sine waves in the feed intake function $f(t)$

made to adjust the DE (7.1), for ad libitum appetite, so that the values of FM will be larger for both the $f(t)$ and $W(t)$, and the total. The approach to modification will be through the residuals using those of the Ross Ltd. fryer males as an example (Fig. 7.12).

Figure 7.12 shows the variation of the residuals of the fryer males about a non-zero mean of 0.0586 kg/wk. It appears quite clear that in these residuals there is a deterministic negative sinewave, with a period, $T$, of about 15 weeks, and an amplitude, $a$, and about 0.4 kg/wk and a time lag, $t_0$, of about 2 weeks superimposed on the stochastic variation of the experimental measurements. Assuming this statement we can start modification of Eq. (7.1) through its solution, namely

$$q^*(t) = (C-D)[1-\exp(-t/t^*)] + D,$$

by subtracting the sinewave $a\sin[2\pi(t-t_0)/T]$ to get the new $q^*(t)$, namely

$$q^*(t) = (C-D)[1-\exp(-t/t^*)] + D - a\sin[2\pi(t-t_0)/T]. \tag{7.26}$$

Since $q^*(t)=dF/dt$, the integral of this modified $q^*(t)$ between the limits, $t-1$ and $t$, yields $f(t)=F(t)-F(t-1)$, as

$$f(t)=(A/T_0)\{1-t^*(1-DT_0/A)[\exp(1/t^*)-1]\exp(-t/t^*)\}$$
$$-[a\sin(\pi/T)T/\pi]\sin[2\pi(t-t_0-1/2)/T] \quad (7.27)$$

and if the lower limit, $t-1$, is taken as zero, the corrected cumulative food consumed is

$$F(t)=(A/T_0)\{t-t^*(1-DT_0/A)[1-\exp(-t/t^*)]\}$$
$$+(aT/2\pi)\{\cos[2\pi(t-t_0)/T]-\cos(2\pi t_0/T)\}. \quad (7.28)$$

Equations (7.27) and (7.28) replace the equations for $f(t)$ and $F(t)$ in Table 7.4 for nonlinear regression to determine the new set of ad libitum feeding and growth parameters $A$, $(AB)$, $T_0$, $t^*$ together with the period of the sinewave, $T$ (or the angular frequency $\omega=2\pi/T$ radians/wk). In addition to the i.c. $D$ there are two i.c., namely $a$ and $t_0$.

The meaning of adding the sinewave component to the main effect of the feeding function, is best approached through the new DE to replace DE (7.1). This differential equation is found by eliminating the initial conditions $D$, $a$, and $t_0$ from Eq. (7.26) and the derivatives of it up to $d^3q^*/dt^3$, with the result that DE (7.29) replaces DE (7.1) as a more appropriate description of how an animal feeds ad libitum

$$d^3q^*/dt^3+(1/t^*)d^2q^*/dt^2+\omega dq^*/dt+(\omega^2/t^*)q^*=\omega^2A/T_0t^*. \quad (7.29)$$

A differential equation exactly describes the fundamental properties of a class of curves, any one of which can be selected by specifying the i.c. However the real world phenomena may show instabilities appearing as oscillations, about the theoretical function, not accounted for in the differential equation. Figures 7.10 and 7.11 illustrate this kind of behaviour of the $(t,f)$ data relative to the theoretical description.

Suppose we let $\Delta(t)$ represent the instability of the food intake as a function of time not accounted for by DE (7.1) and write

$$\Delta(t)=dq^*/dt+(1/t^*)q^*-A/T_0t^*\neq 0,$$

indicating that the actual appetite of the animal does not satisfy the appetency balance required by DE (7.1).

Taking the first and second derivatives of $\Delta(t)$, DE (7.29) can be rewritten as

$$d^2\Delta/dt^2+\omega^2\Delta=0,$$

which imposes an exact description of $\Delta(t)$ as a sinewave with angular frequency, $\omega$. Now suppose this balance equation is again not met in the actual phenomenon and its instability is designated as $\Delta_1(t)=d^2\Delta/dt^2+\omega_1^2\Delta\neq 0$ for all $t$. If $\Delta_1(t)$ is also a sinewave correction applied to $\Delta$ and having an angular frequency $\omega_1$, then $\Delta_1(t)$ is a solution of

$$d^2\Delta_1/dt^2+\omega_1^2\Delta_1=0,$$

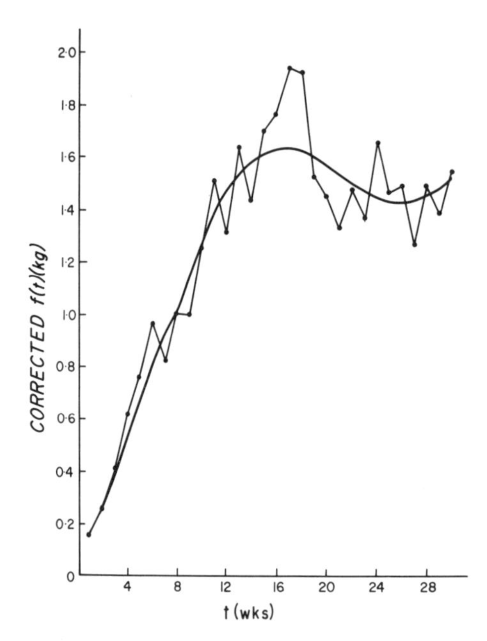


Fig. 7.13. Results of adding a sine wave to the theoretical differential equation for ad libitum feed intake to improve the fit of the theoretical $f(t)$ to the data

with initial condition $b$, and lag (or lead) time $t_{01}$, which is the recorrected version of DE (7.29). This procedure of correcting the corrections with sinewaves can continue until we get the nth correction, $\Delta_n(t)$ which has mean zero and is distributed normally with experimental variance $s^2$. This means that

$$d^2\Delta_{n-1}/dt^2 + \omega_{n-1}\Delta_{n-1} = 0$$

is the final version of DE (7.29) for accurately describing the phenomenon of food intake.

The above may be a quantitative paraphrase of the idea expressed by Mayer and Thomas (1967) that animals have the ability to make short, medium, and long-term corrections of over and under consumption of food during previous periods of time. A set of sinewave corrections with frequencies $\omega < \omega_1 < \omega_2$ etc. may be symbolic of the short, medium, and long term efforts of the Ross fryer males to correct their ad libitum food consumption to the solution of DE (7.1) with partial success as shown in Fig. 7.10.

Equations (7.27) and (7.28), substituted for their equivalents in Table 7.4, were used in nonlinear regression on the fryer male $[t, f(t), W(t)]$ data with the following results.

| $f(t)$ | $A$ KG | $(AB)$ | $T_0$ Wks | $t^*$ Wks | $\omega$ rad/wk | $CV$ Total[a] | $FM$ Total[a] |
|---|---|---|---|---|---|---|---|
| Uncorrected | 5.785 | 0.567 | 3.745 | 6.605 | — | 0.145 | 1.725 |
| Corrected | 5.560 | 0.590 | 3.240 | 9.045 | 0.256 | 0.759 | 2.706 |

[a] For all the $(t, f, W)$ data

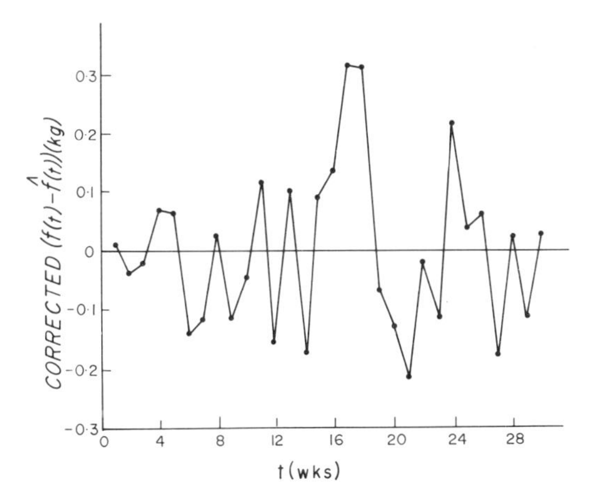


Fig. 7.14. Residuals remaining after correcting food intake for the sine wave

Comparing the fits of the corrected $f(t)$ with the uncorrected, we found the probability that the runs $P(R)$ are random, rose from 0.195 to 0.848, the $CV$ decreased from 0.163 to 0.1056, and most importantly the FM rose from 3.811 to 20.78 indicating the sinewave correction was a large step in the right direction towards getting a better description of the $[t, f(t)]$ data. The improvement is shown in Fig. 7.13 and the deviations, which we now call $\Delta_1$, are illustrated in Fig. 7.14. This correction procedure could be specified with higher frequency sine waves, but I was deterred by the possibility that the instabilities in the $f(t)$ data might be artifacts of the averaging procedures to get $f$ as suggested by Brody. To justify going on with the correction procedure, experiment BG 54 ALF treatment of the fryer males should be repeated under the same nutritional and environmental conditions as reported here. The same genotype and sex of chicken should be used in order to show that the variation of $[t, f(t)]$ data are repeatable and therefore most likely a characteristic of the animal.

The improved description of the actual weekly food consumed also leads to an improved description of the $[t, W(t)]$ data illustrated by Fig. 7.15 in comparison to the way the original $W(t)$ function for the ALF treatment tracked the data in Fig. 7.4 (Figs. 7.4 and 7.13 are drawn to the same scale). The better fit is also indicated by the increase of FM from 10.15 to 21.15 due chiefly to the reduction of $CV$ from 0.0493 to 0.0237 with the probability $P(R)$ that the runs are random remaining about the same.

Accepting DE (7.29) as a replacement for DE (7.1) has no effect on the two predictions previously made for the theory. The FM of the theoretical functions and the statistical tests show that the corrected food intake function and the theoretical live weight functions, are acceptable in describing the BG 54 experiment. The theory is robust enough for use in study of the effects of dietary and other environmental variables on the feeding and growth curves of animals as consequences of the effects on the feeding and growth parameters $A$, $(AB)$, $T_0$, $t^*$, and the frequencies $\omega_i$ of sinewave components of the appetency where they are known to apply.

There is no doubt that competing theories exist and I should like the reader to understand that I do not regard the theory, presented in these sections of this

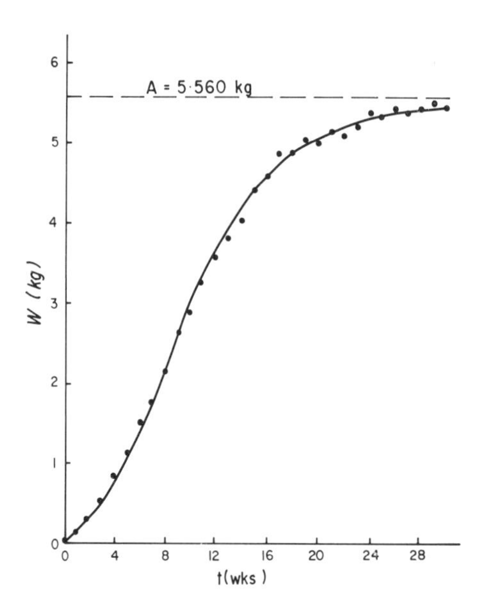


Fig. 7.15. Shows how adding the sine wave to the theoretical feeding function also improves the fit of the theoretical function $W(t)$ to the $(t, W)$ data

chapter, with such high esteem that I am automatically biased against all competitors. But like all inventors of theories I shall consider competing theories in the light of what I know about my theory and its origin and its level of falseness in an experimental test, and raise legitimate questions about the competitors. The most important question I can raise is: "Has the theory been subjected to some strenuous experimental test not included in the experimental situations from which the theory was intuitively derived?"

## 7.5 Alternative Ad Libitum Feeding and Growth Functions

As fas as can be judged from the literature there is no alternative to the controlled growth DE (7.3), which provides the response function when the initial conditions and the feeding function, $q(t)$, are given. However alternatives to the ad libitum growth and/or feeding functions may be specified by the experimenter to describe his data.

As was seen in Chap. 5, the trace of the feeding and growth of an animal, as a space curve, can be described by the parametric equations $W=g(t)$, $F=h(t)$, and $t=t$. Here functions $g(t)$ and $h(t)$ may be selected so that, in the opinion of the experimenter, they adequately describe the apparent geometrical properties of the $(t, W)$ and $(t, F)$ data, when the data are plotted as projections of the trace on the $Wt$ and $Ft$ planes respectively. The usual practice is to fit the selected functions to the data using regression techniques and declare them suitable if the rsd or $CV$ of fit is small and the residuals appear uncorrelated with the independent variable. These are the usual statistical criteria. However, the concept of FM was developed as a more general criterion for judging the relative merits of competing functions (Chap. 7.2). The FM is more general in that it involves not only the rsd (or $CV$) and the probability that the runs of signs of the residuals are random, but more importantly it contains the number of characteristic parameters in the selected functions. All other considerations of goodness of fit being equal, the function containing the smallest number of characteristic parameters will have the highest FM and will be the best theoretical descriptor of the experimental data. Use of the FM requires the experimenter to know the characteristic parameters and their number in the functions $g(t)$ and $h(t)$ he has selected.

In general there have been no alternatives suggested for my theoretical function $F(t)$ representing $h(t)$ (Table 7.4), but there are many $W=g(t)$ functions which possess points of inflection (POI) and have asymptotes denoted as the mature live weights A (Table 1.1, Chap. 1). The logistics function and Richard's generalisation of it have been popular as descriptors of growth data (Fitzhugh 1976). Here we shall consider $W=g(t)$ as the logistic function, the differential equation of which is

$$dW/dt=aW(A-W), \tag{7.30}$$

with the initial condition $W(t)=W_0$ when $t=0$.

The solution of DE (7.30), namely

$$W=A/\{1+[(A-W_0)/W_0]\exp(-Aat)\}, \tag{7.31}$$

can replace $W(t)$ (Table 7.4) to be used with $F(t)$ or $f(t)=F(t)-F(t-1)$ (Table 7.4) to fit $(t, F, W)$ or $[t, f(t), W]$ data. In this case DE (7.1) is accepted, but DE (7.30) replaces DE (7.2) with the consequence that the growth parameters becomes $A$, $T_0$, $t^*$ and $a$ with $A$ as the only parameters these DE's have in common. The growth efficiency factor $(AB)$ has no meaning in this case.

Equation (7.31) does fit the requirement that it be asymptotic to Brody's equation, namely $W=A\{1-\exp[-k(t-t^*)]\}$. If $t$ is considered large enough that $[(A-W_0)/W_0]\exp(-Aat)$ is small compared to unity, Eq. (7.31) asymptotically becomes

$$W=A\{1-[A-W_0)/W_0\exp(-Aat)]\},$$

which is analogous to Brody's equation, if we set

$$Aa=k$$

and set

$$t^*=\ln[(A-W_0)/W_0]k. \tag{7.32}$$

In this formulation this $t^*$ is not a characteristic parameter because it is a function of the i.c. $W_0$ and is fundamentally different from the $t^*$ in DE (7.1). Also it should be noted that the ad libitum discriminant $\alpha$ has no meaning in this case. Using the logistic function [Eq. (7.31)] with $f(t)$ (Table 7.4) we can proceed with the regression using the Ross fryer male ALF data, but without assuming $Aa$ to be Brody's $k$ or that the $t^*$ given by Eq. (7.32) is any way related to Brody's $t^*$ in the $f(t)$ equation. The results of this regression compared to the results using the theoretical functions with $\alpha=1$, are summarized here.

| | $A$ | $(AB)$ | $T_0$ | $t^*$ | $a$ | $CV$ | $FM$ |
|---|---|---|---|---|---|---|---|
| Theory | 5.785 | 0.567 | 3.745 | 6.605 | — | 0.145 | 1.725 |
| Alternative | 5.663 | — | 3.503 | 7.163 | 0.130 | 0.253 | 0.505 |

Even though this alternative to the theory gives fair estimates of $A$, $T_0$, and $t^*$, it must be rejected because the value of its FM is less than one third the value for the theoretical solutions of DE's (7.1) and (7.2) with $\alpha = 1$. It is interesting to note that this alternative has the same number of characteristic parameters, namely four, so that the rejection is due chiefly to the $CV$ and $[1 - P(R)]$ being larger.

A more interesting alternative involves the logistic function [Eq. (7.31) and its DE (7.30)]. The reader may recall the reference I made to Eq. (1.18), namely $dW/dt = (dW/dF)(dF/dt)$ (Chap. 1.5). This equation is useful in setting up this alternative to the theory. The DE of the logistics function can be rewritten as

$$dW/dt = (Aa)(1 - W/A)W. \tag{7.33}$$

Supposing $(Aa)$ is Brody's $k$ and therefore equal to $(AB)/T_0$, DE (7.33) becomes

$$dW/dt = (AB)(1 - W/A)W/T_0. \tag{7.34}$$

Here the reader will recall from Chap. 2 that the product $(AB)(1 - W/A)$ on the right-hand side is none other than the true efficiency of growth, $dW/dF$, from Spillman's hypothesis that live weight follows the law of diminishing returns in the cumulative food consumed, $F$, domain. A consequence of the above expression of the growth rate and $dW/dt = (dW/dF)(dF/dt)$ is that the ad libitum food intake $q^*(t)$ is proportional to the live weight, $W$, or

$$q^*(t) = W/T_0. \tag{7.35}$$

This formula for the food intake quantitatively expresses the quantitative statement made by some animal scientists "that animals eat according to their weight." But how "they eat according to their weight" is not stated.

If Eq. (7.35) is substituted in (7.34) and the terms rearranged, my DE (7.2) is quite neatly derived. However, DE (7.1) for the ad libitum food intake must be replaced by DE (7.36) derived by taking the derivative of Eq. (7.35) and using DE (7.34) with $W/T_0 = q^*$ substituted in it.

$$dq^*/dt + [(AB)/A]q^{*2} = [(AB)/T_0]q^*. \tag{7.36}$$

The initial condition is $q^*(0) = W_0/T_0$ at $t = 0$. The solution is obviously Eq. (7.35) with the solution of DE (7.34) with i.c. $W(0) = W_0$ at $t = 0$ substituted in it. Here it is seen that the increase of the appetite is driven by the variable appetency, $[(AB)T_0]q^*$, and at the same time impeded by the variable resistance, $A/(AB)q^*$. It should be noted that this natural internal resistance to the growth of appetite is the same as the natural internal resistance to the live weight equivalent of the food intake, $A/(AB)q^*(t)$, going to no growth in DE (7.2) which is now acceptable in this alternative to the theory. Here we see that DE's (7.36) and (7.2) are the theoretical bases for this second alternative.

Differential Eq. (7.36) is not new because Newton used it to study the properties of bodies moving through fluids the resistance of which to the motion was proportional to the motion. Differential Eq. (7.36) differs from Newton's only in that it has a variable driving function rather than a constant on the right-hand side.

In order to apply this alternative to the Ross fryer male data expressions for $F(t)$ and $f(t) = F(t) - F(t-1)$ must be found for use with the law of diminishing returns, namely

$$W(t)=(A-W_0)\{1-\exp[-(AB)F(t)/A]\}+W_0 \tag{7.37}$$

to obtain the regression equations for the male fryer data. The cumulative food $F(t)$ is found by integrating (7.35) between the limits zero and $t$ with the condition that $F(t)=0$ at $t=0$. The integration involves integrating the logistic function, namely Eq. (7.31) between zero and $t$, but with $(AB)/T_0$ substituted for $Aa$. The indefinite integral of the logistic function can be found in tables of mathematical functions. The result is

$$F(t)=(A/T_0)[t-[T_0/(AB)]\ln\langle\{W_0+(A-W_0)\times\exp[-(AB)t/T_0]\}/A\rangle] \tag{7.38}$$

and since $f(t)=F(t)-F(t-1)$ it is given by

$$f(t)=(A/T_0)\left\langle 1-[T_0/(AB)]\times\ln\frac{A-(A-W_0)\{1-\exp[-(AB)t/T_0]\}}{A-(A-W_0)\{1-\exp[-(AB)(t-1)/T_0]\}}\right\rangle. \tag{7.39}$$

Equations (7.37), (7.38), and (7.39) satisfy the three basic requirements for theoretical feeding and growth functions, namely that, in the limit when $t$ is large, (1) $q^*(t)$ and $f(t)$ should asymptotically approach the mature food intake $A/T_0$; (2) $F(t)$ should asymptotically approach a straight line the slope of which is the mature food intake $A/T_0$, and (3) the live weight should asymptotically approach the mature live weight $A$.

This alternative, even though it accepts Eq. (7.2) completely, denies the theory proposed in Chap. 7.1 in at least two important ways. Firstly the ad libitum feeding discriminant, $\alpha$, does not exist, and secondly the growth phase plane, GPP, cannot exist, with the consequence that the $T_0$ in this alternative cannot be the slope of anything like the Taylor diagonal. However, the alternative is attractive because it proposes three characteristic growth parameters, namely $A$, $(AB)$, and $T_0$, and is therefore a more parsimonious description of feeding and growth data.

Equations (7.37), (7.38), and (7.39) were applied to the Ross male fryer feeding and growth data and the results, compared with the theoretical, are summarised here.

| | $A$ | $(AB)$ | $T_0$ | $t^*$ | $CV$ | $FM$ (total) |
|---|---|---|---|---|---|---|
| Theory ($\alpha=1$) | 5.785 | 0.567 | 3.745 | 6.605 | 0.145 | 1.725 |
| Alternative | 2.738 | 1.235 | 1.305 | — | 0.486 | 0.658 |

The reduction of the parameters from four to three in the calculation of FM was not sufficient to overcome the almost fourfold increase in the $CV$ of fit of the alternative functions. Also the estimates of $A$, $(AB)$, and $T_0$ are very low in the light of the experience of growing heavy birds like the Ross fryers. The low estimates of the characteristic parameters and the reduced FM are the bases for rejecting this alternative. Rejection of the two alternatives to the theory indicates that the logistic function is not useful as a descriptor of $(t, f, W)$ data in spite of its popularity in studies of $(t, W)$ data.

There are many other alternative theories than the two just discussed, but they must be similarly tested on the BG 54 data in Appendix A.

Lozano (1977) performed a long term experiment on rats from which it was concluded that the equation of diminishing increments did not apply to weight versus cumulative food consumed. Since this equation is a cornerstone of the theory it is important to examine this work in some detail.

The experiment consisted of growing, over an extensive period, two strains of albino rats having known characteristics of medium and low adult weights. The experiment was carried on from weaning at 21 days of age to 197 days with the mean cumulative food consumed, $F$, and mean weight, $W$, per rat measured and reported at various ages. When each population reached six months of age, 10 rats were randomly selected from each group and reared to one year of age. At this age the mean live weight per rat in each group was measured and reported, but the cumulative food was not.

In the analysis of the data Lozano felt there was reason to believe that the animals at weaning would asymptotically reach the mature live weight, $A$, measured at one year of age. This belief led to the use of a technique suggested by Riggs (1963, Sect. 6-13, p. 142) to modify Eq. (7.37) to

$$W(F) = A - a\exp(-bF) - c\exp(-dF). \tag{7.40}$$

This equation states that instead of one growth efficiency factor, $(AB)$, there are two growth efficiency parameters $b$ and $d$, but more importantly it implies that no matter how the animals start at weaning they will arrive at the measured live weight, $A$, at one year of age. The equation also suggests that it is the solution of a second order differential equation involving $b$ and $d$ which needs two i.c. to solve it for $W(F)$. However Lozano specified only that $W(F) = A$ at one year of age and appeared not to have made an attempt to find a theoretical cumulative food consumed, $F(t)$, function.

I used the data from weaning at 21 days of age to 197 days of age of Lozano's rats $\alpha$ on diet $\times$ (Lozano, Table 2) in a nonlinear regression with my theoretical $F(t)$ and $W(t)$ equations (Table 7.4) to obtain values of $A$, $(AB)$, $T_0$, and $t^*$, and to obtain the FM of fit of my theoretical functions to this data. The results of this study are,

$A = 0.380$ Kg; $(AB) = 0.322$; $T_0 = 2.260$ wks; and

$t^* = 3.650$ wks.

The fit of $F(t)$ to the $(t, F)$ data had a $CV = 0.0246$ and a $FM = 10.90$; and the fit of $W(t)$ to the $(t, W)$ data had an $CV = 0.0166$ and $FM = 46.41$. The total fit of the $F(t)$ and $W(t)$ to the combined $(t, F, W)$ data had $CV = 0.0198$ and $FM = 14.63$. These results show that my theoretical functions are good descriptors of Lozano's $(t, F, W)$ data for these rats without the addition of a second growth efficiency factor and without requiring these rats to achieve a given weight at one year of age.

I will only briefly mention three competing theories in which I cannot find any means of comparing them to my theory. These theories attempt more general descriptions of live weight, efficiency, ad libitum intake, body composition, and production of some economic product. They may involve many linear differential equations, constraint equations and initial conditions for solving the various dependent variables as functions of time. The best I can do is present some references and make some remarks about them and leave the matter in the reader's hands for his judgement in the light of my study of the feeding and growth of animals ranging from steers to mice.

Sandland and McGilchrist (1979) gave a widely ranging stochastic analysis of time serial measurements of human height, sunflower height, and duration of pregnancy of Afghan Pika. The mathematical basis of their approach was through the use of the relative growth rate as some function of growth, namely $(1/W)dW/dt=f(W)$ [suggested by Fisher (1946) and expanded by Radford (1967)]. I do not favour the use of relative growth rate in any study of the growth of living organisms. The organism is an open system and the question of whether the variation of the relative growth rate is more influenced by variation of the relative efficiency $(1/W)dW/dF$, or the relative food intake $(1/W)dF/dt$ is ignored. Most often $dF/dt$ is not considered either in those cases in which $dF/dt$ was not measured, where it might have been, or in the case of plants where $dF/dt$ may be unmeasurable. In as much as their study is of techniques for handling data quantitatively, there seems no way to test their approach decisively. I think it can only be said that if the techniques satisfy the worker there can be nothing wrong in using them.

C.Z. Roux (1974, 1976, 1979) and Meissner et al. (1975) have done extensive theoretical work on describing various kinds of data, like body weight, composition etc. of sheep. They have relied heavily on the application of the allometric formulation to many paired biological variables, such as graphing the natural logarithm of body weight versus the natural logarithm of cumulative consumption of metabolizable energy (*ME*). From such formulations they derive some differential equations and statistical matrix equations that help them in analysing their data. They seem to favour transforming their variables so that the graphs are straight lines and the breaks in them can be related to some change of internal physiological status or change of phase of growth (Sandland and McGilchrist 1979) They also tend to rely heavily on Richards's (1959) generalised growth function which includes the logistics, Gompertz, and von Bertallanffy functions as special cases. Here again I do not see how the theory or theories can be decisively challenged and I do not see in their published work what they plan in this respect. It is quite possible to compare their theory with mine as applied to some set of long term $(t, F, W)$ data with some agreement between us on some criterion of goodness of fit.

I shall close this section by referring very briefly to the large amount of work done recently by a group of computer oriented scientists in the Division of Animal Physiology in the Commonwealth Science and Industry Research Organisation (C.S.I.R.O.) in Australia. This work by Baldwin, Black, Faichney, and Graham can be obtained from the C.S.I.R.O., Division of Animal Physiology, Sydney. The capacity of the modern computer, for programming and data handling, invites the use of it in major attempts to whiten the box using data from many experiments on growth of parts of an animal relative to the whole. This approach requires a very large set of D.E.s with rate constants assumed and/or experimentally determined to adequately describe growth and production of animals.

In whitening the box they have introduced a set of interconnected storage units representing organs, blood etc. which utilise the nitrogen and energy input to the system for growth in ways determined by the specified interconnections between the storage units. The growth of the stimulated animal is the day by day sum of the contents in the storage units. None of this work seems to have had any connection with the fundamental theoretical study of the cell by Goodwin (1963) or the work of Morowitz (1968) on the flow of energy in biology.

To me such programs appear to be gigantic accounting schemes, involving the costly portions of the feed, namely nitrogen and energy, and the production of animal tissues of economic value. It also seems to me the user can get out of the program only what information is already built into them from very limited experimental work on various animals. I cannot fathom how such programs can be decisively challenged experimentally. The reader can find entrance to this work through the references in a paper by Baldwin and Black (1979).

I should also remark that whitening the box, by the techniques used in these programs, can be done in an infinity of ways because each compartment may be treated as a set of subcompartments with additional DE's ad infinitum. I have seen nothing in the literature discussing such programs which suggest; (1) that day by day data generated by any such program will be a solution of Eq. (7.2), if the program is fed data which are a solution of Eq. (7.1); (2) that the data generated will be a solution of Eq. (7.3), if the input are any data other than a solution of Eq. (7.1); or (3) there exists any such quantity as the ad libitum feeding and growth discriminant $\alpha$. Also, serious study of these programs has shown no way of obtaining more fundamental biotheoretical meanings of parameters $(AB)$, $T_0$, and $t^*$ as lumped metabolic parameters of the system as a whole.

## 7.6 Rehabilitation from Controlled Feeding Stress

We have seen in Chap. 7.2 how an animal can be control fed to reach point $(q_0, W_0)$ in the controlled growth region of the GPP during a period of time (Fig. 7.16). From this point, $P$, there are many ways of feeding in order to bring the animal back to its ad libitum feeding and growth condition. Of these many ways the simplest to treat theoretically and possibly the simplest experimentally are members of the family of straight lines generated by rotating line PB through angle, $\theta$, from zero radians for line PQ to arctan $(A-W_0)(A/T_0-q_0)$ radians for line PS. The general equation for the family is

$$W(t)=m[q(t)-q_0]+W_0,\ m=\tan(\theta), \tag{7.41}$$

of which lines PQ, PR, PB, PT, and PS are particular members. These lines represent degrees of severity in rehabilitating the animal, in that line PQ represents instantaneous rehabilitation from controlled food intake, $q_0$, to ad libitum food intake, $q_H$, by offering the animal unlimited food supply immediately. The other lines represent programming the food supplied in a controlled fashion to reach ad libitum food intakes, $q_R$, $q_T$ or the mature food intake $q=A/T_0$ during increasing periods of time, thereby reducing the severity of the rehabilitation procedure and allowing the animal more time to adjust its physiological status back to what it would normally be under ad libitum feeding conditions.

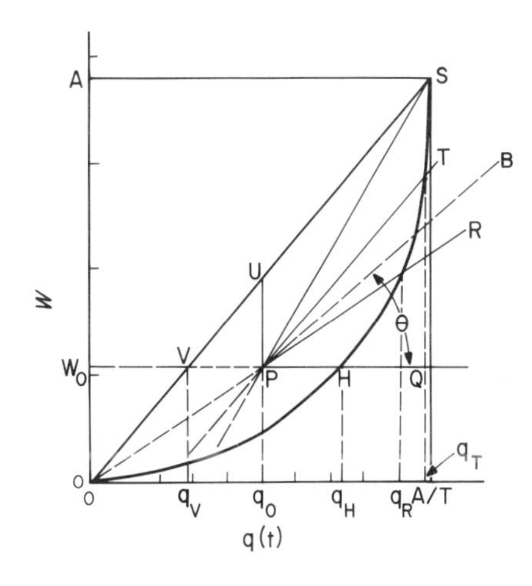


Fig. 7.16. A family of straight lines through point $P$ in the controlled growth region. Line $PB$ generates the family as $\theta$ is increased from zero to $\pi$ radians

We have already seen how the feeding programs, $q(t)$, can be found by substituting Eq. (7.41) in the controlled feeding DE (7.3) and solving it, with initial conditions $W(0)=W_0$ at $t=0$, to get the general differential equation for the required feeding programs, namely

$$dq(t)/dt+(AB)(1/T_0-1/m)q(t)=(AB)(q_0-W_0/m)/T_0\,. \tag{7.42}$$

The solution of this equation with initial conditions $q(0)=q_0$ at $t=0$ is

$$q(t)=[(T_0q_0-W_0)/(m-T_0)]\{1-\exp[-(AB)(1/T_0-1/m)t]\}+q_0\,. \tag{7.43}$$

This equation states the feeding programs as functions of time for $m$ greater than zero excepting $m=T_0$. The equality $m=T_0$ yields an indeterminate 0/0 value for $q(t)$ regardless of time $t$. The indeterminancy of $q(t)$ can be avoided by setting $m=T_0$ in DE (7.42) and $q(t)$ becomes the linear function of time already encountered in the LF1 and LF2 treatments during the planning of the BG 54 experiment (Table 7.2). The response to the rehabilitation feeding programs is found by substituting Eq. (7.43) in Eq. (7.41), with the result

$$W(t)=m[(T_0q_0-W_0)/(m-T_0)]\{1-\exp[-(AB)(1/T_0-1/m)t]\}+W_0\,. \tag{7.44}$$

These equations show that $q(t)$ and $W(t)$ share a common geometrical property when $m$ has a given value, i.e. in both equations when $0<m<T_0$, the quantities $(m-T_0)$ and $(1/T_0-1/m)$ in the factor and exponent respectively are $<0$, which makes $q(t)$ and $W(t)$ increasing exponential functions of time from $q_0$ and $W_0$. When $m$ has a value of $T_0<m\leqq(A-W_0)/(A/T_0-q_0)$ the quantities $(m-T_0)$ and $(1/m-1/T_0)$ are greater than zero so that $q(t)$ and $W(t)$ follow the equation of diminishing increments as functions of time. We can now see why $m=T_0$ in Eq. (7.43) gives an indeterminate condition in both $q(t)$ and $W(t)$, for when $m<T_0$ both $q(t)$ and $W(t)$ curve upwards and when $m>T_0$ they curve downward. Therefore when $m$ passes through $T_0$, either from above or below, the curves cannot continuously change their opposing characters from curving upwards or downwards or vice versa.

Figure 7.16 shows three lines PR, PT, and PS which we can use to graphically demonstrate the feeding and growing of the animal during rehabilitation along these lines. The slope $m$ of line PR is chosen so that PR passes through the origin (0,0) of the GPP also, so that $m = W_0/q_0$. Substituting this in the equations for $q(t)$ and $W(t)$, we get, after some algebraic reduction,

$$q(t) = q_0 \exp[(AB)(T_0 q_0 - W_0)t/T_0 W_0],$$

and

$$W(t) = W_0 \exp[(AB)(T_0 q_0 - W_0)t/T_0 W_0],$$

which are both rising exponential functions of time with doubling time, $t_2$,

$$t_2 = T_0 W_0 \ln 2/(AB)(T_0 q_0 - W_0).$$

When $m = T_0$ line PT is parallel to the Taylor diagonal of the GPP with

$$q(t) = (AB)(q_0 - W_0/T_0)t/T_0 + q_0$$

and

$$W(t) = (AB)(q_0 - W_0/T_0)t + W_0.$$

For line PS $m = (A - W_0)/(A/T_0 - q_0)$ and the least severe rehabilitation feeding program and the response to it are

$$q(t) = (A/T_0 - q_0)\langle 1 - \exp\{-(AB)[(q_0 - W_0/T_0)/(A - W_0)]t\}\rangle + q_0\,,$$

and

$$W(t) = (A - W_0)\langle 1 - \exp\{-(AB)[(q_0 - W_0/T_0)/(A - W_0)]t\}\rangle + W_0\,,$$

which are forms of the equation of diminishing increments in time, $t$. Here it is seen that both the feeding and growth asymptotically approach the mature food intake at a low rate depending on where P is in the controlled region of the GPP. The animal's ad libitum physiological status is regained from what it was at point P.

This gradual return to normal ad libitium feeding and growth is in contrast to what must happen when rehabilitation is carried out along lines PQ, PR, and PT (Fig. 7.16). I have no theoretical approach to the consequences of instantaneous rehabilitation along line PQ, but plotting $q(t)$ and $W(t)$ for lines, PR, PT, PS and the constant intake line PU together with the ad libitum feeding and growth curves, for some chosen animals, may suggest some ideas about possible events as the rehabilitation nears completion.

Figure 7.17 is a plot of the calculated feeding regimes, and the responses to them, using $A = 3.81$ kg, $(AB) = 0.382$, $C = A/T_0 = 0.989$ kg of food/wk and $t^* = 4.19$ wks for the Emmans broiler female chickens (Table 4.1, Chap. 4).

Here it is shown how rehabilitating the chickens from point P ($q_0 = 0.5$, $W_0 = 1.15$ at 8 wks of age) along lines PR and PT, may produce critical situations as the animal nears complete rehabilitation. For example, when the animals food intake for line PR reaches its ad libitum value at R (Fig. 7.17a), the live weight at R″ (Fig. 7.17b) is about 1 kg less than what it should have attained for the animal to be considered rehabilitated to ad libitum growth as well as feeding ad libitum. In as much as the animal must be fed above ad libitum, i.e. about 1.45 kg/wk for about four weeks longer to point R (Fig. 7.17a) for the live weight to reach its ad libitum value at point R (Fig. 7.17b), it is illogical to think the experimenter could

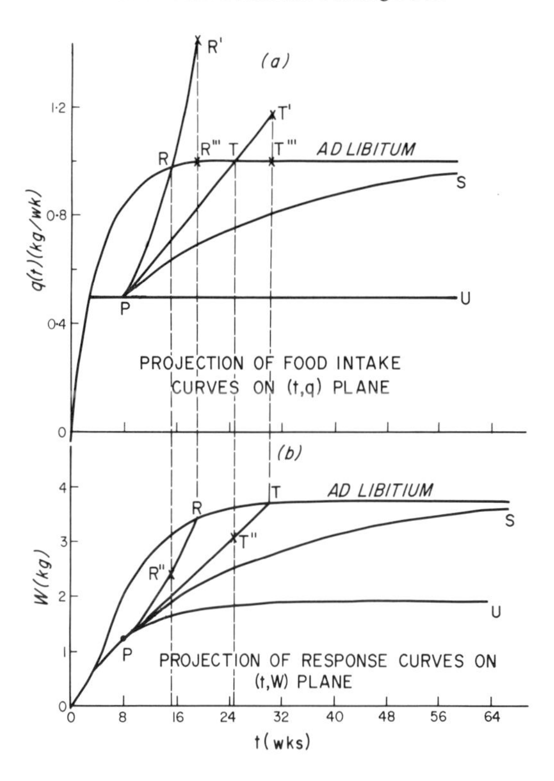


Fig. 7.17 (a, b). Food intake and response curves (a) and (b) respectively for rehabilitation along lines *PR*, *PT*, and *PS* and the nonrehabilitation response curve along the constant food intake into *PU* in Fig. 7.16

then make unlimited food available and the animal's food intake would fall from R′ immediately to the ad libitum value at R‴ (Fig. 7.17 a) corresponding to the ad libitum live weight at R (Fig. 7.17 b). This same situation applies to rehabilitation along line PT, but the adjustment of the food intake from T′ to T‴ (ad libitum) is much less when the animal reaches its ad libitum live weight at T. Rehabilitation along line PS presents no such logical difficulties because both the ad libitum feeding and growth are approached asymptotically, i.e. most gently.

It might be conjectured that, in the case of rehabilitation along line PR, the experimenter should stop controlled feeding at point R (Fig. 7.17 a) and allow the animal free access to food while it is regaining the 1 kg deficit in live weight to attain its normal physiological status. This could be done in a similar manner in rehabilitation along line PT. Figure 7.18 shows the conjectures on how the animal may regain its normal physiological status when it is allowed free access to food when the controlled feeding reaches points R, or T, on the ad libitum feeding curve. The live weights at R″, or T″, would probably follow the dashed curves and asymptotically approach the ad libitum live weight (Fig. 7.18 b), because live weight may have some "biophysical inertia" delaying its change. However the food intake may show overconsumption and follow some path like the dashed curves (Fig. 7.18 a) to a peak followed by an asymptotic decline to ad libitum feeding at about the time the live weight reaches its ad libitum value. Because the rehabilitation along line PT is less drastic, I have indicated a lower peaking of the overconsumption of food.

In the context of the situation of live weight recovery, described in Fig. 7.18 b, the phenomenon of "growth spurt" may be more readily observed than in some

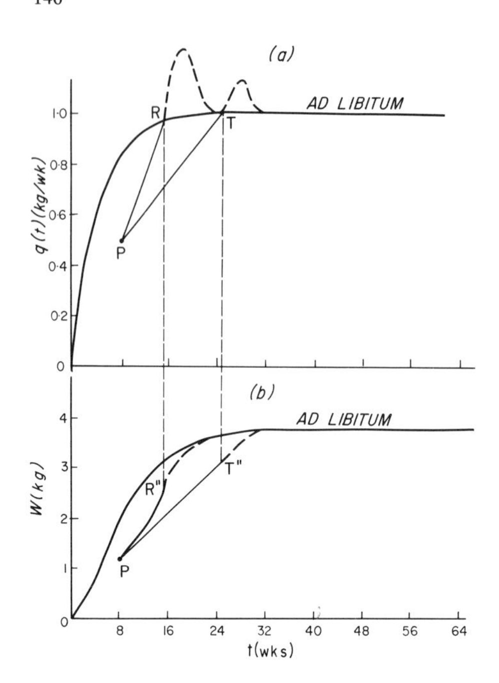


Fig. 7.18. (a) Shows the possible over consumption of food of rehabilitated animals as (b) their live weights return to normal from points $R''$ and $T''$

of the experimental data presented to support the idea of "growth spurt." The dashed curves in Fig. 7.18 b illustrate my notion that for some physiological reason the animal does not dally in its attempt to return to normal live weight and physiological status. The physiological reason may be the secretion of growth hormone at higher levels than usual during the stress of undernourishment (Griffiths 1974, p. 360).

Because my theory is sufficienty robust, I ventured into the field of rehabilitation to get some possible, or rather plausible, idea of what the phrase "the animals were rehabilitated" (quite often found in the literature on inanition) might mean quantitatively. I recently found some experimental support for Fig. 7.18, especially the food intakes, in the work of Pym and Dillon (1974) (Fig. 7.19). They reared two breeds of chicks by ad libitum feeding until 10 weeks of age when one quarter of the flock of each breed was allocated to each of the four treatments, namely, $A$ – ad libitum, $B$ – 80% of $A$, $C$ – 60% of $A$ and $D$ – 40% of $A$. Figure 7.19 shows that when all pens were allowed free access to food at 19 weeks, the food intakes in the following 2 weeks were greater than ad libitum and reached peaks in direct relation to the severity of the preceding restrictions. Subsequently the food intakes diminished towards the ad libitum level while the weights increased asymptotically towards ad libitum live weights. These transients in food intakes took about 10 or 15 weeks to die away. This example is of a rehabilitation quite similar to instantaneous rehabilitation along line PQ ($m=0$) in Fig. 7.16 and may be apropos to Fig. 7.17 and 7.18.

Some controlled rehabilitation experiments based on the theory of controlled growth in the growth phase plane could go far to test the usefulness of the theory and explore the critical region of adapting back to ad libitum feeding and growth

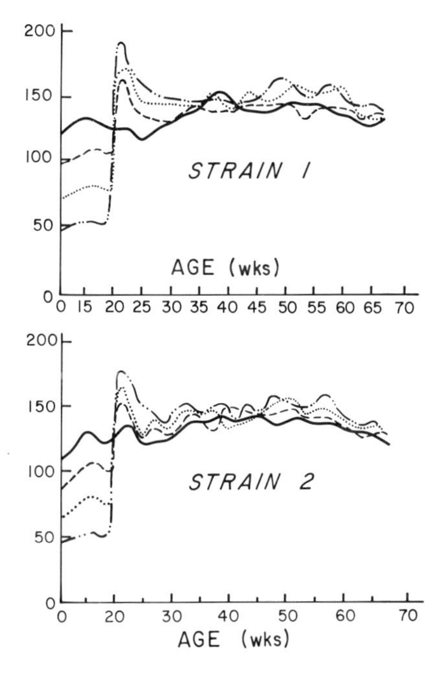


Fig. 7.19. Responses of food intake of two strains of chicks to instantaneous rehabilitation from various levels of restriction from 10 weeks of age, showing over consumption of food in direct relation to the level of restriction. [Reproduced from Fig. 1 of Pym and Dillon (1974) with permission of the authors and of British Poultry Science Ltd]

after more or less severe restriction of feeding. This is a new area of research important to understanding how animals adapt to generalised feeding conditions.

The ideas developed here concerning the response to feeding so that the $[q(t), W(t)]$ curve is a straight line in GPP will be used in Chap. 7.7 to explore the possibility that the Trace of an animal's growth is an optimum growth pattern for ad libitum feeding.

## 7.7 The Ad Libitum Growth Curve as an Optimum

Generally, the optimisation of an activity is thought to be a process of finding those conditions for which some property of the activity such as return or profit is maximised or cost is minimised. Here the idea of an optimum growth curve does not involve economics and hence may need some clarification. Just what is maximised or minimised in the case of an optimum growth curve? In this section the question asked is: "does the ad libitum feeding and growth curve place a lower limit on the time or food consumed for an animal to grow between two arbitrarily chosen points on its trace?" In the context of optimisation problems, the reader must be aware of the general theorem which states that in the process of adjusting the parameters or constraints of any system one and only one quantity, associated with the system, can be minimised or maximised. In the beginning of this section the question of the optimal property of the ad libitum feeding and growth curve, i.e. the trace, was stated in terms of minimum time or food; it cannot be both. Later it will be shown that time is the property to be minimised with the cumulative food consumed being a minimum because of its direct dependence on time. This question can be given a general meaning for all animals (except man) by discussing it

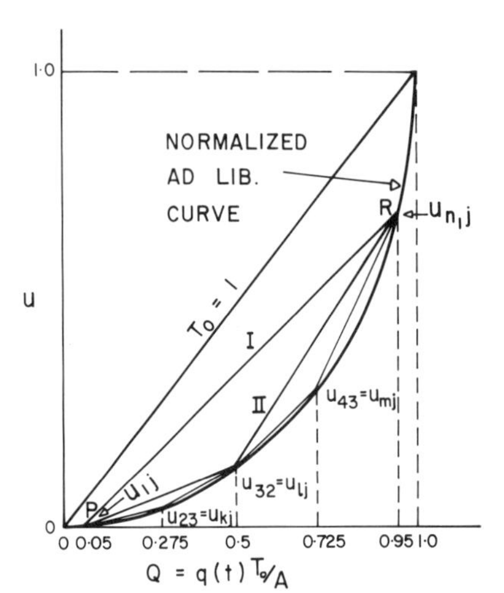


Fig. 7.20. The normalised GPP for any animal with $\alpha \leqq 1$ with inscribed polygons *I*, *II*, and *III* which are members of a sequence of polygons which approaches the normalised ad libitum growth phase curve as a limit as the number of sides of the polygon increases without limit

in the context of the normalised growth and feeding function in the normalised GPP, namely

$$U = 1 - (1 - Q)^{\alpha} \exp(\alpha Q), \tag{7.45}$$

(Fig. 7.20). Here $U = W/A$, $Q = q^*/C = q^* T_0/A$; later we shall use the normalised time $T = t/t^*$ and the normalised cumulative food consumed $\phi$ as the integral of $Q$ with respect to $T$ between the limits $T = 0$ and $T = T'$ when $Q$ has reached some preselected value. Equation (7.45) was derived and discussed in Chap. 5 [Eq. (5.8)], and it describes a limiting border of the controlled growth region of the GPP. The border is limiting because it can only be approached from the interior of the controlled growth region.

A possible way of growing an animal from point P to point R in Fig. 7.20 is along a sequence of inscribed chords of the ad libitum growth phase curve, which then becomes the limit of the broken line sequence as the number of segments increases without limit. $Q$ equal to 0.05 at point P and equal to 0.95 at point R in Fig. 7.20 were chosen only to illustrate the method of choosing a sequence of chords in the approach to the normalised growth phase curve. In the numerical proof the values of $Q$ at P and R are chosen to be 0.1 and 0.999 respectively. In the following, a sequence of chords will be called a polygon. Let $j$ = the ordinal number of a polygon and $i$ = the ordinal number of the apices of the polygon on the normalised growth phase curve starting with $i = 1$ at point P and becoming $n_j + 1$ at R, where $n_j$ is the number of chords in polygon $j$. Let $Q_{ij}$ = the $i^{\text{th}}$ ad libitum food intake on the $j^{\text{th}}$ polygon, then by Eq. (7.45) the coordinates of the $i$-th apex on the $j$-th polygon are $(Q_{ij}, U_{ij})$ where $U_{ij}$ is

$$U_{ij} = 1 - (1 - Q_{ij})^{\alpha} \exp(\alpha Q_{ij}).$$

The coordinates of the next apex on the $j$-th polygon are $(Q_{i+1j}, U_{i+1j})$ and the slope $m_{ij}$ of the chord from the $i$-th apex to the $i+1$ apex is

$$m_{ij} = (U_{i+1j} - U_{ij})/(Q_{i+1j} - Q_{ij}).$$

Since it was agreed to count point P as the first apex on all the polygons, then $m_{ij}$ is the slope of the $i$-th chord of the $j$-th polygon. The $j$ index can be dropped temporarily for added clarity of notation in the following discussion.

The equation of the $i$-th chord is

$$u = m_i(q - Q_i) + U_i. \tag{7.46}$$

Here $u$ and $q$ denote normalised variables on the chord in the controlled growth region of the normalised GPP. We have already seen how the controlled normalised feeding regime, $q$, and the normalised response, $u$, can be respectively derived from the normalised form of DE (7.3), namely $du/dT + \alpha u = \alpha q$, by using Eq. (7.46) to get

$$q = [(Q_i - U_i)/(m_i - 1)]\{1 - \exp[-\alpha(1 - 1/m_i)T]\} + Q_i, \tag{7.47}$$

and

$$u = [(Q_i - U_i)/(1 - 1/m_i)]\{1 - \exp[-\alpha(1 - 1/m_i)T]\} + U_i. \tag{7.48}$$

In the previous section we pointed out a possible problem here if $m_i$ were exactly unity. But in the context of these polygons with increasing numbers of sides, it is possible for $m_i$ to approximate unity quite closely, i.e. $m_i = 1 + \varepsilon_i$, where $\varepsilon_i$ is arbitrarily small. In this case the exponent $\alpha(1 - 1/m_i)T$ of the exponential is near zero in the order of $\alpha\varepsilon_i T/(1 + \varepsilon_i)$, so the factor $\{1 - \exp[-\alpha\varepsilon_i T/(1 + \varepsilon_i)]\}$ will approximate $\alpha\varepsilon_i T/(1 + \varepsilon_i)$ and the factor $1/(1 - 1/m_i)$ becomes $(1 + \varepsilon_i)/\varepsilon_i$. Substituting these quantities in Eqs. (7.47) and (7.48), in the limit they will become

$$q = (Q_i - U_i)\alpha T + Q_i,$$

$$u = (Q_i - U_i)\alpha T + U_i,$$

respectively. These equations show that $m_i$ near unity will cause no problem in calculating the normalised total time $T$ and normalised cumulative food consumed $\phi$ along any polygon from P to R.

Here we are starting the $i$-th controlled feeding regime from the $i$-th apex, where $T = 0$, to arrive at the $i + 1$ apex where $T = T_i$. As we have seen in the previous section, this is a rehabilitation process which starts from a point on the normalised GPP curve to an adjacent point on the same curve. Figure 7.18 and the discussion of rehabilitation imply that we must choose between two values of the time, $T_i$, to go from one apex to the next along the $i$-th chord. If we choose the time for $q$ to go from $Q_i$ to $Q_{1+1}$, then $u < U_{i+1}$ and the endpoint falls short of the desired value of $U_{i+1}$ and if we choose the time for $u$ to go from $U_i$ to $U_{i+1}$, then $q > Q_{ii}{}^{+}{}_{1}$ and the endpoint exceeds the desired value of $Q_{i+1}$.

However, Eqs. (7.47) and (7.48) both show that the times, $T_i$, for $q$ to go from $Q_i$ to $Q_{i+1}$ and for $u$ to go from $U_i$ to $U_{i+1}$ are equal and, given by

$$T_i = [\alpha(1 - 1/m_i)]^{-1}\ln[(Q_i - U_i)/(Q_{i+1} - U_{i+1})]. \tag{7.49}$$

When the animal is control fed from a normal state to an adjacent normal state, choice of time is unnecessary.

The cumulative food $\phi_i$, required for controlled growth of the animal between these points, is the integral of $q$ [Eq. (7.47)] between the limits $T = 0$ and $T = T_i$, namely

$$\phi_i = (a_i + Q_i)T_i - a_i/b_i[1 - \exp(-b_i T_i)], \tag{7.50}$$

where to simplify the algebra

$$a_i = (Q_i - U_i)/(m_i - 1),$$

and

$$b_i = \alpha(1 - 1/m_i).$$

Here it is seen that the time $T$ is the quantity naturally minimised by ad libitum feeding and $\phi$ is minimised by association.

Equations (7.49) and (7.50), with the definitions of $a_i$ and $b_i$, are the times it takes and the cumulative food required respectively to control feed the animal along the $i$-th chord of any polygon from point $(Q_i, U_i)$ to point $(Q_{i+1}, U_{i+1})$ on the ad libitum growth phase curve. I propose now to restore the polygon index, $j$, and calculate the total normalised time, $T_j$, and the total normalised cumulative food consumed, $\phi_j$, to grow an animal along any polygon $j$ from point P to point R, where

$$T_j = \sum_{i=1}^{n_j} T_{ij} \tag{7.51}$$

and

$$\phi_j = \sum_{i=1}^{n_j} \phi_{ij}\,. \tag{7.52}$$

Here $n_j$ is the number of chords in polygon $j$.

I shall use (7.51) and (7.52) to calculate $T_j$ and $\phi_j$ respectively along any polygon, $j$, from point P to point R on the ad libitum growth phase curve and show that, for a chosen value of $\alpha < 1$, $T_j > T$, and $\phi_j > \phi$. Here $T$ and $\phi$ are respectively the ad libitum normalised time and normalised food consumed to grow the animal from P to R along the normalised growth phase curve (Fig. 7.20) and are given respectively by

$$T = \ln[(1 - Q_{ij})/(1 - Q_{n_j j})]\,,$$

and

$$\phi = T - (1 - Q_{ij})[1 - \exp(-T)]\,.$$

This problem is a straight forward arithmetic type and can be made easier with the proper choice of calculating the coordinates of the apices $(Q_{ij}, U_{ij})$ of the $j$-th polygon. Figure 7.20 shows the first three polygons, I, II, and III with sides numbering 1, 2, and 4 respectively, and apices numbering respectively 2, 3, and 5, counting P as the first and R as the last, from which it may be inferred that by dividing the difference between $Q$ at P and $Q$ at R by powers of two, a simple sequence of polygons can be generated.

This choice, for sequentially creating the polygons, gives the coordinates $Q_{ij}$ and $U_{ij}$ of the $i$-th apex on the $j$-th polygon (counting $j = 1$ for chord PR) as shown here

$$Q_{ij} = Q_{ij} + (Q_{n_j j} - Q_{ij})(i - 1)/n_j\,,$$

Table 7.10. Formulae for the normalized time, $T_j$, and food consumed $\phi_j$ to grow an animal, with a known value of $\alpha$, along polygon $j$, for comparison with $T$ and $\phi$ when the animal grows along its Growth Phase Curve

---

Equations for the Growth Phase Curve, $T$ and $\phi$.

$U=1-(1-Q)^\alpha \exp(\alpha Q)$; $\quad T=\ln[(1-Q_P)/(1-Q_R)]$

$\phi=T-(1-Q_P)[1-\exp(-T)]$

Definition of polygon $j$ the segments of which increase geometrically as $j$ increases.
Number of segments $=n_j=2^{(j-1)}$
Coordinates of the apices of polygon $j$

$Q_{ij}=Q_P)+(Q_R-Q_P)(i-1)/n_j, \quad i=1,2,\ldots,n_j+1$

$U_{ij}=1-(1-Q_{ij})^\alpha \exp(\alpha Q_{ij})$

Equations for $T_{ij}$ and $\phi_{ij}$ to grow an animal along segment $i$ of polygon $j$.

Slope of segment $=m_{ij}=(U_{i+1j}-U_{ij})/(Q_{i+1j}-Q_{ij})$

$T_{ij}=(1/b_{ij})\ln[(Q_{ij}-U_{ij})/(Q_{i+1j}-U_{i+1j})]$

$\phi_{ij}=(a_{ij}+Q_{ij})T_{ij}-(a_{ij}/b_{ij})[1-\exp(-b_{ij}T_{ij})]$

Here $a_{ij}=(Q_{ij}-U_{ij})/(m_{ij}-1)$; $b_{ij}=\alpha(1-1/m_{ij})$

Equations for $T_j$ and $\phi_j$

$T_j=\sum_{i=1}^{n_j} T_{ij}$; $\quad \phi_j=\sum_{i=1}^{n_j} \phi_{ij}$

---

where $n_j=2^{(j-1)}$,

$$U_{ij}=1-(1-Q_{ij})^\alpha \exp(\alpha Q_{ij}), \qquad i=1,2,\ldots,n_j+1.$$

The slope of the $i$-th chord of the $j$-th polygon is

$$m_{ij}=(U_{i+1j}-U_{ij})/(Q_{i+1j}-Q_{ij}).$$

It remains to choose a value for $\alpha$. This can be done by choosing $\alpha$ for some particular animal from Table 5.2 (or 5.3), Chap. 5. However, for generality I shall choose the mean value of $\alpha=0.443$ for the animals listed in Table 5.3 and known to have been fed ad libitum on a nutritious diet and under good management.

For easy reference, Table 7.10 contains all the formulae required to calculate $T_j$ and $\phi_j$, for any polygon $j$ connecting points P and R, and the ad libitum formulae for $T$ and $\phi$, and Table 7.11 contains the results of calculating $T_j$ and $\phi_j$ with $j=1,2,\ldots,10$. Here it is seen that $T_j$ drops precipitously for the first few polygons after which the successive differences decrease more gradually. Between polygons 8 and 9 the difference is about 75 in 73,000. It seems clear that $T=6.8024\ldots$ is the lower limit of the sequence as the number of chords increase without limit. The slowing of the differences between $T_j$ and $T$, namely $T_j-T$, is best seen by plotting $\ln(T_j-T)$ versus the polygon number, $j$. Figure 7.21 illustrates this and implies another idea which might have been guessed beforehand. The natural logarithm of the difference for $j>7$ levels off with increasing rapidity and appears as though it may go to negative infinity when $T_j=T$, so slowly that $T$ as the lower limit of $T_j$ may not be reached exactly as $2^{(j-1)}$ approaches infinity. This means that the points on the ad libitum growth phase curve do not belong to the totality of points of the

Table 7.11. The sequence of times $T_j$ and associated food consumed $\phi_j$ as the number of chords, equal to $2^{(j-1)}$, is increased for comparison to time $T$ and $\phi$ when the animal naturally grows and feeds along its biotrace. $\Omega$ at P=0.1; $\Omega$ at R=0.999; $\alpha$=0.443

| Polygon No. $j$ | Number of chords $2^{(j-1)}$ | $T_j$ | $\phi_j$ |
|---|---|---|---|
| 1 | 1 | 24.802 | 14.195 |
| 2 | 2 | 10.067 | 7.8772 |
| 3 | 4 | 8.2785 | 6.9385 |
| 4 | 8 | 7.6843 | 6.5827 |
| 5 | 16 | 7.4398 | 6.4181 |
| 6 | 32 | 7.3310 | 6.3378 |
| 7 | 64 | 7.2821 | 6.2992 |
| 8 | 128 | 7.2609 | 6.2817 |
| 9 | 256 | 7.2534 | 6.2745 |
| 10 | 512 | 7.2494 | 6.2719 |
| Ad libitum | Infinite | 6.8024 | 5.9034 |

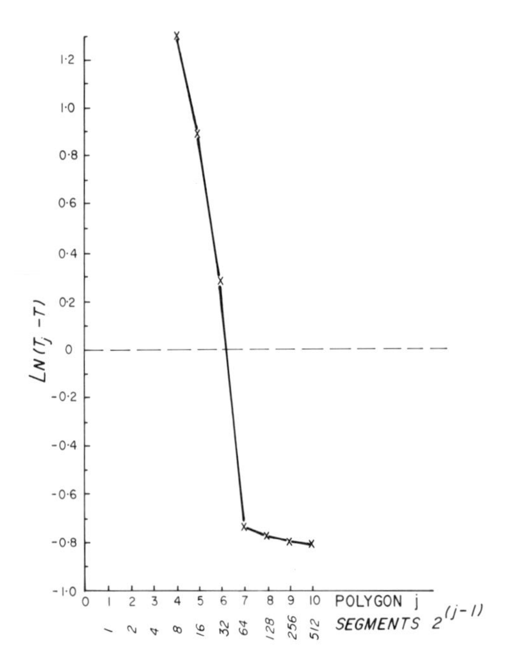


Fig. 7.21. Showing the slowing of the changes of $\ln(T_j - T)$ for each doubling of the number of sides in the polygons beyond $j=7$

GPP which the ad libitum growth phase curve bounds. With respect to the ad libitum growth phase curve the GPP is an open region. I said previously that this result may have been anticipated because the totality of points comprising the solutions of a given differential equation cannot contain the totality of points comprising the solutions of a differential equation which is independent of the given one. Therefore the controlled growth region of the GPP must be open with respect to the ad libitum feeding and growth curve bounding it.

Here it has been shown that feeding ad libitum is a biologically optimal process for the animal to grow from infancy to maturity in a minimum of time. However in the practical world ad libitum is sometimes considered economically wasteful because too large a fraction of the food seems to be going to unwanted fat storage in the animal as it ages. A quite real consequence of this economic attitude is the ever present possibility that animals in a long and costly experiment may be unintentionally control fed for economic reasons when they were intended to be fed ad libitum.

## 7.8 Determination of Fraps' Productive Energy (*PE*) of a Foodstuff; a Critical Evaluation

This subject is of interest here because the determination of *PE* is achieved by feeding an animal at two widely differing levels of intake for a period of time (Fraps and Carlyle 1939, Fraps 1946). Animal scientists generally call this procedure feeding on two different planes of nutrition, but most often there is no specification of how the animals are actually fed on either plane. Furthermore the method of determining (*PE*) is a curious combination of the energy balance during growth and the live weight variation with food consumed.

Fraps' concept of the productive energy of a foodstuff can be expressed as that of the food (corrected for maintenance) which goes directly to producing increased body energy stores. This idea gave use to the following equation with terms which could be experimentally measured.

$$(PE)(F-gW)=E. \qquad (7.53)$$

Here $F$ is the cumulative food consumption of the foodstuff by the animal freely fed (i.e. ad libitum) for a period of time, $t$; $W$ is the mean live weight during this period and $E$ is the mean gain in body energy over time period, $t$; $g$ is the food required to maintain one gram of the mean live weight; and (*PE*) is therefore the productive energy per gram of corrected foodstuff consumed to produce the body weight gain of energy $E$.

This equation is reasonable and involves variables $F$, $W$, and $E$ which can be experimentally determined for economically useful food materials for commercial animals. But it also involves two unknown factors for a given foodstuff, namely (*PE*) and $g$. In order to eliminate $g$ Fraps saw that, feeding the foodstuff at a much lower level than ad libitum for the same period of time, $t$, used in ad libitum feeding, Eq. (7.53) could be written, in terms of the new experimental variables $F'$, $W'$, $E'$, as

$$(PE)(F'-gW')=E'. \qquad (7.54)$$

Fraps solved (7.53) and (7.54) simultaneously to eliminate $g$ and obtain the productive energy (*PE*) of the foodstuff, namely

$$(PE)=(E'W-EW')/(F'W-FW'). \qquad (7.55)$$

The "truth" of Eq. (7.55) was so convincing, that Fraps and other nutritionists determined the (*PE*) for many foodstuffs. At about that time, linear programming came into use for determining the least cost blending of animal diets to meet the many requirements nutritionists had developed for the "optimum growth" of commercial species and breeds within these species that were commercially valuable. Since energy is one of the major requirements it was natural to turn to the Fraps $(PE)$ values for the commercial foodstuffs as quantities for insertion into the linear programming matrix. This use of $(PE)$ lasted for a number of years, only to fade out of the literature and linear programming.

I have never learned why the concept faded. Some have said "$g$ in Fraps' equation could not be expected to remain constant between ad libitum and limited feeding," but proof of this assertion was not offered. Davidson, McDonald, and Williams (1957) experimentally found erratic results for $(PE)$ on some foodstuffs and concluded Eq. (7.53) was arithmetically unsound but again no reason was offered.

The reader may think I am flogging a "dead horse," but in the context of the theory it could be illuminating to examine Fraps' concept in relation to a suggestion by G.C. Emmans (personal communication, 1974) that Fraps' $(PE)$ may be related to $(AB)$.

Differential equations (7.2) and (7.3) can be respectively rearranged to get

$$dW/dt=(AB)[q^*(t)-q^*(t)W/A], \tag{7.56}$$

and

$$dW/dt=(AB)[q(t)-W/T_0). \tag{7.57}$$

Integrating these equations over the same time interval, zero to $t$, with the same initial conditions, namely at $t=0$, $W=W_0$, we get, using $q^*(t)dt=dF$, the results

$$W(t)-W_0=(AB)\left[F-(1/A)\int_0^F WdF\right] \tag{7.58}$$

and

$$W_1(t)-W_0=(AB)\left[F_1-(1/T_0)\int_0^t Wdt\right]. \tag{7.59}$$

Here $W_1(t)$ and $F_1$ signify respectively the different weight attained and food consumed during time $t$ under limited feeding. Since $\int_0^F WdF$ and $\int_0^t Wdt$ are respectively the areas under the $W(F)$ and $W(t)$ curves between the limits $F=0$ to $F=F$, and $t=0$ to $t=t$, these areas can be written as $F\bar{W}_F$ and $t\bar{W}_t$, where $\bar{W}_F$ is the mean live weight over the ad libitum food consumed interval in time $t$ and $\bar{W}_t$ is the mean over the same time interval during controlled feeding. Equations (7.58) and (7.59) become respectively

$$W(t)-W_0=(AB)[F-(F/A)\bar{W}_F], \tag{7.60}$$

and

$$W_1(t)-W_0=(AB)[F_1-(t/T_0)\bar{W}_t]. \tag{7.61}$$

From the interpretations of DE's (7.2) and (7.3) given in Chap. 7.1, it can be seen that $[F-(F/A)\bar{W}_F]$ and $[F_1-(t/T_0)\bar{W}_t]$ are respectively the cumulative foods consumed $F$ and $F_1$ corrected for those portions, namely $(F/A)\bar{W}_F$ and $(t/T_0)\bar{W}_t$, not going to the growth from the initial live weight $W_0$. This means Fraps moved in the right direction but was hampered by the simple arithmetic concepts he used.

If an energy factor, $e$, with units kcal. per unit gain of live weight, is assumed and used with Eqs. (7.60) and (7.61) we get

$$E=[e(AB)][F-(F/A)\bar{W}_F] \qquad (7.62)$$

and

$$E'=[e(AB)][F_1-(t/T_0)\bar{W}_t]. \qquad (7.63)$$

Here $E$ and $E'$ are respectively the whole body energy gains under the two feeding regimes.

Emmans' conjecture appears well founded since $(PE)$ does resemble $[e(AB)]$ in the comparison of Fraps' Eqs. (7.53) and (7.54) with Eqs. (7.62) and (7.63). However Fraps' assumption of a $g$ factor constant for both ad libitum and limited feeding is unfounded because $F/A$ cannot be equal to $t/T_0$. Furthermore the body energy stores, $E$, of an animal on a partial starvation or some controlled feeding regime are less than the stores, $E$, of an animal ad libitum fed for the same period of time, hence $e$ for controlled feeding cannot be the same value of the $e$ factor for ad libitum feeding.

I cannot recommend use of Eqs. (7.62) and (7.63) for determining what might be termed a "productive energy" of some foodstuff, because the introduction of the energy factor, $e$, is just as ad hoc as Fraps' introduction of the factor, $g$, into Eq. (7.53) even though Eqs. (7.62) and (7.63) have a better theoretical foundation.

## 7.9 Some Concluding Remarks

The first six chapters of this book laid the foundation for the theory of feeding and growth advanced and tested in this chapter. It should be noted here that though the concepts of the relative growth rate, the allometric relations, the material and energy balance relations, and the metabolic sizes of animals are considered fundamental to animal science, they have not been useful in or deriveable from this theory. However some interesting and important results have emerged from this theory, namely: (1) there appears to be no species effects on the DE.s of feeding and growth; (2) among the parameters the efficiency factor $(AB)$ and the ad libitum feeding discriminant $\alpha$ appear uniform over the species from mice to cattle and (3) the normalised ad libitum feeding and growth functions in the following equations

$$u=1-\exp[-Z(T)], \qquad (7.64)$$

$$Z(T)=\alpha[T-1+\exp(-T)], \qquad (7.65)$$

$$T=T, \qquad (7.66)$$

describe the biotraces of all animals excepting the primates.

The wide variety of breeds within some economic species is a remarkable fact of life. Britain has a plethora of breeds of cattle, sheep, poultry, dogs etc., all produced by people, generally called animal breeders, with some knowledge of the science of genetics but with a great deal more practical knowledge and patience in culling their flocks for animals the progeny of which will sell at a handsome reward for the breeder.

The existence of the wide variety of cattle in Britain has raised some important economic questions, important enough for the Agriculture Research Council to support a very large multibreed experiment designed by St. Clair S. Taylor at the Animal Breeding Research Organisation in Edinburgh (Taylor 1972, 1975). The above equations and the theory could prove useful in the study of the data from this experiment to discover clearly the fundamental differences among the breeds.

I have studied ad libitum feeding and growth data from two British breeds of Jersey and Friesian cattle and have shown that the major difference in their feeding and growth performance is their mature weight $A$, that of the Friesian being larger by a factor of 1.5 (Table 4.1, Chap. 4, data from Monteiro, 1974). It is of interest to note in Table 4.1 that the value of the efficiency factor ($AB$) for cattle has changed little from 1924 (Spillman and Lang) to 1974 (Monteiro). Examination of Figs. 1 and 2 of Theissen's (1979) report on some recent results from the Multibreed experiment suggests that the major differences in the growth and feeding parameters overall will be found in the mature weights over the breeds from the smallest to the largest.

## References


Baldwin RL, Black JL (1979) A computer model for evaluating effects of nutritional and physiological status on the growth of mammalian organs and tissues. Anim Res Lab Tech Pap No 6, CSIRO, Aust

Box GEP, Jenkins GM (1970) Time series analysis forecasting and control. Holden-Day, San Francisco

Brody S (1974) Bioenergetics and growth. First published: Reinhold, New York (Reprinted: Hafner Press, New York)

Davidson J, McDonald I, Williams RB (1957) The utilisation of dietary energy by poultry. 1. A study of the algebraic method for determining the productive energy of poultry feeds. J Sci Food Agric 8:173–182

Fisher RG (1946) Statistical methods for research workers. Oliver and Boyd, Edinburgh

Fitzhugh HA (1976) Analysis of growth curves and strategies for altering their shape. J Anim Sci 42:1036–1051

Fraps GS (1946) Composition and productive energy of poultry feeds and rations. Bull 678 Tex Agric Exp Stn

Fraps GS, Carlyle EC (1939) The utilization of the energy of feed by growing chickens. Bull 571 Agric Exp Stn

Goodwin B (1963) Temporal organisation in the cell: a dynamic theory of cellular control processes. Academic Press, London New York

Griffiths M (1974) Introduction to human physiology. MacMillan, New York

Hogg RV, Craig AT (1970) Introduction to mathematical statistics. MacMillan, New York

Lozano SC (1977) The effect of diet and genetic constitution on growth curves and efficiency in rats. Anim Prod 25:381–388

Mayer J, Thomas D (1967) Regulation of food intake and obesity. Science 156:328–337

Meissner HH, Roux CZ, Hofmeyer HS (1975) Voluntary feed intake, growth, body composition and efficiency in sheep: breed and sex differences. Agroanimalia 7:105–114

Morowitz HJ (1968) Energy flow in biology. Academic Press, London New York
Pym RAE, Dillon JF (1974) Restricted food intake and reproductive performance of broiler breeder pullets. Br Poult Sci 15:245–259
Radford PJ (1967) Growth analysis formulae – their use and abuse. Crop Sci 7:171–175
Richards FJ (1959) A flexible growth function for empirical use. J Exp Bot 10:290–300
Riggs DS (1963) Mathematical approach to physiological problems. MIT Press, Cambridge
Roux CZ (1974) The relationship between growth and feed intake. Agroanimalia 6:49–52
Roux CZ (1976) Model for the description of growth and production. Agroamalia 8:83–94
Roux CZ (in press) A dynamic stochastic model for animal growth
Sandland RL, McGilchrist CA (1979) Stochastic growth curve analysis. Biometrics 35:258–271
Spillman WJ, Lang E (1924) The law of diminishing increment. World, Yonkers
Taylor ST CS (1972) ARC animal breeding research organisation report, Edinburgh
Taylor ST CS (1975) ARC animal breeding reserach organisation report, Edinburgh
Thiessen RB (1979) ARC animal breeding research organisation report, Edinburgh
Titus HW, Full MA, Hendricks WA (1934) Growth of chickens as a function of the feed consumed. J Agric Res 48:817–835

# Chapter 8 A General Euclidean Vector Representation of Mixtures

Thus far I have discussed the traces of growing animals without referring to the constitution of the food consumed but assuming the food to be nutritions, constant in composition, and either unrestricted or controlled in relation to the animal. The word "constitution" here can have the narrow meaning of the collection of a few or many ingredients to form meals, with or without heat, or a wider meaning of the collection of the chemicals of the food. Further, I shall use the word "food" to denote a material, however constituted, which some animal will consume and use in its metabolic processes without the appearance of dietary pathologies. A food is therefore a mixture of many constituents, which are in turn mixtures of constituents, so that one unit weight of the food can be represented by the symbol, $(x_i)$, $i=1,2,...,N$, where $x_i$ is the weight fraction of the $i^{\text{th}}$ constituent of the food composed of $N$ separately recognisable constituents and where, for various reasons, $N$ can be any integer, small or large. This concept of a mixture can be generalised to matters other than food. If on mixing $N$ ingredients, two or more of the ingredients undergo chemical reaction to form other constituents, then the degrees of freedom for creating sets of mixtures from the original ingredients are changed. Without analysis to discover the number and identities of the new ingredients, the symbol $(x_i)$, $i=1,2,...,N$, has no meaning.

Often it is necessary to compare various mixtures of the same constituents. This tedious process can be aided visually (and by mental imagery) if each mixture can be represented as a point in a geometrical diagram or space. Differences can then be seen at a glance. There are undoubtedly many geometries which can serve this purpose. One type of geometry useful in nutrition was proposed by Moon and Spencer (1974) who show that the present calculations of nutritionists can be geometrically represented as linear operations on foods as vectors, but with vector addition, and multiplication of a vector by a scalar as the only defined arithmetical operations. Their geometry is an affine geometry which has no metric, in other words, the distance between two points is not defined, neither is the product of two vectors defined. Given nutrition as it stands today, namely the preparation of healthful foods, it may be convenient to consider the nutrition space of the food ingredients as an affine space. The approach of Moon and Spencer to the geometry of nutrition can be generalised to any number of ingredients.

I shall propose here another geometry in which a mixture of $N$ constituents, i.e. ingredients $(x_i)$, $i=1,2,...,N$, will be represented as a vector of $N-1$ components, $[y_j], j=1,2,...,N-1$, in a hyper Euclidean space in which not only the operations of vector addition, and multiplication of a vector by a scalar are defined, but the product of two vectors and the distance between two points of the space are also defined. I shall use $(x_i)$ to represent a collection of unique but identifiable fractions

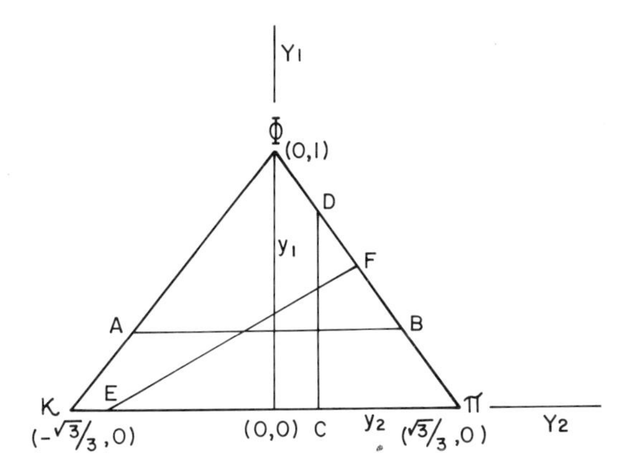


Fig. 8.1 The nutrition space as an equilateral triangle of unit heigth when $N=3$

of a unit mass of the mixture, so that

$$\sum_{i=1}^{N} x_i = 1,\ 0 \leq x_i \leq 1, \tag{8.1}$$

and use the square bracketed symbol $[y_j]$ to represent the vector, having $N-1$ components, assigned to the mixture $(x_i)$. The unit mass may be 1 g, 100 g, or 1 kg as the user of these ideas may desire.

Here I shall develop the transformation of $(x_i)$ to $[y_j]$ and back to $(x_i)$, and show that $[y_j]$ can be represented as a vector in Euclidean space (hyperspace if $N>4$) with $N-1$ coordinate axes $Y_j$ which are mutually perpendicular. It will be shown in Chap. 8.2 that the vector space, representing the totality of mixtures $(x_i)$ of $N$ ingredients, is a regular hypersolid which reduces to familiar geometric figures such as a straight line of unit length for $N=2$, an equilateral triangle (Fig. 8.1) when $N=3$, and a regular tetrahedron (Fig. 8.2) when $N=4$. The vector space has some useful properties which are discussed in Chap. 8.3 and are used in describing proteins as mixtures of amino acids (Chap. 8.4) and the musculature of animals as mixtures of a set of specific muscle masses (Chap. 8.5).

## 8.1 Transformation of Mixture $(x_i)$ to Vector $[y_j]$

Using the properties of $(x_i)$, given in Eq. (8.1), and the null vector [0], all components of which are zero, I shall show how $[y_j]$ can be sequentially built up, component by component from [0], as a starting point, to establish the relation between the vector $[y_j]$ and the mixture $(x_i)$. We shall use the notation $(x_i)_k$ and $[y_j]_k$ to distinguish between two or more mixtures and their vector representations.

Firstly, choose a $y_1$ which satisfies $0 \leqq y_1 \leqq 1$, then the vector $[y_1, 0, \ldots 0]$ corresponds to mixture

$$(y_1,\ (1-y_1)/(N-1), \ldots),$$

the constituents of which satisfy the requirements of Eq. (8.1). Secondly, choose a $y_2$ which with $y_1$ satisfies $0 \leqq (1-y_1)/(N-1)+y_2 \leqq 1$, then the vector $[y_1,\ y_2,$

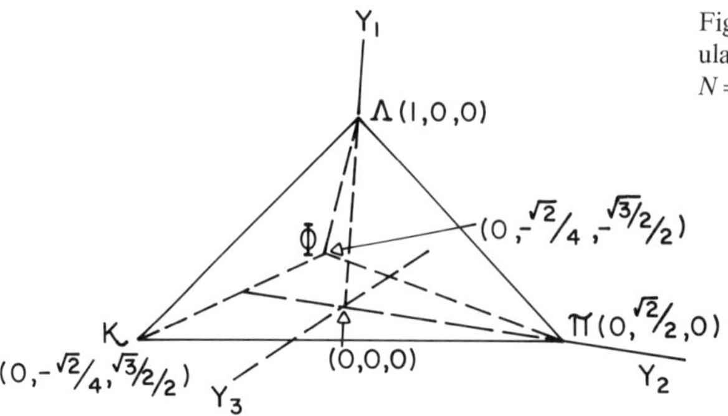


Fig. 8.2. The nutrition space as a regular tetrahedron of unit height when $N=4$

$0,\ldots,0]$ corresponds to mixture

$$(y_1, (1-y_1)/(N-1)+y_2, (1-y_1)/(N-1)-y_2/(N-2), \ldots),$$

which satisfies the requirements of Eq. (8.1). Thirdly, select $y_3$ to satisfy, with the chosen values of $y_1$ and $y_2$,

$$0 \leqq (1-y_1)/(N-1)-y_2/(N-2)+y_3 \leqq 1,$$

vector $[y_1, y_2, y_3, 0, \ldots, 0]$ corresponds to mixture

$$(y_1, (1-y_1)/(N-1)+y_2, (1-y_1)/(N-1)-y_2/(N-2)+y_3,$$

$$(1-y_1)/(N-1)-y_2/(N-2)-y_3/(N-3), \ldots).$$

This process is continued until the zeros of the null vector are exhausted. By inference from the three preceding steps, namely

$$x_i = x_1, \tag{8.2}$$

$$x_2 = 1/(N-1) - y_1/(N-1) + y_2,$$

and

$$x_3 = 1/(N-1) - y_1/(N-1) - y_2/(N-2) + y_3,$$

it follows that the $i^{\text{th}}$ constituent of the mixture is given by

$$x_i = 1/(N-1) - \sum_{j=1}^{i-1} y_j/(N-j) + y_i, \; 1 < i \leq N-1 \tag{8.3}$$

with $x_N$ given by

$$x_N = 1 - \sum_{i=1}^{N-1} x_i \, . \tag{8.4}$$

Because Eq. (8.3) is recursive in the $y$'s it can be readily used to solve for the $y$'s in terms of the $x$'s, with the following results.

$$y_1 = x_1 \tag{8.5}$$

$$y_j = x_j - \left(1 - \sum_{k=1}^{j-1} x_k\right)/(N-j+1), \; j = 2, 3, \ldots, N-1. \tag{8.6}$$

Equations (8.5) and (8.6) represent the transformation of mixture $(x_i)$ to the vector $[y_j]$ and Eqs. (8.2), (8.3), and (8.4) represent the reverse transformation of $[y_j]$ to $(x_i)$.

## 8.2 Geometric Representation of $[y_i]$ in a Euclidean Space

Since I wish to plot $[y_j]$ in a Euclidean hyperspace of $N-1$ dimensions, I construct $N-1$ Cartesian coordinate axes $Y_j$, which have as their origin the null vector [0], and which are all mutually perpendicular. In order to plot $[y_j]$ in this hypergeometric space I must have a set of scale factors $s_j$ expressed as units of $y_j$, which means that the point on the $Y_j$ axis corresponding to $y_j$ is at $y_j/s_j$ units of $Y_j$ from the origin of axes [0]. For example $y_1$ is marked at the point $y_1/s_1$ on the $Y_1$ axis, $y_2$ is marked at the point of $y_2/s_2$ on the $Y_2$ axis and so on.

The relative values of the $N-1$ scale factors, $s_j$, determine the geometrical shape of the surface and volume of the Euclidean space which contains the $[y_j]$ points corresponding to all the possible mixtures $(x_i)$ of the same $N$ ingredients. It is obvious that the order of the ingredients in the lists under study should remain constant during application of the techniques discussed here. I shall designate the volume, in the hyperspace, as the mixture space.

Including the scale factors, the transformation of $(x_i)$ to $[y_j]$ becomes

$$y_1 = x_1 s_1, \tag{8.7}$$

$$y_j = \left[x_j - \left(1 - \sum_{k=1}^{j-1} x_k\right)/(N-j+1)\right] s_j, \; j = 2, 3, \ldots, N-1 \tag{8.8}$$

and the reverse transformation becomes

$$x_1 = y_1/s_1, \tag{8.9}$$

$$x_i = 1/(N-1) - \sum_{k=1}^{i-1} (y_k/s_k)/(N-k) + y_i/s_i, \; i = 2, 3, \ldots, N-1, \tag{8.10}$$

$$x_N = 1 - \sum_{i=1}^{N-1} x_i. \tag{8.11}$$

A vector is defined as the distance from the point of origin to its terminus and the direction of that distance relative to the coordinate axes. Since the hyperspace is assumed Euclidean, then the vector $[y_j]$ has length

$$r = \left[\sum_{j=1}^{N-1} y_j^2\right]^{\frac{1}{2}} \tag{8.12}$$

with its direction given by the direction cosines

$$l_j = y_j/r. \tag{8.13}$$

The scalar product of two vectors, $[y_j]_i$ and $[y_j]_k$, is defined as the dot $(\cdot)$ product, namely

$$[y_j]_i \cdot [y_j]_k = \sum_{j=1}^{N-1} y_{ij} y_{kj} = d_{ik}.$$

Here $i$ and $k$ are the identification indices of the two vectors, and $d_{ik}$ is a pure number scalar obtained by adding the products of the corresponding components. The ratio of $d_{ik}$ and the product of the lengths, $r_i \times r_k$ of the vectors, defines the cosine of the angle $\theta$ (in units of radians) between the vectors as shown here

$$\cos \theta_{ik} = d_{ik}/r_i r_k = \sum_{j=1}^{N-1} y_{ij} y_{kj}/r_i r_k = \sum_{j=1}^{N-1} l_{ij} l_{kj},$$

or

$$\theta_{ik} = \arcos(d_{ik}/r_i r_k). \tag{8.14}$$

Here, it is seen that $\cos\theta$ is also the sum of the products of the corresponding direction cosines of the two vectors. If $\theta$ is desired in angular degrees it can be multiplied by $180/\pi$ degrees per radian.

These basic concepts make possible the extension of the propositions of Cartesian analytic geometry to the hypergeometry of the mixture space.

Some useful propositions are given in the following statements.

1. The difference between vectors $[y_j]_k$ and $[y_j]_m$ is $[y_{jk} - y_{jm}]$. Considering the vector difference as originating at $[y_j]_m$ and terminating at $[y_j]_k$, the length, $d_{km}$, of the difference is

$$d_{km} = \left[ \sum_{j=1}^{N-1} (y_{jk} - y_{jm})^2 \right]^{\frac{1}{2}}$$

and its direction cosines, $l_{km}$, are

$$l_{jkm} = (y_{jk} - y_{jm})/d_{km}, j = 1, \ldots, N-1.$$

Another relation between $d_{km}$ and the angle, $\theta_{km}$, between the vectors, is given by the law of cosines, namely

$$d_{km}^2 = r_k^2 + r_m^2 - r_k r_m \cos(\theta_{km})$$

2. The equation of a hyperstraight line through the termini of the two vectors is

$$(y_1 - y_{1m})/(y_{1k} - y_{1m}) = (y_2 - y_{2m})/(y_{2k} - y_{2m}) = \cdots$$
$$= (y_{N-1} - y_{N-1m})/(y_{N-1k} - y_{N-1m}).$$

Stated in an alternative form with the direction cosines, $l_{jkm}$, the equation of the hyper straight line is

$$(y_1 - y_{1m})/l_{1km} = (y_2 - y_{2m})/l_{2km} = \cdots = (y_{N-1} - y_{N-1m})/l_{N-1km}.$$

3. The general equation of a hyperplane is

$$\sum_{j=1}^{N-1} a_j y_j + c = 0,$$

and in intercept form the equation is

$$\sum_{j=1}^{N-1} y_j/b_j = 1.$$

Here the $b_j$ are the intercepts of the hyperplane on the coordinate axes $y_j$.

4. The equation of a hypersphere of radius, $R$, with its centre at the terminus $[y_j]_k$ is

$$\sum_{j=1}^{N-1} (y_j - y_{jk})^2 = R^2.$$

5. The equation of a hyperellipsoid, with semiaxes $a_j$ and centred on the terminus of $[y_j]_k$ is

$$\sum_{j=1}^{N-1} (y_j - y_{jk})^2/a_j^2 = 1.$$

In using theorems 1 to 5 and others the reader may devise by extending the Euclidean geometry of two and three dimensions to the Euclidean space of $N-1$ dimensions, it must be remembered that the vectors $[y_j]$ must be in the mixture space of the hypersolid representing all the possible mixtures, $(x_i)$, of the same $N$ ingredients. If they are not, then on transforming from $[y_j]$ to $(x_i)$ some of the $x_i$ will be greater than unity and less than zero.

Theorems 4 and 5 are useful in considering the experimental errors $(\varepsilon_i)$ incurred during the analysis of the mixture. A mixture need only be analyzed for $N-1$ of its components, with the error in the Nth component given by $\left(\sum_{i=1}^{N-1} \varepsilon_i^2\right)^{\frac{1}{2}}$. Transforming the mixture $(x_i + \varepsilon_i)$ to its corresponding vector $[y_j + e_j]$ yields an error hyperellipsoidal surface with variables $1_j$ defined by

$$\sum_{j=1}^{N-1} 1_j^2/e_j^2 = 1$$

with its centre at the terminus of $[y_j]$. If the transformation of the mixture error $(\varepsilon_j)$ yields the vector error $[e_j]$ in which $e_j = e_k = e$, $j$ and $k = 1, \ldots, N-1$, then the error hyperellipsoid becomes the error hypersphere

$$\sum_{j=1}^{N-1} 1_j^2 = e^2$$

at the terminus of $[y_j]$.

There are $N$ mixtures, $I_i, i = 1, 2, \ldots, N$, composed of only one of the pure ingredients, all the other ingredients being at zero level, for example $I_1$ is $(1, 0, 0, \ldots, 0)$, $I_2$ is $(0, 1, 0, 0, \ldots, 0)$ and so on. Each of these mixtures corresponds to a point in the mixture space. These points are the extreme points, $P_i$, of the mixture space which contains the points representing all the mixtures $(x_i)$ made from the $N$ pure ingredients. The coordinates of the extreme points, $P_i$, corresponding to the mixtures $I_i$ are important to know. Using the transformation Eqs. (8.7) and (8.8) the co-ordinates of the $P_i$ were found to be those shown in Table 8.1. Another important point in the mixture space (which I shall designate $P_c$) is located in the interior of the hypervolume and corresponds to the mixturc $(1/N)$ in which the levels of the ingredients are all equal to $1/N$. The vector coordinates of $P_c$ are $[s_1/N, 0, \ldots, 0]$, so that $P_c$ lies on the $Y_1$ axis at a distance of $s_1/N$ from the origin of coordinates, and a distance of $(1 - 1/N)\, s_1$ is designated here as $r_{c1}$.

Table 8.1. Co-ordinates of points $P_i$ corresponding to pure ingredients $I_i$

| $P_i$ | $P_1$ | $P_2$ | $P_3$ | . | . | $P_k$ | . | . | $P_N$ |
|---|---|---|---|---|---|---|---|---|---|
| $y_j$ | | | | | | | | | |
| $y_1$ | $s_1$ | 0 | 0 | . | . | 0 | . | . | 0 |
| $y_2$ | 0 | $(N-2)s_2/(N-1)$ | $-s_2/(N-1)$ | . | . | $-s_2/(N-1)$ | . | . | $-s_2/(N-1)$ |
| $y_3$ | 0 | 0 | $(N-3)s_3/(N-2)$ | . | . | $-s_3/(N-2)$ | . | . | $-s_3/(N-2)$ |
| . | . | . | 0 | . | . | . | . | . | . |
| $y_{k-1}$ | . | . | . | . | . | $-s_{k-1}/(N-k+2)$ | . | . | $-s_{k-1}/(N-k+2)$ |
| $y_k$ | 0 | 0 | 0 | . | . | $(N-k)s_k/(N-k+1)$ | . | . | $-s_k/(N-k+1)$ |
| . | . | . | . | . | . | 0 | . | . | . |
| . | . | . | . | . | . | . | . | . | . |
| $y_{N-1}$ | 0 | 0 | 0 | . | . | 0 | . | . | $-s_{N-1}/2$ |

$s_j = [(N-1)(N-j+1)/N(N-j)]^{1/2}$

I think the reader can see the utility of determining the scale factors, $s_j$, so that the mixture space will be a regular hypersolid, i.e. extreme points $P_i$ are equidistant from each other and are points on circumscribed hypersphere of radius $r$ the centre of which is the interior point $P_c$. I have found that requiring the $P_i$ to lie on the circumscribed hypersphere is the easiest way to find suitable values of $s_j$.

Since the space is Euclidean, the distance $r_{cj}$ between points $P_c$ and $P_j$ can be calculated using

$$r_{cj}=\left[\sum_{i=1}^{N-1}(y_{ci}-y_{ji})^2\right]^{\frac{1}{2}}, \tag{8.15}$$

where $y_{ci}$ is the ith coordinate of $P_c$ and $y_{ji}$ is the $i$th coordinate of $P_j$ in Table 8.1. In terms of the coordinates of $P_j$ Eq. (8.15) becomes

$$r_{cj}=\left[\sum_{i=1}^{j-1}s_i^2/(N-i+1)^2+s_j^2(N-j)^2/(N-j+1)^2\right]^{\frac{1}{2}}. \tag{8.16}$$

The points $P_j$, $i \cup 1$ will lie on the hypersphere if

$$r_{cj}=r_{c1}=s_1(1-1/N)=s_1(N-1)/N \tag{8.17}$$

From Table 8.1 and Eq. (8.15), the distance $r_{c2}$ between $P_c$ and $P_2$ is

$$r_{c2}=[s_1^2/N^2+s_2^2(N-2)^2/(N-1)^2]^{1/2}.$$

Equating this to the right hand side of Eq. (8.17) and squaring both sides leads to

$$s_2=s_1[(N-1)(N-1)/N(N-2)]^{1/2}$$

after some algebraic re-arrangement of the terms of the equation. Performing the same operation with $r_{c3}=r_{c1}$ we get

$$s_3=s_1[(N-1)(N-2)/N(N-3)]^{1/2}.$$

By inference the general formula for $s_j$ is

$$s_j=s_1[(N-1)(N-j+1)/N(N-j)]^{1/2},\ j=1,2,\ldots,N-1. \tag{8.18}$$

These scale values put all the $P_i$ on the circumscribed hypersphere centred at $P_c$, but are the distances $r_{ik}$ between $P_i$ and $P_k$, $k \neq i$, all equal? This question can be answered generally using Table 8.1 and Eq. (8.15) with index $i$ substituted for $c$. It will serve our purpose to show $r_{12}=r_{13}=r_{23}$ and inferring $r_{ij}=r_{12}$. Applying Eq. (8.15) to the coordinates of $P_i$ in Table 8.1 yields for the distance $r_{12}$

$$r_{12}=[s_1^2+s_2^2(N-2)^2/(N-1)^2]^{1/2},$$

which, on using the preceding value of $s_2$ and some algebraic reduction, becomes

$$r_{12}=s_1[2(N-1)/N]^{1/2}.$$

By the same procedure

$$r_{13}=[s_1^2+s_2^2/(N-1)^2+(N-3)^2s_3/(N-2)^2]^{1/2},$$

which, on using the values of $s_2$ and $s_3$ and more algebraic reduction, becomes

$$r_{13}=s_1[2(N-1)/N)]^{1/2}.$$

The distance $r_{23}$ is, from Table 8.1 and Eq. (8.15),

$$r_{23}=\{[s_2(N-2)/(N-1)+s_2/(N-1)]^2+(N-3)^2 s_3/(N-2)^2\}^{1/2},$$

which, by the methods used before, becomes

$$r_{23}=s_1[2(N-1)/N]^{1/2}=r_{12}=r_{13}.$$

Through tedious algebra, it can be shown that $r_{ij}=r_{kl}=s_1[2(N-1)/N]^{1/2}$, and all the extreme points $P_i$ will be equidistant from each other and have $P_c$ as their centre. Therefore, the mixture space is a regular hypersolid of $N-1$ dimensions, the volume of which depends only on the value of $s_1$.

Choosing $s_1=1$, the mixture space becomes a regular convex space with $N$ vertices and $N$ altitudes of unit length and the distance between any pair of vertices is $[2(N-1)/N]^{1/2}$.

The final forms of the transformation equations for $(x_i)$ to $[y_j]$ are

$$y_1=x_1 \tag{8.19}$$

$$y_j=\left[x_j-\left(1-\sum_{k=1}^{j-1} x_k\right)/(N-j+1)\right]s_j,\ j=2, 3, \ldots, N-1, \tag{8.20}$$

$$s_j=\{(N-1)\ (N-j+1)/N\ (N-j)\}^{\frac{1}{2}},\ j=1, 2, 3, \ldots, N-1 \tag{8.21}$$

and for $[y_j]$ to $(x_i)$ are

$$x_1=y_1, \tag{8.22}$$

$$x_i=1/(N-1)-\sum_{k=1}^{i-1}(y_k/s_k)/(N-k)+y_i/s_i,\ i=2, 3, \ldots, N-1, \tag{8.23}$$

$$x_N=1-\sum_{i=1}^{N-1} x_i. \tag{8.24}$$

## 8.3 Some Useful Geometric Properties of the Mixture Space

In research, mixtures are prepared as part of an experiment, but since the number of degrees of freedom, $N$, for preparing a mixture may be large, the experimenter is thus faced with very large sets of mixtures from which to choose or to work with. The geometry of the mixture space can possibly be of some help. For example he could choose a set of mixtures, the points of which in the mixture space would lie on a straight line. A simple example is one in which the desired set of mixtures corresponds to points along the length of the vector $[y_j]$ of a chosen mixture $(x_i)$. Let $p$ be a positive fraction between zero and one, and since the vector is a line from the origin to the terminus of the vector, then the vector $[py_j]$ lies along the original vector but with the length given by

$$pr=\left[\sum_{j=1}^{N-1}(py_j)^2\right]^{\frac{1}{2}}.$$

Allowing $p$ to take values from 0 to 1 in equal or unequal steps gives a set of co-linear vectors $[p_k y_j]$, where $k$ indexes the chosen values of $p$, which can be retransformed to the desired mixtures $(x_i)_k$. This idea can be generalised to finding mixtures whose points in the mixture space lie on any line which lies wholly within the mixture space.

Holding one ingredient, say $x_m$, constant and varying the levels of the others yields a hyperplane in the mixture space which is parallel to the extreme hyperplane corresponding to the mixture with $x_m = 0$. Any hyperplane which is parallel to the hyperplane for the mixture in which $x_m = 0$ is a hyperplane for which $x_m$ satisfies $0 < x_m < 1$, but when $x_m = 1$ all the other ingredients vanish and the hyperplane degenerates to the extreme point $P_m$ of the mixture space.

The experimenter may wish to experiment with a set of mixtures $(x_i)_n$, $n = 1, \ldots, N-1$ whose points in the mixture space line on the hyperplane

$$\sum_{j=1}^{N-1} a_j y_j = d, \tag{8.25}$$

where he has chosen $a_j$ and $d$ so that the hyperplane sections the regular hypersolid. He may then choose as many points as he desires, say $[y_j]_n$, which satisfy Eq. (8.25) and lie in the section or on its boundary. He could then transform the $n$ vectors to $n$ mixtures $(x_i)_n$ and carry on. If he had chosen $(x_i)_n$ in the beginning, the vectors $[y_j]_n$ would not be related to each other in any predictable fashion.

In biology the mixtures I am considering could be foods, in which case nutritionists would be interested in the response $R$ of an animal consuming the mixtures, where $R$ is some feature of interest in the interaction of the animal and the food. Hopefully $R$ would be a fundamental parameter associated with the interaction. Because I am now discussing foods as mixtures I shall name the regular hypersolid, nutrition space, instead of mixture space. If now the nutritionist has found $R$ for foods the points of which lie along hyperlines, planes, or regions in the nutrition space; he is in a position to investigate $R$ as a function of the $y_j$, stated as

$$R = \phi(y_j).$$

Usually $\phi(y_j)$ is expressed as a polynomial in the $y_j$ and multiple linear regression is used on the $(y_j, R)$ data points. If in the analysis the polynomial has only linear terms, which are significant at some chosen level of confidence, it can be concluded that the effects of the ingredients are additive with minor, if any, interactions, and function $\phi(y_j)$ is a hyperplane. However, if the analysis shows the polynomial is of higher degree than unity then $\phi(y_j)$ is a hypersurface with curvature in various directions thus indicating interactions. If the curvature is positive at some point the ingredients are said to be synergistic and if negative the ingredients are antagonistic. These ideas hold because the $y_j$ are linear in the $x_i$ and are orthogonal variables.

Because foods are mixtures of mixtures, the preceding general approach to nutrition may be cumbersome, but it should not be forgotten in the following simple approach. Chemists and nutritionists have shown that the nutritious portions of all foods can be reduced to at least three ingredients; namely, protein, edible fats and oils, and carbohydrates as sugars and starch, for monogastric animals. For herbivores a fourth nutritious portion, namely cellulosic fibre, must be considered.

For monogastrics the mixtures of interest are one gram of nutrient made up of a fraction of fat, $x_1 = \phi$, a fraction of protein $x_2 = \pi$ and a fraction of carbohydrate $x_3 = \kappa$ with

$$\phi + \pi + \kappa = 1$$

Since $N = 3$ the nutrition space becomes two dimensional and reduces to an equilateral triangle of unit height with the pure ingredients at the vertices (0,1), $(\sqrt{3}/3, 0)$ and $(-\sqrt{3}/3, 0)$ (Fig. 8.1). It is arbitrary how the triangle is oriented, but I shall use the orientation shown in Fig. 8.1 where the vector co-ordinates are denoted by $y_1$ and $y_2$. Application of the transformation equations to the mixtures $(\phi, \pi, \kappa)$ gives the vector coordinates of any ration in the diet space as

$$y_1 = \phi, \tag{8.26}$$

$$y_2 = (\sqrt{3}/3)\,(2\pi + \phi - 1). \tag{8.27}$$

These equations and Fig. 8.1 show that protein fraction $\pi$, in the diets along line AB, is the only variable with the fat fraction $\phi$ being held constant, so that the response, $R$, of an animal to diets along AB is a function only of $y_2$ or $\pi$, i.e. $R = f(\pi)$. In diets along AB, $\pi + \kappa = 1 - \phi$ is constant, i.e. a change in $\pi$ is counterbalanced by an equal and opposite change in $\kappa$. Nutritionists often experiment with diets along such lines of constant fat fraction but rationalise this by assuming the diets along AB are isocaloric. Here it is seen that the geometry of the nutrition space makes possible nutrition experiments before introduction of the concept of dietary energy.

Responses of animals may be found to diets along the line CD, parallel to the $y_1$ axis, with constant $y_2$, say equal to $c$. By Eqs. (8.26) and (8.27) the diets would be composed in such a way that

$$c = (\sqrt{3}/3)\,(2\pi + \phi - 1)$$

and solving this for $\pi$ gives

$$\pi = (c\sqrt{3} + 1 - \phi)/2.$$

Choosing levels of $\phi$ and calculating $\pi$ as above, the levels of $\kappa$ are $1 - \phi - \pi$. In this case the response $R$ is best considered a function of $y_1$ (or $\phi$), since $\phi$ is the only independent variable. By using a grid of lines like AB and CD, on which to choose rations, the response surface $R$ becomes the function

$$R = f(y_1, y_2),$$

where $y_1$ and $y_2$ are orthogonal variables and range over the nutrition space. Transforming the variables $y_1$ and $y_2$ back to $\pi$, $\kappa$, and $\phi$ gives the response surface in terms of the food variables.

Sometimes the response $R$, especially if it is something called metabolizable energy, is thought to be a linear combination of the nutrients such as

$$R = a\phi + b\pi + c\kappa + \varepsilon \tag{8.28}$$

and truly represents the response surface with error term $\varepsilon$. So convinced, the experimenter can select three sets of $(\phi, \pi, \kappa)$, as widely separated as practicable, and replicate $R$, of each ration, as many times as desirable and fit Eq. (8.28) by linear regression to get the values of the metabolizeable energies of each ingredient $a, b$, and $c$ and their errors. Equation (8.28) implies that $\pi=\phi=\kappa=0$ is a point in the diet space. But this violates the rule that the sum of the ingredients should be unity for a unit amount of ration. This situation can be avoided by writing Eq. (8.28) as

$$R=d_0+d_1 y_1+d_2 y_2+e. \tag{8.29}$$

If the three widely separated rations $(\phi, \pi, \kappa)$ are selected and converted to three $(y_1, y_2)$ points in the nutrition space, namely the equilateral traingle (Fig. 8.1) and $R$ measured at these points then linear regression can be used to obtain values of $d_0$, $d_1$, $d_2$ and the rsd $e$. Using Eqs. (8.26) and (8.27) to transform Eq. (8.29) back to the form of Eq. (8.28), the metabolizeable energies $a, b, c$ of each ingredient becomes

$$a=d_0+d_1$$
$$b=d_0+d_2 \sqrt{3}/3$$
$$c=d_0-d_2 \sqrt{3}/3.$$

Equation (8.29) is an acknowledgement that a diet free of nutrition is not represented by a point in the nutrients space, and that the constant, $a$, is the metabolizable energy of a diet for which $y_1=y_2$ or $\pi=50\%$, $\kappa=50\%$, and $\phi=0\%$ of the total nutrients. Concepts like these will be found useful in Chap. 9 where responses of animals to various diets are studied.

Interactions between $\pi$, $\kappa$, and $\phi$ do not appear because of the prior judgement that there are none in the specification of Eq. (8.28) and only three diets. A case for the flatness of $R$ can better be made by selecting three or more diets with $\pi$, $\kappa$, and $\phi$ suitably arranged among the three already chosen, and replicating $R$ again and then fitting

$$R=a+by_1+cy_2+dy_1y_2+ey_1^2+fy_2^2$$

to determine the co-efficients $a$ to $f$ inclusive and applying the $t$-test to determine the significance of the difference of $d, e$, and $f$ from zero at some chosen level of confidence. It is quite possible that the general response surface given by $R=f(y_1, y_2)$, for some response of an animal to various diets, may not be linear in the coefficients so that nonlinear regression techniques must be used to describe the surface.

Another simple case of interest to the experimenter is the response of animals to diets along line EF (Fig. 8.1) in nutrition space which has

$$y_1=my_2+b \tag{8.30}$$

as its equation, where $m$ is the slope of the line relative to the $y_2$ co-ordinate axis and $b$ is the intercept of the line on the $y_1$ axis. Using Eqs. (8.26) and (8.27) to transform $y_1$ and $y_2$ to terms in $\pi$ and $\phi$ gives

$$\phi=(2m\sqrt{3}/(3-m\sqrt{3}))\,\pi+(3b-m\sqrt{3})/(3-m\sqrt{3}) \tag{8.31}$$

as the specification of those diets which lie along the line in the nutrition space given by Eq. (8.30). I shall have occasion to refer to equations of this type in Chap. 9

when I examine the protein to energy ratio of a diet as a variable in the nutrition space.

Herbivores are distinguished from monogastrics by their ability to use cellulosic fibre as a nutrient as well as protein, starches, and fats. Here then $N=4$, and letting $\lambda$ represent the fibre fraction of the nutrients in the diet we get

$$\lambda+\pi+\kappa+\phi=1$$

with the nutrition space reducing to the regular tetrahedron shown in Fig. 8.2. The theorems of solid analytic geometry can be used to get rectangular arrays of diet points from which to deduce the form of the general response "surface" given by

$$R=f(y_1, y_2, y_3).$$

If $R$ is thought (or found by experiment) to represent a "plane" then $R$ is given by

$$R=a+by_1+cy_2+dy_3,$$

indicating again that there are no interactions. Also if $R$ is some measure of dietary energy for herbivores, $y_1$, $y_2$, and $y_3$ can be transformed, as was done previously, by the following equations

$$y_1=\lambda, \tag{8.32}$$

$$y_2=(3\sqrt{2}/4)\pi+(\sqrt{2}/4)\lambda-\sqrt{2}/4, \tag{8.33}$$

$$y_3=(\sqrt{3/2})\kappa+(\sqrt{3/2}/2)\lambda+(\sqrt{3/2}/2)\pi-\sqrt{3/2}/2 \tag{8.34}$$

to get $R=(a+b)\lambda+(a+c\sqrt{2}/2)\pi+(a-c\sqrt{2}/4)\kappa+(a-c\sqrt{2}/4-d\sqrt{3/2}/2)\phi$, (8.35)

the coefficients of which are the dietary energies per unit of the pure dietary component $\lambda$, $\pi$, $\kappa$, and $\phi$.

For carnivores, the components of the diet are protein and fat ($N=2$), and the nutrition space becomes a line, $y_1$, of unit length. The response $R$ then becomes a curve

$$R=f(y_1),\ 0\leqq y_1\leqq 1$$

where $y_1$ can be either $\pi$ or $\phi$.

I shall close this section by recalling the beginning of this chapter where there was a general discussion of a food as a mixture of ingredients (foodstuffs) which are themselves mixtures. Much work has been done to determine the metabolisable energy ($ME$) of individual foodstuffs of economic importance for animal production. For each species there are long lists of foodstuffs with their individual $ME$'s and chemical analyses used in least cost linear programming to formulate diets (see for example, Scott et al. 1969, McDonald et al. 1973). In the use of these tables of data it is generally assumed that the $ME$'s of the foodstuffs are additive, i.e., the $ME$ of the ration is given by

$$ME=\sum_{i=1}^{N}\varepsilon_i x_i, \tag{8.36}$$

where $\varepsilon_i$ is the metabolisable energy per unit of the $i^{th}$ foodstuff and $x_i$ is the fraction of the $i^{th}$ foodstuff per unit of the ration, and $N$ is the number of foodstuffs available to the ration formulators.

Some of my associates and I, realising the empirical nature of our science, think that Eq. (8.36) is an assumption which could be subjected to experimental verification. Equation (8.36) expresses a necessary condition for the existence of the *ME* of a diet as an additive property without interactions among the $(x_i)$. But given the *ME* of *N* rations made up of various fractions of the available foodstuffs, can we show experimentally that no interactions exist? If this were shown at a sufficient level of confidence, then Eq. (8.36) would not only be necessary but also sufficient for expressing the *ME* of a diet from *N*, or any subset, of the foodstuffs. This seems trivial because a theorem of algebra tells us that given *N* diets $(ME_j, j=1,2,...,N)$ with fractions of foodstuffs $x_{ij}$, we can get the $\varepsilon_i$ by simultaneous solution of the set of equations.

$$ME_j = \sum_{i=1}^{N} \varepsilon_i x_{ij}$$

provided the determinant of $(x_{ij})$ is not at or near zero. But this already assumes Eq. (8.36) necessary and sufficient. There is the objection I have expressed previously to equations such as Eq. (8.36), because it is assumed that the point $(x_i)$, where all the $x_i=0$, is in the diet space when in fact it is not. To experimentally prove that Eq. (8.36) is sufficient, I suggest the following procedure. Begin with the Euclidean nutrition space of $N-1$ dimensions and generate a grid of $[y_j]$ points, beginning at the origin of co-ordinates, by stepping off points at equal intervals along each co-ordinate axis in the positive direction of $y_1$ and in the positive and negative directions of the other co-ordinates. In this way a large number, *M*, of points $[y_j]_k, j=1,2,...,N-1$, and $k=1,2,...,M$ where $M>N$ can be obtained and transformed to diets $(x_i)_k$. Depending on how far the stepping off of points on the coordinate axes was carried; some of the *M* diets will have some fractions of ingredients $(x_i)_k$ which will violate the restriction $0 \leq x_{ik} \leq 1$. These diets are deleted and $N+$ diets, $(x_i)_j, j=1,2,...,N+$, can be selected from the rest for experimental determination of the $ME_j$ with replication. Since the $[y_j]_k, k=1,2,...,N-1$ nutrition space points are known for each selected diet the fit of

$$ME_j = a + \sum_{i=1}^{N-1} b_i y_{ij} + \sum_{i,k=1}^{N-1} c_{ik} y_{ij} y_{kj} + \text{etc.} \tag{8.37}$$

to the $([y_i]_j, ME_j), i=1,2,...,N-1$, data points can be examined for the existence of the interaction factors $c_{ik}$ etc. If the interaction factors, $c_{ik}$, are found negligible, transforming the $[y_i]_j$ vectors back to $(x_i)_j$ substituting in Eq. (8.37) will give the $ME_i$ per unit of $x_i$, which is denoted by $\varepsilon_i$, as the set of equations

$$\varepsilon_i = f_i(a, b_i, N) \pm sd(\varepsilon_i), \; i=1,2,...,N$$

such as shown in Eqs. (8.28) for $N=3$ and Eq. (8.35) for $N=4$. These values of $\varepsilon_i$ could then be compared to the tabular values of $\varepsilon_i$ and if the null hypothesis applies at a sufficient level of confidence then Eq. (8.36) could be judged a sufficient condition also, but only for the foodstuffs and species of animal used in the experiment.

If some of the interaction coefficients in Eq. (8.37) were judged significant, they would show which foodstuffs interact, and at what intensity, the interaction being synegistic or antagonistic depending on whether the co-efficients are positive or negative.

## 8.4 Proteins as Mixtures of Amino Acids

One half of the 22 amino acids which are building blocks for proteins are considered essential for some metabolic function of the animal, the actual number depending on the species concerned and on the investigators. For example, Tables 8.2–8.5 give lists of 10 to 13 amino acids considered by various authors essential for proper metabolic function. The list in Table 8.2 is considered essential for the proper metabolic functioning of humans. I think the list in Table 8.3 applies to animals generally while Tables 8.4 and 8.5 were compiled for rats and chickens, respectively. All the lists contain common amino acids. In Table 8.3 arginine, glycine, and histidine are substituted for tryptophan in Table 8.2, whereas in Table 8.4 arginine, glycine, and histidine are added to the list in Table 8.2. In Table 8.5 histidine and arginine are substituted for cystine and tyrosine in the list in Table 8.2 and tyrosine, cystine, and hydroxylsine are included as being less than essential but not nonessential. There are eight or nine amino acids common to Tables 8.2 through 8.5.

I present these lists of essential amino acids to show the possibility that the list may differ not only between species of animals but also between breeds and, in the extreme, may differ between individuals within breeds within species. However, all these differences can be incorporated into the general geometry of the proteins as vectors in a protein nutrition space.

Since proteins are polymers which are broken down by digestion into upwards of 22 molecular constitutents called amino acids, the differences between the pro-

Table 8.2. Ten amino acids essential according to the Food and Agriculture Organisation of the United Nations (1957)

| | |
|---|---|
| Isoleucine | Methionine |
| Leucine | Cystine[a] |
| Lysine | Threonine |
| Phenylalanine | Tryptophan |
| Tyrosine | Valine |

[a] Difference between total sulphur amino acid and methionine

Table 8.3. Twelve amino acids essential according to the U.S. Feed grains Council (1973)

| | |
|---|---|
| Isoleucine | Cystine |
| Leucine | Threonine |
| Lysine | Valine |
| Phenylalanine | Arginine |
| Tyrosine | Glycine |
| Methionine | Histidine |

Table 8.4. Thirteen amino acids considered essential by Block and Mitchell (1946)

| | |
|---|---|
| Isoleucine | Cystine |
| Leucine | Threonine |
| Lysine | Tryptophan |
| Phenylalanine | Valine |
| Tyrosine | Arginine |
| Methionine | Glycine |
| | Histidine |

Table 8.5. Thirteen amino acids considered essential by Scott et al. (1969). The last three can be synthesised by the chick, but only from limited substrates

| | |
|---|---|
| Isoleucine | Threonine |
| Leucine | Tryptophan |
| Lysine | Histidine |
| Phenylalanine | Valine |
| Methionine | Arginine |
| | Tyrosine |
| | Cystine |
| | Hydroxylysine |

teins are the differences of the concentrations ($z_i$) of the individual amino acids. A protein, therefore, can be seen as a mixture ($z_i$), $i=1,2,\ldots,22$, where $z_i$ can be expressed as grams of the $i^{\text{th}}$ amino acid in 100 grams of protein, or as milligrams of nitrogren in 1 gram of total protein. Regardless of how the $z_i$ are expressed, they can be reduced to frations ($x_i$) of 1 unit of protein by

$$x_i = z_i / \sum_{i=1}^{22} z_i ,$$

so that $x_i$ satisfy the inequality $0 \leqq x_i \leqq 1$, and the protein can then be represented by the vector $[y_j]$, $j=1,2,\ldots,21$, in a Euclidean hyperspace of 21 dimensions. Each protein will be represented by a point in the protein nutrition space. The geometrical properties of this space are the same as those discussed in Chap. 8.3 and we can talk generally about mixtures of amino acids whose points lie on lines, in planes, or on hypersurfaces of various dimensions up to $N=21$, and we can express biological responses, $R$, of animals to dietary proteins by the general formula

$$R = f(y_j).$$

Nutritionists, adhering to Liebig's law of the minimum, have found that the relative levels of 10–13 of the 22 amino acids account for most of the differences between the responses, $R$, of animals consuming the proteins, with 9–12 of the amino acids appearing to be neutral. The 10–13 metabolically active amino acids are called essential amino acids and the neutral ones called nonessential. The dimensions of the protein nutrition space can accordingly be reduced to the set of essential amino acids minus unity. I shall refer to this reduced space as the amino acid nutrition space.

I have chosen the list of ten essential amino acids in Table 8.2 because the FAO Nutritional Study, Bulletin Number 16 (1957) presents the levels of each of these amino acids in the proteins of 30 foodstuffs and mixtures of them. The Bulletin also proposes a "provisional" set of levels $z_i$ of these amino acids which are based on the minimal requirements for each of them in human nutrition as estimated in 1957. To this group of amino acid levels I have added, in Table 8.6, the levels of these amino acids occuring in the crystalline amino acid mix used by Scott and his colleagues at the University of Illinois (Klain et al. 1960, Velu et al. 1971). Table 8.6 shows the levels $z_i$ of the amino acids, listed in Table 8.2, for various natural sources of protein along with the FAO provisional pattern and the crystalline amino acid mix of Scott et al. (1969). The mixtures $(x_i)_k$ representing each of these sources of nitrogen are found by

$$x_{ik} = z_{ik} / \sum_{i=1}^{10} z_{ik} , \tag{8.38}$$

where $k$ is the index number of the amino acid source. Transforming these mixtures to vectors $[y_i]_k$ gives an amino acid nutrition space of nine dimensions containing the vectors. The values of $x_{ik}$, of the amino acid sources in Table 8.6, are shown in Table 8.7 for level by level comparison of the sources, if that is desired.

Since the vectors represent the mixtures as wholes, the differences between the mixtures can be quickly seen by computing the lengths, $r_k$, of the vectors and the angles, $\theta_{kl}$, in degrees between the vectors taken in pairs using Eqs. (8.12) and

Table 8.6. Essential amino acid levels ($z_i$, g/kg N) in various proteins and in the crystalline amino acid mixes recommended by F.A.O. or developed at the University of Illinois

| $z_i$ | | | | Milk | | | | | Beef | | | |
|---|---|---|---|---|---|---|---|---|---|---|---|---|
| | F.A.O. | Illionois | Egg | Human | Cow | Casein | Wheat | Soy | Liver | Kidney | Heart | Muscle |
| Isoleucine | 270 | 3.11[a] | 428 | 411 | 407 | 402 | 261 | 333 | 327 | 304 | 317 | 332 |
| Leucine | 306 | 6.22 | 565 | 573 | 630 | 628 | 426 | 484 | 577 | 542 | 558 | 515 |
| Lysine | 207 | 6.17 | 396 | 402 | 496 | 497 | 107 | 395 | 468 | 453 | 573 | 540 |
| Phenylalanine | 180 | 2.59 | 368 | 297 | 311 | 334 | 308 | 309 | 315 | 294 | 282 | 256 |
| Tyrosine | 180 | 2.33 | 274 | 355 | 323 | 367 | 192 | 201 | 234 | 232 | 232 | 212 |
| Methionine | 144 | 1.82 | 196 | 140 | 154 | 190 | 123 | 86 | 147 | 128 | 149 | 154 |
| Cystine | 162 | 1.82 | 147 | 134 | 57 | 25 | 100 | 111 | 79 | 80 | 68 | 83 |
| Threonine | 180 | 3.37 | 310 | 290 | 292 | 272 | 151 | 247 | 302 | 278 | 288 | 275 |
| Tryptophan | 90 | 0.778 | 106 | 106 | 90 | 85 | 60 | 86 | 94 | 92 | 81 | 75 |
| Valine | 270 | 4.25 | 460 | 420 | 440 | 448 | 264 | 382 | 393 | 365 | 360 | 345 |
| | 2016 | 32.458 | 3250 | 3128 | 3200 | 3248 | 1992 | 2580 | 2936 | 2768 | 2908 | 2787 |

[a] Gram per 100 grams of protein; others are mg per gram nitrogen of the proteins

Table 8.7. Essential amino acid levels, $x_i$, calculated from the $z_i$ in Table 8.6

| $x_i$ | | | | Milk | | | | | Beef | | | |
|---|---|---|---|---|---|---|---|---|---|---|---|---|
| | F.A.O. | Illinois | Egg | Human | Cow | Casein | Wheat | Soy | Liver | Kidney | Heart | Muscle |
| Isoleucine | 0.134 | 0.096 | 0.132 | 0.131 | 0.127 | 0.124 | 0.131 | 0.129 | 0.111 | 0.110 | 0.109 | 0.119 |
| Leucine | 0.152 | 0.192 | 0.174 | 0.183 | 0.197 | 0.193 | 0.214 | 0.188 | 0.197 | 0.196 | 0.192 | 0.185 |
| Lysine | 0.134 | 0.190 | 0.122 | 0.129 | 0.155 | 0.153 | 0.154 | 0.153 | 0.159 | 0.164 | 0.197 | 0.194 |
| Phenylalanine | 0.089 | 0.080 | 0.113 | 0.095 | 0.097 | 0.104 | 0.155 | 0.120 | 0.107 | 0.106 | 0.097 | 0.092 |
| Tyrosine | 0.089 | 0.072 | 0.084 | 0.113 | 0.101 | 0.113 | 0.096 | 0.078 | 0.080 | 0.084 | 0.080 | 0.076 |
| Methionine | 0.071 | 0.056 | 0.060 | 0.045 | 0.078 | 0.058 | 0.062 | 0.033 | 0.050 | 0.076 | 0.051 | 0.055 |
| Cystine | 0.063 | 0.056 | 0.045 | 0.043 | 0.018 | 0.008 | 0.050 | 0.043 | 0.027 | 0.029 | 0.023 | 0.030 |
| Threonine | 0.089 | 0.104 | 0.095 | 0.093 | 0.091 | 0.084 | 0.076 | 0.096 | 0.103 | 0.100 | 0.099 | 0.099 |
| Tryptophan | 0.045 | 0.024 | 0.033 | 0.034 | 0.028 | 0.026 | 0.030 | 0.033 | 0.032 | 0.033 | 0.028 | 0.027 |
| Valine | 0.134 | 0.131 | 0.142 | 0.134 | 0.133 | 0.138 | 0.133 | 0.127 | 0.134 | 0.132 | 0.124 | 0.124 |

Table 8.8. The lengths ($r_k$) and angles ($\theta_{kl}$) between the vectors of the amino acid sources listed in Table 8.6 taken in three groups

| Source | $\theta_{kl}$, angular degrees | | | | | | $r_k$ |
|---|---|---|---|---|---|---|---|
| F.A.O. | 0 | | | | | | 0.166 |
| Illinois | 26 | 0 | | | | | 0.186 |
| Beef muscle | 22 | 10 | 0 | | | | 0.202 |
| heart | 23 | 12 | 6 | 0 | | | 0.198 |
| liver | 21 | 16 | 11 | 6 | 0 | | 0.191 |
| kidney | 22 | 15 | 11 | 5 | 2 | 0 | 0.191 |
| F.A.O. | 0 | | | | | | 0.166 |
| Illinois | 26 | 0 | | | | | 0.186 |
| Cow's milk | 18 | 19 | 0 | | | | 0.204 |
| Casein | 20 | 23 | 6 | 0 | | | 0.204 |
| Soya | 19 | 20 | 12 | 17 | 0 | | 0.196 |
| Wheat | 35 | 47 | 24 | 34 | 31 | 0 | 0.207 |
| F.A.O. | 0 | | | | | | 0.166 |
| Illinois | 26 | 0 | | | | | 0.186 |
| Human milk | 15 | 26 | 0 | | | | 0.187 |
| Egg | 13 | 26 | 12 | 0 | | | 0.181 |

Table 8.9. Angles between the vector representations of all the sources of essential amino acids in Table 8.6 and the F.A.O.provisional pattern and the crystalline amino acid mix of the University of Illinois (Scott et al. 1969)

| Source | F.A.O. | Illinois | Source | F.A.O. | Illinois |
|---|---|---|---|---|---|
| F.A.O. | 0 | 0 | F.A.O. | 0 | 26 |
| Illinois | 26 | 0 | Illinois | 26 | 0 |
| Egg | 13 | 26 | Soy | 19 | 20 |
| Human milk | 15 | 26 | Beef liver | 21 | 16 |
| Cow's milk | 18 | 19 | Beef kidney | 22 | 15 |
| Casein | 20 | 23 | Beef heart | 23 | 12 |
| Wheat | 35 | 47 | Beef muscle | 22 | 10 |

(8.14). I wrote a computer programm to take the $z_{ik}$ in Table 8.6 as data and calculate $r_k$ and $\theta_{kl}$ using Eq. (8.38) to get the $(x_i)_k$ Table 8.7 and the transformation Eqs. (8.19), (8.20), and (8.21) to calculate the vector $[y_i]_k$ and Eqs. (8.12) and (8.14) to get the $r_k$ and $\theta_{kl}$. Into this program I wrote a subroutine to transform the vectors back to $(x_i)_k$, using Eqs. (8.22), (8.23), and (8.24), as a check routine which I could call as desired. Because of limited computer capacity I wrote the program to operate on six mixtures of ten constituents each, so that I had to apply it to the data in Table 8.6 in three parts, with the results shown in Table 8.8. Since the angles are between pairs of vectors, the table need only be triangular in order to list all possible pairs. The angles in the columns are angles of the sources relative to the source with 0 degrees in its row. I used the FAO and Illinois patterns in each run in order to compare the natural sources with two artificial sources arrived at through application of biological and nutritional principles. Table 8.9 summarizes the angular data in Table 8.8 by referring the various sources to the FAO and Illinois patterns only.

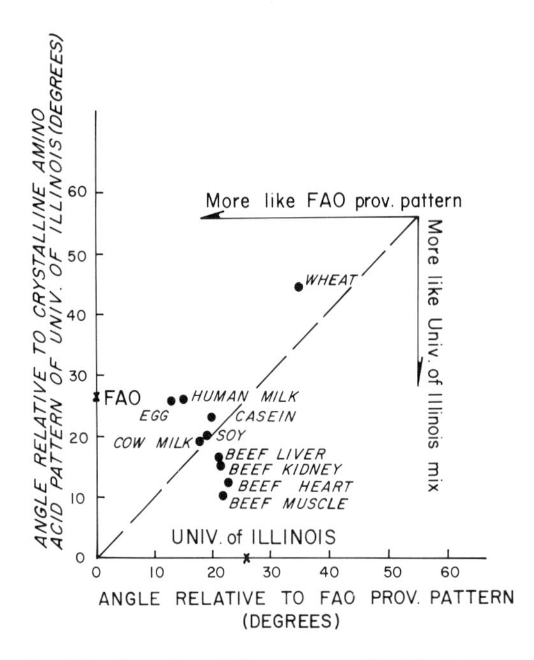


Fig. 8.3. Angles of sources of essential amino acids in Table 8.6 relative to the FAO provisional pattern and the pattern in the crystalline amino acid mix used at University of Illinois

The lengths of the vectors (Table 8.8) are not very different from each other, as might have been expected, but the angles show some interesting things; for example, the beef proteins are closer to the Illinois pattern than to the FAO pattern, and egg and human milk proteins are the other way around while wheat protein is far from all the sources. Figure 8.3 illustrates these relations in clear and quick

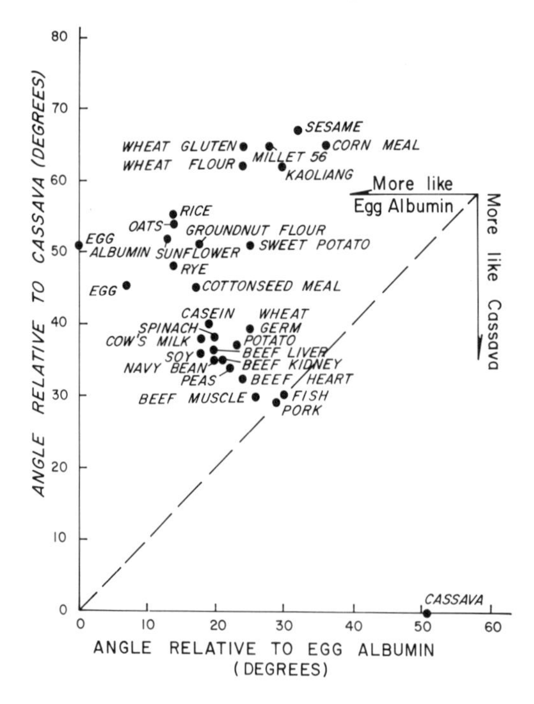


Fig. 8.4. Angles of proteins listed in Table 5, F.A.O. Bulletin 16, relative to egg albumin and cassava proteins

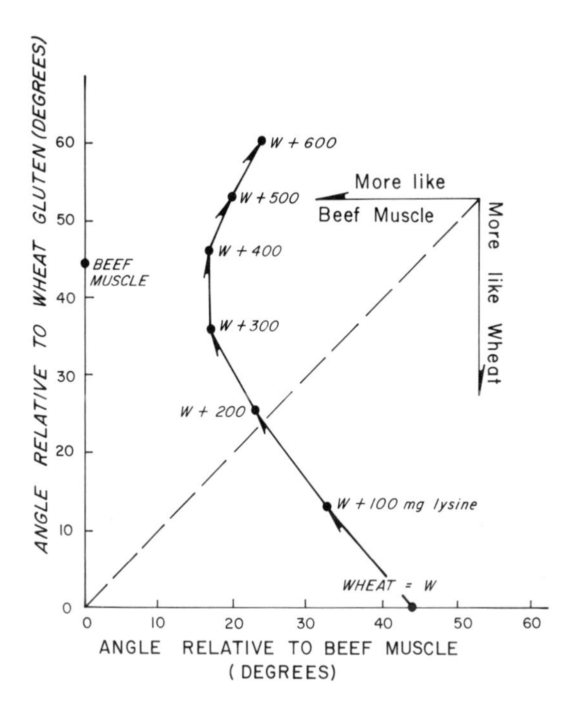


Fig. 8.5. The effect of adding lysine to wheat gluten on the angle between the new vector and the vector for beef muscle

fashion. Two more points of interest from Fig. 8.3 and Table 8.9 are, egg and human milk proteins are more like the FAO provisional pattern of essential amino acids developed from minimal human essential amino acid requirements, and the beef, liver, kidney, heart, and muscle proteins are more like the University of Illinois crystalline amino acid mix developed to meet the requirements of chickens for these essential amino acids.

Encouraged by these results I applied the program, in five steps of six sources each, to the data for 30 natural foodstuffs in Table 5 of FAO Bulletin Number 16. Figure 8.4 pictures the results using the angles relative to egg albumin and cassava protein as the co-ordinates. Here it is seen that practically all the natural proteins are closer to egg albumin than to cassava which has an angle of 51 degrees relative to egg albumin. It is curious that there are no other natural essential amino acid sources in this group of 30 which are close to cassava. Furthermore it appears that the natural proteins sparsely occupy a small portion of the protein nutrition space.

Because of the biological values, BV, of the sources, wheat gluten and cassava, with large angles relative to the reference source egg albumin, were low, I examined the biological values of the sources in Table 5 of FAO Bulletin 16 for a relation to the angles they made with egg albumin. I found no evidence of a systematic change of biological value with angle; there was only a general drift downward of biological value, BV, as the angle increased. This lack of high correlation was not unexpected, since the angle between the source and the reference is not unique. There is a hypercone of an infinite number of vectors, with the reference as the axis of the cone, all of which have the same angle with the reference.

It is known that adding lysine to wheat gluten improves its BV. I was curious to see how adding lysine to $z_3$ of wheat in Table 8.6, in 100 mg increments from zero to 600 mg total added, changes the angle $\theta$ of the resulting vector relative to the vector for beef muscle. We have already seen that this angle between wheat gluten and beef muscle is 44 degrees. Figure 8.5 shows that the additions of lysine up to 300 mg reduce the angle from 44 to 17 degrees with the next increment causing no change and subsequent increments causing the angle to increase. The effect of adding lysine to wheat (Table 8.6) is shown in Table 8.10 where the $x_i$ of the treated wheat can be compared with the $x_i$ of the untreated wheat and beef muscle. Here it can be seen that the $x_i$ of wheat plus lysine become more like those of beef muscle as the lysine level is increased from 0.057 to about 0.178, but thereafter the similarity to beef muscle begins to fade. Figure 8.5 dramatically shows these effects.

The vector of wheat gluten plus lysine approaches that of beef muscle but later recedes from it. The implication is that BV is correlated with the angle $\theta$, i.e. BV=40 of wheat gluten may be increased to a value near BV=76 of beef muscle, but as far as I know sufficient experimental work has not been done to elucidate the functional relation and I am not sure that BV as it is presently defined is a proper measure of the value of a protein to a growing animal (see Chap. 9).

The ideas in this section lead me to think that the idea of the "limiting amino acid" is an unduly restrictive concept in that it encouraged an experimentally inefficient stepwise approach to finding the vector representation of an "ideal" protein in a diet to achieve stated changes of body weight and composition.

The same remark can be made about the limitations of the concept of the "limiting nutrient" in a food. In experimentation with mixtures there may be an optimum blend of the differential concept of "limiting factor" and the integrated concept of vector representation, optimum in the sense that the least number of experiments are needed to explore the response of an animal to mixtures of amino acids, or food ingredients.

## 8.5 Musculature of Animals as a Mixture of Specific Muscle Masses

Berg and Butterfield (1976) defined a set of nine standard muscle groups which could be used to specify the musculature of an animal as the set of nine weights

Table 8.10. The modifications of the $x_i$ for wheat (Table 8.6) by the addition of Lysine in 100 mg steps to $z_3$ (lysine in Table 8.6) for comparison to the $x_i$ of beef muscle

| $x_i$ | Basal lysine | Lysine +100mg | +200 | +300 | +400 | +500 | +600 | Beef muscle |
|---|---|---|---|---|---|---|---|---|
| Isoleucine | 0.131 | 0.125 | 0.119 | 0.114 | 0.109 | 0.105 | 0.101 | 0.119 |
| Leucine | 0.214 | 0.204 | 0.194 | 0.186 | 0.178 | 0.171 | 0.164 | 0.185 |
| Lysine | 0.054 | 0.099 | 0.140 | 0.178 | 0.212 | 0.244 | 0.273 | 0.194 |
| Phenylalanine | 0.155 | 0.147 | 0.141 | 0.134 | 0.129 | 0.127 | 0.119 | 0.092 |
| Methionine | 0.062 | 0.059 | 0.056 | 0.054 | 0.051 | 0.049 | 0.047 | 0.055 |
| Cystine | 0.050 | 0.048 | 0.046 | 0.044 | 0.072 | 0.070 | 0.039 | 0.030 |
| Threonine | 0.076 | 0.072 | 0.069 | 0.066 | 0.063 | 0.061 | 0.058 | 0.099 |
| Tryptophan | 0.030 | 0.029 | 0.027 | 0.026 | 0.025 | 0.024 | 0.023 | 0.027 |
| Valine | 0.133 | 0.126 | 0.120 | 0.115 | 0.110 | 0.106 | 0.102 | 0.124 |

Table 8.11. The nine standard muscle groups in the order used to obtain the vector representing the groups as muscle mixture (Berg and Butterfield 1976)

1. Primal pelvic limb
2. Distal pelvic limb
3. Spinal
4. Abdominal
5. Proximal thoracic limb
6. Distal thoracic limb
7. Thorax to thoracic limb
8. Neck to thoracic limb
9. Neck and thorax

Table 8.12. Angle, $\theta_{ij}$, between vectors for muscle masses of species and breeds of animals

| | | | | | | | | | | | | | | | |
|---|---|---|---|---|---|---|---|---|---|---|---|---|---|---|---|
| Steers | 0 | | | | | | | | | | | | | | |
| Barrows | 11 | 0 | | | | | | | | | | | | | |
| Bulls $w=77.6$ kg | 5 | 11 | 0 | | | | | | | | | | | | |
| Rams | 12 | 5 | 11 | 0 | | | | | | | | | | | |
| Big buffalo | 10 | 18 | 8 | 17 | 0 | | | | | | | | | | |
| Crossbred bulls | 8 | 13 | 10 | 13 | 11 | 0 | | | | | | | | | |
| Small buffalo | 10 | 16 | 9 | 15 | 7 | 8 | 0 | | | | | | | | |
| Banteng steer | 9 | 13 | 12 | 13 | 13 | 4 | 8 | 0 | | | | | | | |
| Angus × Charolais bull | 9 | 15 | 5 | 14 | 6 | 13 | 11 | 15 | 0 | | | | | | |
| Big bull moose | 13 | 18 | 13 | 18 | 13 | 14 | 8 | 12 | 14 | 0 | | | | | |
| Small bull moose | 13 | 18 | 14 | 18 | 14 | 12 | 7 | 10 | 16 | 4 | 0 | | | | |
| Heifer | 2 | 11 | 7 | 13 | 11 | 7 | 10 | 9 | 11 | 13 | 13 | 0 | | | |
| Doe | 10 | 11 | 10 | 10 | 14 | 7 | 9 | 6 | 14 | 12 | 10 | 10 | 0 | | |
| Mature buck | 16 | 17 | 12 | 13 | 12 | 15 | 13 | 16 | 11 | 17 | 17 | 18 | 14 | 0 | |
| Bison bull | 10 | 14 | 8 | 12 | 7 | 9 | 7 | 10 | 8 | 11 | 11 | 11 | 10 | 10 | 0 |

of the muscle groups as percentages of the live weight (Table 8.11). Chapter 7 of their book contains sets of these percentages for a wide range of animal sizes and species. I used these data to compute the vector $[y_i], j=1,2,...,8$ for each of the animals and the angles $\theta_{k1}$ between them in pairs (Table 8.12).

Figure 8.6 shows the angles of each animal relative to the big buffalo and the barrow. Most of the animals, excepting the rams, are more like the big buffalo than the barrow, but there appear to be three groups of animals which are more like each other than either the big buffalo or barrow.

1. The crossbred bulls, steers, bulls of average weight of 77.6 kg, heifers, Banteng steer, and female deer are more like each other in musculature than either the big buffalo or barrows.
2. The big bull moose, small bull moose and mature buck deer are more alike than the big buffalo or the barrow or the animals in item 1.
3. The Angus × Charolais bull, small buffalo and Bison bulls are more like each other and the big buffalo than they are like the others.

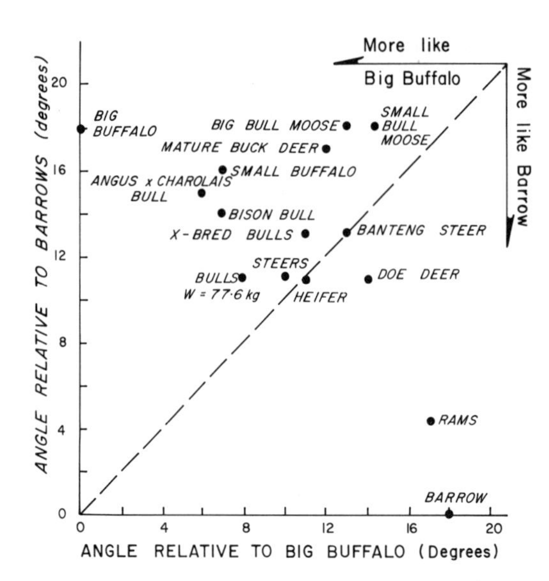


Fig. 8.6. Angles between vectors representing the musculature of animals and the reference vectors of the musculatures of big buffalos and barrows, showing three possible groupings

If economic values could be assigned to each of the muscles (Table 8.11) for each species, an economic vector could be assigned to each species and the angles $\theta_{ij}$ would then express relative economic values between the species $i$ and $j$. A figure like Fig. 8.6 would then show the economic groupings relative to the least and most valuable species.

The vector concept may be a tool useful to comparative anatomists. There are undoubtedly many mixtures of biological substances which can be usefully represented as vectors in multidimensional Euclidean spaces.

## References

Berg RT, Butterfield RM (1976) New concepts of cattle growth. Univ Press, Sydney

Block RJ, Mitchell HH (1946) The correlation of the amino acid composition of proteins with their nutritive value. Nutr Abstr Rev 16:249–278

Klain CJ, Greene DE, Scott HM, Johnson BC (1960) The protein requirement of the growing chick determined with amino acid mixtures. J Nutr 71:209-212

McDonald P, Edwards RA, Greenhalgh JFD (1973) Animal nutrition, 2nd edn. Longman, New York

Moon P, Spencer DE (1974) A geometry of nutrition. J Nutr 104:1535–1542

Scott ML, Nesheim MC, Young RJ (1969) Nutrition of the chicken. Scott and associates, Ithaca

Velu JG, Baker DH, Scott HM (1971) Protein and energy utilisation by chicks fed graded levels of a balanced mixture of crystalline amino acids. J Nutr 101:1249–1256

# Chapter 9 The Effects of Diet Composition on the Growth Parameters

Chapter 7 showed the theory of feeding and growth of animals was sufficiently robust to be used as a guide in experimentation on the various ways the trace can be shifted about in the $t, F, W$ space with the consequent shifting of the traditional growth curve in the $t, W$ plane. According to the theory the trace depends solely on the values of the characteristic parameters $A$, $(AB)$, $T_0$, and $t^*$ appearing in the DE's for ad libitum or control fed animals. Shifting the trace by various treatments of the animal is therefore manifested by changes of the characteristic parameters caused by the treatments. It is well known that the growth curve can be manipulated through nutrition, i.e. by varying the composition of the food fed. Here we seek the manner in which the trace is manipulated by changes in diet composition.

The primary aim of this chapter is to find how composition affects the growth parameters. Nutritional concepts such as "balanced" diets and proteins, biological value and requirements may be considered, but only to the extent necessary to achieve our aim. Much work involving growth of animals, nitrogen balance, blood plasma analysis, biochemistry, and so forth has been done in bringing nutrition to its present state of development (Morrison 1954, McCollum 1957, Mitchell 1962, Albanese 1963, Hafez and Dyer 1969, McDonald et al. 1973). The reader undoubtedly knows that the references given here are a small sample of the voluminous literature devoted to nutrition. However, too little of the work can be used to estimate parameters of growth, since the animals were used mostly as technical indicators of some physiological state while consuming diets or components of diets said to be deficient in one respect or another.

The ubiquitous nature of food makes difficult its definition in terms which give general acceptability. Chapter 9.1 will review briefly the general geometrical approach to specifying a unit of weight of a food as a mixture advanced in Chap. 8. Chapter 9.2 will briefly review this approach to a few dietary entities that nutritionists use to specify foods. Chapter 9.3 will further restrict the discussion to foods usually fed to monogastrics and add another dimension to the geometry of the diet space, namely the response of one or more of the growth parameters of a growing animal to the diet consumed.

Thus far I have not mentioned the energy of the food chiefly because the objective of the preceding chapters was quantitative description of how the mass of food eaten is related to the mass of the animal. Animals consuming food, air and water are simultaneously consuming the potential chemical energy of the food, which is utilised by the animal to augment or sustain the chemical potential energy of its live weight and to maintain itself as an active ongoing system. Chapter 9.4 examines metabolisable energy in relation to the geometrical representation of foods while the remaining sections are devoted to the growth parameters as responses to alteration of diet composition in one respect or another.

Chapters 9.5 and 9.6 discuss the effects on the growth parameters of alteration of the dietary energy and protein levels respectively. Chapters 9.6 and 9.7 suggest new definitions of the growth promoting abilities for proteins and single amino acids. Chapter 9.8 will briefly discuss the concept of requirement and its inadequacy in the light of the discussions in Chaps. 9.5–9.7.

## 9.1 Diet as a Mixture

A diet, or some fixed preparation of a food, is a very complex mixture of ingredients which are themselves complex in chemical composition. The activity of the digestive tract breaks down food into simpler chemical and physical structures which are transported across the walls of the tract to the metabolic apparatus for use in growth, warmth, production, and daily life by the animal.

Description of a diet as a mixture of ingredients, i.e., a recipe, cannot serve as a scientific base for nutrition studies. Over the years methods have been developed to analyse a diet chemically for three major nutrients; namely protein, carbohydrate, and fat. Chemical analysis of a food can never be complete because of the numerous microconstituents like vitamins, cofactors, and minerals (many known, others still unknown) which are vital for effective use of the food by the animal. Accepting the incompleteness, the analysis of a unit mass of the diet for fractions of protein, nitrogen free extract (NFE), ether extract, water, ash, and fibre may be a basis for study of the effect of diet composition on the growth parameters. Such an analysis is called the proximate analysis, thus acknowledging its incompleteness. This line of enquiry on food composition will be continued later, but meanwhile it should be noted that the proximate analysis of a diet yields six components, three of which, namely protein, nitrogen free extract, and ether extract, provide material and energy for growth, maintenance, and production in monogastrics. Ruminants can use fibre in which case the number of nutritious components becomes four, with the total number of components being seven. Also protein can be considered as a mix of its constituent amino acids. One unit of protein can be hydrolysed into fractions of about 22 amino acids. Hence, a diet can be considered a mixture of at least 27 components.

Diets of various compositions differ primarily in the levels ($x_i$) of identifiable components, where $x_i$ is the amount of the $i^{th}$ component as a fraction of one unit of the diet and $i$ is counted from 1 to $N$, where $N$ is the number of components chemically identifiable. Let $x_{ij}$ symbolize the level of the $i^{th}$ component in the $j^{th}$ diet. One way to compare the chemical analyses of two diets $j$ and $k$ is to note the differences between levels of the same component; namely to note whether $x_{ij} \lesseqgtr x_{ik}$, and relate these differences to the differences between some measured performance characteristic of animals feeding on the two diets. Such a procedure can be tedious and unrewarding with the possibility that individual differences may be overemphasised in the biological context. Furthermore the individual component comparison ignores two fundamental properties of any, say the $j^{th}$, diet composition. Firstly, the sum of all the levels $x_{ij}$ of the $j^{th}$ diet must equal unity as indicated by Eq. (9.1) and secondly, the animal performs on the whole diet.

$$\sum_{i=1}^{N} x_{ij}=1, \quad 0 \leqq x_{ij} \leqq 1. \tag{9.1}$$

The preceding discussion of diets as mixtures of a few or many components introduces the general notion of comparative study of different mixtures of the same components before some animal performance rating is assigned to them. Chapter 8 proposed a method of representing a mixture $(x_i)$ by a vector $[y_j]$ in an $(N-1)$ dimensional Euclidean hyperspace $(N>4)$ and develops the geometry and some of the vector algebra of this space.

Dropping the diet index for convenience and taking scale factor $s_1$ as unity, Eqs. (9.2), (9.3), and (9.4) are the equations for transforming a mixture $(x_i)$ of $N$ ingredients to the vector $[y_j]$, $j=1,2,...,N-1$.

$$y_1=x_1 \tag{9.2}$$

$$y_j=\left[x_j-(1-\sum_{i=1}^{j-1} x_i)/(N-j+1)\right] s_j \tag{9.3}$$

$$s_j=[(N-1)(N-j+1)/N(N-j)]^{1/2}, \quad j=2,3,\ldots,N-1. \tag{9.4}$$

Since the vector $y_j$ is in a Euclidean space of $N-1$ dimensions, the usual notions of distance between points and angular direction in two or three dimensional space carry over unchanged to Euclidean spaces of higher dimensions where $N-1>3$. Hence, regardless of number of dimensions, it is proper to refer to vector $C$ as being close to vector $A$ or vector $B$ or to neither.

The vector $[y_j]$ places the mixture composition $(x_i)$ in a regular hypersolid of $N-1$ dimensions with the pure components of the mixture $(x_k=1,\ x_i=0$ for $i \neq k)$ at its $N$ vertices. The $N$ heights of the hypersolid are all equal to unity and intersect at the centre of the solid which corresponds to the mixture composition $x_i=1/N$, $i=1,\ldots N$. All the edges of hypersolid are of equal length with the points along any edge representing mixtures of the pure components terminating the edge. All the points of the equilateral triangular faces of the solid represent mixtures of the pure components at their vertices and so on. In Chap. 8 I named this solid the nutrition space.

When $N=3$ the nutrition space becomes an equilateral triangle with the three heights equal to unity as shown in Fig. 8.1 (Chap. 8) and Eqs. (9.2), (9.3), and (9.4) reduce to

$$y_1=x_1 \tag{9.5}$$

$$y_2=(x_1+2x_2-1)\sqrt{3}/3 \tag{9.6}$$

thereby making it possible to discuss collections of three component mixtures in the context of plane analytical geometry using Cartesian co-ordinates $Y_1$ and $Y_2$ having the origin of co-ordinates at the centre of the side with vertices which correspond to pure components $x_2=1$ and $x_3=1$.

When $N=4$ the nutrition space becomes a regular tetrahedron, as shown in Fig. 8.1 (Chap. 8), and the transformation equation are

$$y_1 = x_1 , \tag{9.7}$$

$$y_2 = (x_1 + 3x_2 - 1)\sqrt{2}/4 , \tag{9.8}$$

$$y_3 = (x_1 + x_2 + 2x_3 - 1)\sqrt{3/2}/2 \tag{9.9}$$

thus opening up the possibility of applying solid analytical geometry to problems involving mixtures of four components. The principles of analytical geometry can be generalised to apply in Euclidean spaces of any dimension. Examples of application of some of these principles to experimental data are presented in the next section.

## 9.2 Nutrient Composition of a Food as a Point in the Nutrition Space

I have already mentioned the development of what is generally known as the proximate analysis of a food. In this analysis the six chemically identifiable components of a diet, crude protein, ether extract (fats and oils), nitrogen free extract (NFE), crude fibre, ash, and water, are quantitatively expressed in percentages which can be converted to fractions of one unit of diet on division by 100 (Atwater and Bryant 1903, Morrison 1954, Mitchell 1962, Fox and Cameron 1968). Since water and ash supply no dietary energy there are four possible nutrients and since monogastrics do not utilise crude fibre the nutrition Space containing their diets is the equilateral triangle briefly discussed in the preceding section. Ruminants, however, can utilise crude fibre thus making their nutrition space the regular tetrahedron also discussed previously. There is no particular reason to stop at $N=4$ in chemically specifying the components of a food. Carpenter and Clegg (1956) have analysed the NFE of feed grains into starch, sugar, and a carbohydrate residue thereby making the nutrition space of monogastrics a four dimensional hypersolid since the sets are then considered to be composed of five nutrients, namely crude protein, fat, sugar, and carbohydrate residue.

The proximate chemical analysis of diets has disadvantages among which is the estimation of percent protein by the formula 6.25 × the percent nitrogen in the diet. In response to the problem of inaccuracy of proximate analysis biochemists and nutritionists have followed one of two routes in producing chemically defined diets. One of these routes is to formulate the diet of known amounts of highly purified natural proteins, fats and oils, and starches and/or sugar as nutrients fortified with the known needs, of the experimental animals, for minerals, vitamins, and co-factors plus an inert material to provide bulk and roughage (Metta and Mitchell 1954, Mitchell 1962, Fox and Cameron 1968). Diets such as these are called semi-purified diets and are used primarily to develop scientific measures of nutritional value and to learn how to distribute these values among the various nutrients.

The other route is essentially the same except that the purified protein is replaced by its nitrogen equivalent of crystalline amino acids with some basic material to neutralize the acidity (Klain et al. 1960, Leveille et al. 1960, Velu et al. 1971). Crystalline amino acid diets are flexible in the ratios of the individual amino acids and hence, in principle, can be made analogous to the various natural proteins which are characterised by the ratios of their various amino acid constituents. This flexibility opens the possibilities of not only making a standard (or reference) protein, the analysis of which is fixed and can be used for performance comparisons with natural proteins (Parks 1973), but also of making amino acid mixtures corresponding to no naturally occuring proteins. The crystalline amino acid proteins can also be used to study dietary deficiencies of individual amino acids such as lysine and methonine and others (Leveille et al. 1960, Velu et al. 1971).

## 9.3 Nutrition Space and Response Surfaces of Monogastrics

Assume a diet contains the following fractions of constituents, $P$ of proteins, $C$ of carbohydrate, $F$ of fat, $a$ of ash or minerals, and $w$ of water, then the fractions of dry nutrient can be expressed as

$$\Pi = P(1-a-w),$$

$$\phi = F/(1-a-w) \text{ and}$$

$$\kappa = C/(1-a-w),$$

where $\Pi+\phi+\kappa=1$, $0 \leqq \Pi, \phi, \kappa \leqq 1$, $N=3$ and the nutrition space becomes the equilateral triangle discussed in the preceeding section and shown in Fig. 9.1. Since the three altitudes are each unity and the sum of the perpendicular distances from any point within the triangle to the sides is unity, the diet space can be covered with a triangular network of lines, of which lines of constant $\phi$ are parallel to the $\kappa\Pi$ side, lines of constant $\kappa$ are parallel to the $\phi\Pi$ side and lines of constant $\Pi$ are parallel to the $\kappa\phi$ side. The intersection of any two of these lines locates a diet in the diet space. These lines are called the triangular co-ordinates of the diets. It should be noted that there is no point in the nutrition space corresponding to $\phi=\Pi=\kappa=0$, i.e. to a diet composition containing no nutrients. This remark will be useful to recall a little later when metabolisable energy is being discussed. Any two of the above fractions $\pi$, $\phi$, and $\kappa$ can be used to determine the location of a diet in the nutrition space. It does not matter which two are chosen. Here $\Pi$ and $\phi$ will be used. Figure 9.1 shows the location of the three diets used by Metta and Mitchell (1954) to determine the metabolisable energies of casein, lard, and starch in rat feeding.

If $\phi$ and $\Pi$ are substituted for $x_1$ and $x_2$ and $y$ and $x$ for $y_1$ and $y_2$ respectively, in Eqs. (9.5) and (9.6) the transformations from triangular to rectangular co-ordinates in the nutrition space, become

$$y = \phi \tag{9.10}$$

$$x = (\phi + 2\Pi - 1)\sqrt{3}/3 \tag{9.11}$$

where the origin is at $\phi=0$, $\Pi=\kappa=1/2$.

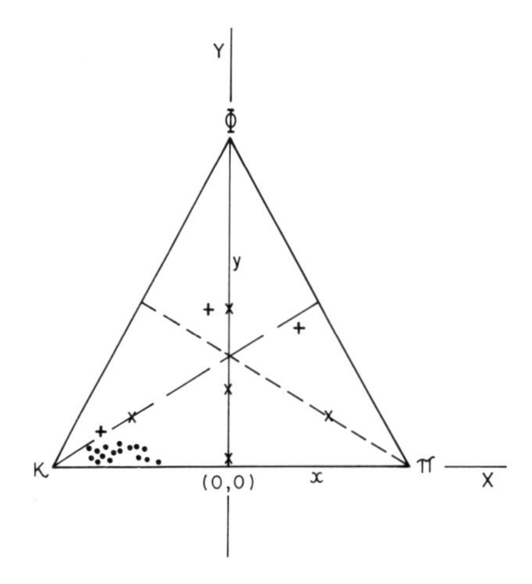


Fig. 9.1. Nutrition space for $N$ equal 3, showing the locations of: + the three diets used by Metta and Mitchell (1954), ● the diets studied by Carpenter and Clegg (1956), × four additional diets which might be used better with Metta and Mitchel's diets to cover the nutrition space

Generally the points of interest in the diet space (Fig. 9.2) may lie on a straight line with slope $m$ and $y$-axis intercept $b$, the equation of which is

$$y = b + mx\,, \tag{9.12}$$

and which by Eqs. (9.10) and (9.11) becomes

$$\phi = (b\sqrt{3} - m)/(\sqrt{3} - m) + 2m\Pi/(\sqrt{3} - m)\,, \tag{9.13}$$

in the triangular co-ordinates $\phi$ and $\Pi$. By keeping $m$ constant and varying $b$, a family of parallel straight lines of slope $m$ can be generated across the diet space. Since the equation of a line normal to the line given by Eq. (9.12) and having the $y$ axis intercept $a$ is

$$y = a - x/m, \tag{9.14}$$

then the equation, in the triangular co-ordinates, of the family of lines normal to the family given by Eq. (9.13) is

$$\phi = (am\sqrt{3} + 1)/(m\sqrt{3} + 1) - 2\Pi/(m\sqrt{3} + 1)\,. \tag{9.15}$$

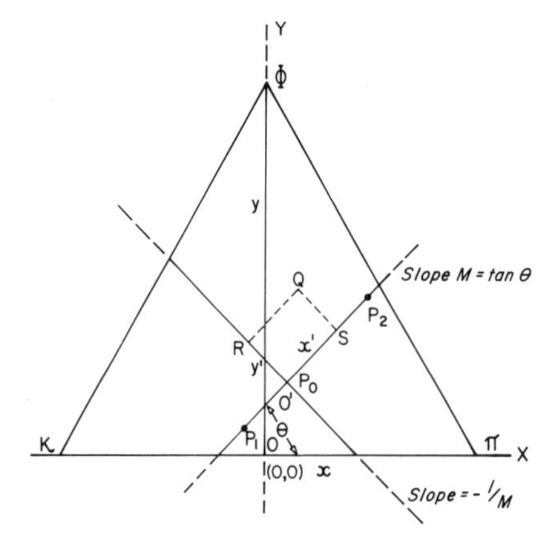


Fig. 9.2. A general set of rectangular axes in the nutrition space, $N=3$, obtained by rotating the transformed $x$ and $y$ axes and translating the origin O to diet point $P_0$

In this manner Eqs. (9.12) and (9.14) can be used to set up a new rectangular co-ordinates system of lines of constant $m$ over the diet space with the diet components satisfying Eqs. (9.13) and (9.15). The preceding mathematics are the same as choosing $P_0$ in Fig. 9.2 as a new origin of co-ordinates and using the translation and rotation theorems of analytical geometry to construct a new system of rectangular co-ordinates over the nutrition space. Any portion of the diet space can be covered with a rectangular grid of experimental diets having the property of a constant $m$ arbitrarily chosen. The parameter $m$ is the slope of the new axis of $x$ relative to the old $x$ axis, namely the $\kappa\Pi$ side of the nutrition space.

Nutritionists usually choose the new $x$ axis as a line of constant specific dietary energy (say $E_0$kcal/g). Energy $E_0$ is considered the sum of the energies contributed by the individual nutrients given by

$$E_0 = \alpha\Pi + \beta\kappa + \gamma\phi.$$

Here $\alpha$, $\beta$, and $\gamma$ are respectively the specific energies of protein, carbohydrate, and fats or oils. Using the transformation Eqs. (9.10) and (9.11) to solve $\phi$ and $\Pi$ in terms of $x$ and $y$ and $\kappa = 1 - \Pi - \phi$, the equation of the line of constant $E_0$ is

$$y = (\alpha + \beta - 2E_0)/(\alpha + \beta - 2\gamma) + [(\sqrt{3}(\alpha - \beta)/(\alpha + \beta - 2\gamma)]x\,. \tag{9.16}$$

This is the equation of a family of parallel straight lines of slope

$$m = \sqrt{3}(\alpha - \beta)/(\alpha + \beta - 2\gamma)\,, \tag{9.17}$$

and $y$ intercept

$$b = (\alpha + \beta - 2E_0)/(\alpha + \beta - 2\gamma)\,, \tag{9.18}$$

which can vary with $E_0$ with fixed values of the nutrient specific energies $\alpha$, $\beta$ and $\gamma$. Proceeding as before, the family of lines normal to this family has a slope of

$$-1/m = (\alpha + \beta - 2\gamma)/\sqrt{3}(\alpha - \beta)\,,$$

and a $y$ intercept, $a$, of arbitrary value.

These families of lines form a new rectangular coordinates system which is rotated (but not yet translated) about the origin 0 in Fig. 9.2 through an angle the tangent of which is $m$ given by Eq. (9.17) and which depends only on the nutrient specific energies. The origin O may now be translated to a new origin O′ (Fig. 9.3) by selecting a particular point $(x_0, y_0)$, corresponding to the diet composition $(\Phi_0, \Pi_0, \kappa_0)$ with specific energy

$$E_0 = \alpha\Pi_0 + \beta\kappa_0 + \gamma\phi_0,$$

thus constructing a new set of coordinate axes such that diets of constant specific dietary energy lie along the new $x$ axis and diets of varying energy lie along the new $y$ axis. Figure 9.3 shows three sets of axes of this nature, the rotations of which are due only to different sets of values of $\alpha$, $\beta$, and $\gamma$.

Nutritionists use many measures of specific dietary energy in units of kcal/g, all of which are deduced from the combustion bomb energy, $E_c$, of the diet and the performance of animals consuming it. The digestible energy, is $E_c$ less the faecal energy; the metabolisable energy $E_m$ is digestible energy less the urinary energy and

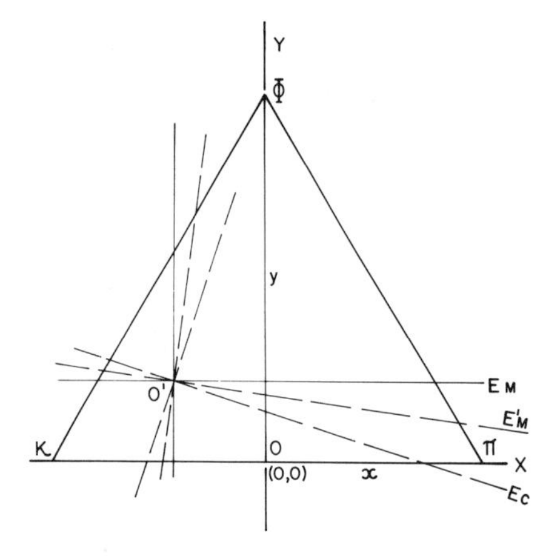


Fig. 9.3. Three sets of rectangular coordinates the $x$ axes of which may be lines of constant dietary energy depending on how the energy is defined. $E_c$ is combustion energy, $E_m$ is metabolisable energy by Atwater and Bryant (1903) and $E'_m$ is metabolisable energy by Metta and Mitchel (1954)

Table 9.1. Specific nutrient energies and slopes of various lines of constant specific dietary energy. [Adapted from Parks (1973) with permission of Journal of Theoretical Biology]

| Energy | $\alpha$ (protein) | $\beta$ (NFE) | $\gamma$ (fat) | $m$ | Angle $\theta$ |
|---|---|---|---|---|---|
| Combustion $E_c$ | | | | | |
| Brody | | | | | |
| Atwater and Bryant (1903) | 5.65 | 4.10 | 9.45 | −0.293 | −16.4 |
| Metta and Mitchell (1954) | 5.831 (casein) | 4.197 (starch) | 9.494 (lard) | −0.316 | −17.6 |
| Metabolisable $E_m$ | | | | | |
| Atwater (1899) | 4 | 4 | 9 | 0 | 0 |
| Metta and Mitchell (1954) | 4.673 ± 0.089 | 3.962 ± 0.016 | 8.770 ± 0.065 | −0.1381 ± 0.0191 | − 7.9 |
| Scott et al. (1969) | 3.65 | 3.65 | 8.95 | 0 | 0 |
| Parks[a] from data by Metta and Mitchell (1954) | 4.683 ± 0.041 | 3.967 ± 0.041 | 8.760 ± 0.100 | | |

[a] See Table 9.8

the productive energy, $PE$, is $E_m$ less the thermal energy loss, commonly called the specific dynamic action, plus energy loss in gas formation (ruminants), during the digestive process. Only two of these measures are of interest here, namely bomb energy $E_c$ and metabolisable energy $E_m$. Experiments have been performed, with necessary refinements, to determine values of $\alpha$, $\beta$, and $\gamma$ for $E_c$ and $E_m$ (Table 9.1).

Table 9.1 also shows the slopes and angles of the constant specific dietary energy families of lines relative to the $\kappa\Pi$ side of the nutrition space. The lines of constant dietary bomb energy, $E_c$, have angles of about −17° and those of constant specific metabolisable energy, $E_m$, have angles of 0° or about −8° with the $\kappa\Pi$ side of the nutrition space. Here it is seen that $E_c$ increases as the diet point moves along a line of constant $E_m$ with increasing $\Pi$, showing that the energetic efficiency of the physiological apparatus for extracting metabolisable energy from the dietary com-

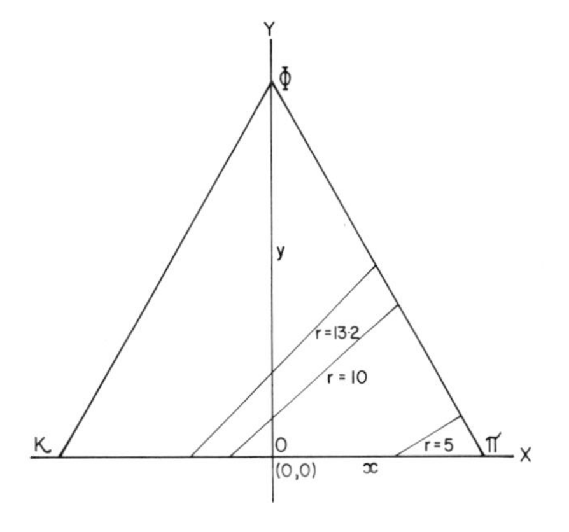


Fig. 9.4. Lines of constant energy-to-protein ration (kcal/$P$) in the nutrition space, $N=3$ showing how they are rotated and displaced as the ratio increases

bustion energy progressively decreases as $\Pi$ increases and the diet becomes richer in protein.

While it has been shown that lines of constant $E_m$ have smaller slopes than lines of constant $E_c$, the exact relation is not known. Atwater and Bryant (1903) have shown that $\alpha=4$, $\beta=4$, and $\gamma=9$ kcal/g are quite serviceable values for the metabolizable energies of these nutrients. As a consequence $\alpha=\beta=4$, the slope $m$ reduces to zero and lines of constant $E_m$ are parallel to the $\kappa\Pi$ side of the nutrition space. Scott et al. (1969) make $\alpha=\beta=3.65$, again ensuring that lines of constant $E_m$ are parallel to $\kappa\Pi$.

More recent work on determining metabolizeable energies of diet ingredients is reported by Haresign and Lewis (eds. 1980) and Chami et al. (1980). The preceding discussion of the geometrical ideas for relating diets of varying composition to each other are as useful in this newer work as they are in the study of older work.

Another type of straight line in the nutrition space, of interest to nutritionists, contains all the points in the nutrition space which have the same dietary energy to protein ratio $r=E_m/\Pi$.

Assuming $\alpha=\beta$ and using $\kappa=1-\Pi-\phi$, the ratio $r$ becomes

$$r=[\alpha+(\gamma-\alpha)\phi]/\Pi.$$

Solving Eqs. (9.10) and (9.11) as before for $\phi$ and $\Pi$ in terms of $y$ and $x$ and substituting them in the preceding relation reduces it to the following straight line

$$y=(r-2\alpha)/[r-2(\gamma-\alpha)]+\{r\sqrt{3}/[r-2(\gamma-\alpha)]\}\,x\,.$$

All the diet formulations on this line have the same energy to protein ratio. Here it is seen that for various values of $r$ the lines change slope and intercept on the $y$ axis. For large values of $r$ the slope approaches $\sqrt{3}$, and the line becomes more parallel to the $\kappa\Phi$ side of the nutrition space. For small values of $r$ the line has a lower slope, becomes shorter in length and crowds into the region of the nutrition space adjacent to the apex $\Pi$ (Fig. 9.4).

Figure 9.4 illustrates how the lines of constant ratio, $r$, of dietary $ME$ to protein, (sometimes referred to as $C/P$ ratio), change position over the diet space as $r$ is varied. The intercepts and slopes of these lines were calculated using $\alpha=\beta=4$ and $\gamma=9$ kcal/g and the values of $r$ shown in the figure. When responses to diets

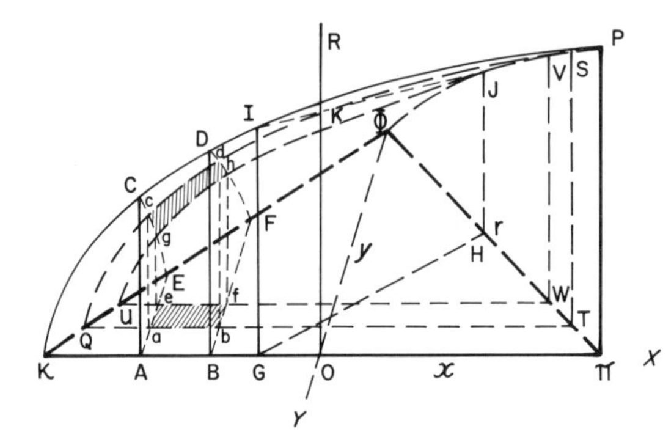


Fig. 9.5. Response surface over the nutrition space, $N=3$, showing a portion of the response surface projected into a rectangle in the nutrition space, and showing various response curves to diets along the lines in the nutrition space

are discussed, it should be noted that the lines of constant energies, $E_c$ or $E_m$, are families of parallel lines, whereas there are no families of lines of constant $r$. This is a major geometric difference between the use of dietary energy and the energy to protein ratio as design characters in the formulation of experimental diets for animals.

Suppose an experimenter has measured responses, $R$, of an animal consuming various diets of known composition $(\Pi, \phi, \kappa)$ and erects lines of length $R$ perpendicular to the plane of the nutrition space at each diet point. The result is a three dimensional figure (Fig. 9.5). A surface passing through the ends of the lines is the response surface described by Eq. (9.19) where the dependent variable $R$ is expressed as a function of the co-ordinates $x$ and $y$ of the diets as independent variables.

$$R = R(x, y). \tag{9.19}$$

The surface, $\kappa P \Phi$, shown in Fig. 9.5, starts along the $\kappa\Phi$ side of the nutrition space, then rises and curves in diminishing increments as $\Pi$ approaches unity, i.e. a pure protein diet. It is apparent that a response surface can have almost any kind of curvature depending on how the animal responds to the various diets. Figure 9.5 illustrates the relation between response curves and a response surface. Response curve $\kappa P$ relates to the responses of the animal to diets free of fat and composed of various combinations of protein and carbohydrate described by $\Pi + \kappa = 1$. A geometrical interpretation is that curve $\kappa P$ is the intersection of the response surface by a plane passing through the $\kappa\Pi$ side of and at right angles to the plane of the nutrition space. This interpretation can be generalised, i.e. a response curve is the intersection of the response surface by a plane passing through any straight line in the nutrition space and at right angles to it.

Lines QT and UW are parallel to the $\kappa\Pi$ side ($x$ axis) and define diets of constant fat levels. Planes erected through these lines and normal to the nutrition space yield response curves QS and UV respectively. Lines AE and BF are parallel to the $y$ axis. Planes passing through these lines in the nutrition space yield response curves CE and DF. Response curves QS and UV and curves CE and DF block out the shaded region *cdhg* in the response surface which corresponds to the shaded rectangle *abfe* in the nutrition space. The more limited the experimental region, i.e. the smaller the rectangle *abfe*, the more limited will be the region *cdhg* of the response surface explored. It is regrettable that much nutrition research has been concerned with responses (growth, feed efficiency, etc.) to diets in limited regions, like *abfe*, of the nutrition space. This may be practical, but is of no great use to animal scientists and biochemists who could use information about the response surface as a whole in so far as it can be established experimentally. More extensive experiments using diets on a broader rectangular grid over the nutrition space can be used to obtain the shape parameters of the response surface. I know of no such experiments published thus far.

Also shown in Fig. 9.5 is a line GH in the nutrition space, along which the diets have constant energy to protein ratio, $r$. A plane passed through line GH and normal to the nutrition space intersects the response surface to give response curve IKJ. It is apparent that there may be no unique response to diets of constant $r$, the response depends on where the diet point is along line GH. For these reasons I find it difficult to understand how the energy to protein ratio can be considered a fundamental parameter in nutrition. If there is a response surface, there is nothing gained in exploring the properties of the surface using $r$ as a dietary variable, and confusion may result from varying shapes of the response curves, such as curve IKJ in Fig. 9.5, as $r$ is varied (Donaldson et al. 1956).

Because of the inadequacies of proximate analysis of diets, there have been extensive experiments on the metabolisable energies of a variety of foodstuffs but relatively few on distributing the metabolisable energy of a diet among the nutrients composing the diet (Scott et al. 1969; Mitchell and Beadles 1930). I have encountered the feeling in some quarters that the concepts of nutrients and the animals requirements of them, need more fundamental clarification. Accepting proximate analysis, the next section will discuss metabolisable energy of diets as a plane response surface over the nutrition space with specific reference to two experiments seeking to distribute metabolisable energy among the nutrients in the diet.

## 9.4 Metabolisable Energy as a Plane Response Surface

Metta and Mitchell (1954), being in doubt about the Atwater values of 4, 4 and 9 kcal/g nutrient for $\alpha$, $\beta$, and $\gamma$ as applied to nutrition of the rat, performed an experiment feeding purified diets of casein (vitamin free), starch and lard to rats with sufficient replication to determine $\alpha$, $\beta$, and $\gamma$ for these known nutrients with a high degree of precision. The diets were three in number, quite widely different in nutrient levels as seen in Table 9.2 and by the plus marks in Fig. 9.1, thus expressing their assumption that metabolisable energy response surface is a plane, and thereby determined by only three points in the nutrition space.

The diets were fed to 12 mature rats individually caged in metabolism cages for separation of faeces and urine. There were four rats on each diet for replication. The diets were fed in three ten-day periods with the diets interchanged from period to period so that each rat was fed each diet. This design, after removing two sick rats in the third period, yielded 136 pieces of experimental data to be used in linear regression [Eq. (9.20)] to determine the specific metabolisable energies of the dietary components. The rats were daily hand fed measured amounts of each ration

Table 9.2. Composition of diets used by Metta and Mitchell

| | Diet | | |
|---|---|---|---|
| | 1 | 2 | 3 |
| Constituents | | | |
| Casein (Vit. free) | 43.0 | 10.0 | 20.0 |
| Corn starch | 10.0 | 73.0 | 30.0 |
| Lard | 38.0 | 8.0 | 41.0 |
| Wheat germ oil | 0.5 | 0.5 | 0.5 |
| Cod liver oil | 1.5 | 1.5 | 1.5 |
| Minerals[a] | 7.0 | 7.0 | 7.0 |
| Vitamins[b] | – | – | – |
| Total | 100.0 | 100.0 | 100.0 |

| | | |
|---|---|---|
| [a]Minerals | 4 | mineral mixture |
| | 1 | sodium chloride |
| | 2 | barium sulfate (roughage) |
| [b]Vitamins | 0.002 | calcium pantothenate |
| | 0.400 | choline chloride |
| | 0.00025 | pyridoxine hydrochloride |
| | 0.00025 | riboflavin |
| | 0.00025 | thiamine hydrochloride |
| | 0.001 | nicotinic acid |
| | 0.005 | P-aminobenzoic acid |
| | 0.010 | inositol |
| | 0.41875 | |

sufficient to maintain their body weights. Body weight maintenance was achieved with few refusals of food.

The diets were analysed and gross bomb energies were determined. The gross energy and dry matter contents of the nutrients were also determined (Table 9.3). From these data the gross energy and the amount of dry nutrients consumed by each rat could be calculated.

At the end of each period the gross energy and nitrogen contents of the faeces and urine of each rat were determined. The metabolisable energy extracted from the food consumed was determined by the difference between the gross energy consumed and the gross energy of faeces plus urine, corrected for nitrogen balance. Tables 9.4–9.6 and contain the results of the experiment expressed as consumption of gross energy, dry casein, $C$, dry starch, $S$, dry lard, $F$, and metabolisable energy, $E_m$, by each rat in each 10-day period.

Metta and Mitchell then fitted

$$E_m = aC + bS + cF, \tag{9.20}$$

to the 136 pieces of data by the method of least squares to obtain expected values and standard errors of the coefficients $a$, $b$, and $c$. The results of this calculation are shown in Table 9.8 and 9.1 ($\alpha = a$, $\beta = b$, and $\gamma = c$) where the slopes of lines of constant dietary energy were discussed. It is readily seen that the expected values of

Table 9.3. Analyses (per 100 g) of the diets in Table 9.2

| | Diet | | |
|---|---|---|---|
| | 1 | 2 | 3 |
| Constituents | | | |
| Gross energy (kcal) | 642.12 | 419.28 | 624.16 |
| Total nitrogen (g) | 6.1713 | 1.4831 | 2.8515 |
| Ether extract (g) | 39.44 | 9.97 | 42.72 |
| Dry matter (g) | 97.16 | 92.46 | 96.88 |

Analyses (per 100 g) of the nutrients in Table 9.2

| Nutrient | Casein | Corn starch | Fats |
|---|---|---|---|
| Dry matter | 93.12 | 90.35 | – |
| Gross energy (kcal/g dry matter) | 5.831 | 4.197 | 9.494 |

Table 9.4. Results of first 10-day experimental period. (Data from Metta and Mitchell 1954)

| Rat | Diet | Gross energy consumed (kcal) | Casein consumed (g) $C$ | Starch consumed (g) $S$ | Fat consumed (g) $F$ | Metabolisable energy extracted[a] (kcal) $E_m$ |
|---|---|---|---|---|---|---|
| 1 | 1 | 430.22 | 26.83 | 6.05 | 26.80 | 379 |
| 2 | 1 | 449.48 | 28.03 | 6.32 | 28.00 | 392 |
| 3 | 1 | 404.53 | 25.23 | 5.69 | 25.30 | 362 |
| 4 | 1 | 404.53 | 25.23 | 5.69 | 25.20 | 362 |
| 5 | 2 | 398.32 | 8.85 | 62.63 | 9.50 | 372 |
| 6 | 2 | 398.32 | 8.85 | 62.63 | 9.50 | 373 |
| 7 | 2 | 398.32 | 8.85 | 62.63 | 9.50 | 373 |
| 8 | 2 | 398.32 | 8.85 | 62.63 | 9.50 | 375 |
| 9 | 3 | 405.71 | 12.11 | 17.61 | 27.95 | 370 |
| 10 | 3 | 436.91 | 13.04 | 18.97 | 30.10 | 399 |
| 11 | 3 | 405.71 | 12.11 | 17.61 | 27.95 | 367 |
| 12 | 3 | 405.71 | 12.11 | 17.61 | 27.95 | 371 |

[a] Corrected to nitrogen balance

$a, b$, and $c$ are significantly different from the Atwater values of 4, 4, 9 kcal/g and that $a$ and $b$ are significantly different from each other signifying that for rats consuming diets composed of casein, starch, and lard, the lines of constant metabolisable energy are not parallel to the $\kappa\Pi$ side of the nutrition space but slope downward at an angle of about 8° (Table 9.1).

I am critical of the use of Eq. (9.20) as the regression equation in this problem. For one thing, it is assumed that the constant term, which normally appears in linear regression, is zero implying $C=S=F=0$ is a point in the nutrition space when in fact it is not. For another, it is assumed that there is more information in Tables 9.4–9.6 than there really is. Also Eq. (9.20) implies that a space of four dimensions is involved in this problem, whereas the previous sections have shown

Table 9.5. Results of second 10-day experimental period. (Data from Metta and Mitchell 1954)

| Rat | Diet | Gross energy consumed (kcal) | Casein consumed (g) $C$ | Starch consumed (g) $S$ | Fat consumed (g) $F$ | Metabolisable energy extracted[a] (kcal) $E_m$ |
|---|---|---|---|---|---|---|
| 1 | 2 | 398.32 | 8.85 | 62.63 | 9.50 | 379 |
| 2 | 2 | 398.32 | 8.85 | 62.63 | 9.50 | 374 |
| 3 | 2 | 398.32 | 8.85 | 62.63 | 9.50 | 378 |
| 4 | 2 | 398.32 | 8.85 | 62.63 | 9.50 | 373 |
| 5 | 3 | 405.71 | 12.11 | 17.61 | 27.95 | 371 |
| 6 | 3 | 436.91 | 13.04 | 18.97 | 30.10 | 401 |
| 7 | 3 | 405.71 | 12.11 | 17.61 | 27.95 | 373 |
| 8 | 3 | 405.71 | 12.11 | 17.61 | 27.95 | 374 |
| 9 | 1 | 403.22 | 26.83 | 6.05 | 26.80 | 384 |
| 10 | 1 | 449.48 | 28.03 | 6.32 | 28.00 | 404 |
| 11 | 1 | 417.38 | 26.03 | 5.87 | 26.00 | 378 |
| 12 | 1 | 404.32 | 25.21 | 5.69 | 25.19 | 362 |

[a] Corrected to nitrogen balance

Table 9.6. Results of third 10-day experimental period. (Data from Metta and Mitchell 1954)

| Rat | Diet | Gross energy consumed (kcal) | Casein consumed (g) $C$ | Starch consumed (g) $S$ | Fat consumed (g) $F$ | Metabolisable energy extracted[a] (kcal) $E_m$ |
|---|---|---|---|---|---|---|
| 1 | 3 | 468.12 | 13.97 | 20.32 | 32.25 | 427 |
| 2 | 3 | 468.12 | 13.97 | 20.32 | 32.25 | 432 |
| 3 | 3 | 468.12 | 13.97 | 20.32 | 32.35 | 427 |
| 4 | 3 | 468.12 | 13.97 | 20.32 | 32.25 | 431 |
| 5 | 1 | 481.59 | 30.03 | 6.77 | 30.00 | 434 |
| 7 | 1 | 441.58 | 27.54 | 6.21 | 27.51 | 397 |
| 9 | 2 | 461.21 | 10.24 | 75.52 | 11.00 | 429 |
| 10 | 2 | 461.21 | 10.24 | 72.52 | 11.00 | 426 |
| 11 | 2 | 461.21 | 10.24 | 72.25 | 11.00 | 430 |
| 12 | 2 | 461.21 | 10.24 | 72.25 | 11.00 | 429 |

[a] Corrected to nitrogen balance

that metabolisable energy for omnivores is a response surface in three dimensions, namely $X$, $Y$, and $E_m$. Hence Eq. (9.20) should be modified to

$$E'_m = a' + b'x + c'y. \qquad (9.21)$$

Here $x$ and $y$ are the Cartesian co-ordinates corresponding to the location of each of the three experimental diet points in the nutrition space using $\Pi = C/(C+S+F)$ and $\Phi = F/(C+S+F)$ and Eqs. (9.10) and (9.11) and $E'_m$ is the metabolisable energy per gram of dry nutrients consumed calculated by $E'_m = E_m/(C+S+F)$.

Using the data in Tables 9.2 and 9.3, $\Pi$ and $\Phi$ for each diet were calculated and transformed to $x$ and $y$, and calculating $E'_m$ from the data in Tables 9.4–9.6 the $(x, y, E'_m)$ data points were found (Table 9.7). Application of Eq. (9.21) to the

Table 9.7. Metabolisable energy per gram of nutrient consumed versus $x$ and $y$ co-ordinates of the diets. (Data by Metta and Mitchell 1954)

| $x$ | $y$ | $E_m$ (kcal/g) | $x$ | $y$ | $E_m$ (kcal/g) |
|---|---|---|---|---|---|
| 0.204 | 0.430 | 6.08 | −0.0623 | 0.462 | 6.12 |
| | | 6.02 | | | 6.13 |
| | | 6.18 | | | 6.07 |
| | | 6.18 | | | 6.14 |
| | | 6.16 | | | 6.16 |
| | | 6.21 | | | 6.17 |
| | | 6.25 | | | 6.18 |
| | | 6.18 | | | 6.12 |
| | | 6.22 | | | 6.19 |
| | | 6.21 | | | 6.12 |
| −0.391 | 0.107 | 4.19 | | | 6.14 |
| | | 4.16 | | | 6.18 |
| | | 4.20 | | | |
| | | 4.19 | | | |
| | | 4.21 | | | |
| | | 4.22 | | | |
| | | 4.22 | | | |
| | | 4.24 | | | |
| | | 4.29 | | | |
| | | 4.23 | | | |
| | | 4.27 | | | |
| | | 4.22 | | | |

Table 9.8. Comparisons of $\alpha$, $\beta$, $\gamma$ in $E=\alpha\pi+\beta K+\gamma\Phi$ determined by using various linear regression formulae and data of Metta and Mitchell

| $\alpha$ | $\beta$ | $\gamma$ | Remarks |
|---|---|---|---|
| $4.673 \pm 0.089$ | $3.962 \pm 0.016$ | $8.770 \pm 0.065$ | Calculated by Metta and Mitchell |
| $4.683 \pm 0.041$ | $3.967 \pm 0.041$ | $8.760 \pm 0.100$ | Using transformed coordinates |

recalculated $(x, y, E_m')$ data points using linear regression gave $a'=4.325$, $b'=0.6199 \pm 0.0714$ and $c'=4.435 \pm 0.1001$ with residual error$=0.0500$. Transforming back to $E_m'$ as a function of $\Pi$, $\kappa$ and $\Phi$, Eq. (9.21) becomes

$$E_m'=\alpha\Pi+\beta\kappa+\gamma\phi \pm 0.050,$$

where $\alpha=a'+b'\sqrt{3}/3$, $\beta=a'-b'\sqrt{3}/3$ and $\gamma=a'+c'$. The values of $\alpha$, $\beta$, and $\gamma$ and their errors shown in Table 9.8 indicate that between the two methods the expected values of the co-efficients are not very different but the standard errors are quite different, which is a reflection of the fact that Metta and Mitchell treated their data as though there was more information in their experiment than was warranted. It seems they thought there were replications of $C$, $S$, and $F$, when in fact there were none (Table 9.7). Except for the coefficient $\alpha$ their standard errors of $\beta$ and $\gamma$ were considerably smaller than those calculated by Eq. (9.21) (Table 9.8).

Carpenter and Clegg (1956) did an experiment on 17 feed grains used in poultry feeds in an effort to relate the metabolisable energy of the grains to their proximate analyses. The analyses were performed in two parts, namely part A, the usual proximate analyses, and part B proximate analyses with the NFE analysed into starch, sugar, and carbohydrate residue. Hence for part A, where $N=3$, the nutrition space is the equilaterial triangle shown in Fig. 9.1 but for part B the nutrition space is a regular hypersolid of four dimensions, because the analysis is for $N=5$ components, namely, $\Pi$ = fraction of crude protein; $\phi$ = fraction of fat; $\Sigma$ = fraction of starch; $G$ = fraction of sugar, and $D$ = carbohydrate residue.

The data for part A can be handled, in the manner used with the Metta and Mitchell data, by Eq. (9.21), but for part B the regression equation is

$$E=a+bx+cy+dz+gv. \tag{9.22}$$

Here $x, y, z$, and $v$ are the Cartesian co-ordinates of a grain analysis in the nutrition space obtainable from the general transformation Eqs. (9.2)–(9.4) with $N=5$. Application of Eq. (9.21) to the part A data yielded $a=0.3457$, $b=0.6572 \pm 4.1623$, and $c=0.62713 \pm 6.0891$, residual error $=0.4389$ and squared correlation coefficient $R^2=0.2857$, a poor fit as Carpenter and Clegg pointed out. However, fitting Eq. (9.22) gave $a=2.7921$, $b=0.0951 \pm 0.9220$, $c=6.4366 \pm 2.8790$, $d=4.0183 \pm 0.7446$, $g=-1.8311 \pm 1.3203$, residual error $=0.1499$ and $R^2=0.9412$, indicating a good fit by accounting for 94% of the total variance in the data. Using the transformation equations and $\Pi+\Phi+\Sigma+G+D=1$, Eq. (9.22) was recast to

$$E_m=\alpha\Pi+\gamma\phi+\beta\Sigma+\delta G+\varepsilon D,$$

where, with usual propogation of error, $\alpha=2.887 \pm 0.922$, $\gamma=7.777 \pm 2.230$, $\beta=3.334 \pm 0.900$, $\delta=-1.130 \pm 1.142$, and $\varepsilon=1.186 \pm 1.142$. This indicates that sugar and carbohydrates appear to add little if anything to the metabolisable energy for chickens, since $\delta$ and $\varepsilon$ are not significantly different from zero. It is interesting to note as Carpenter and Clegg did, that starch is the most potent fraction of the NFE and that the metabolisable energy per gram of crude protein, $\alpha=2.887$, of these feed grains is considerably less than Atwater's value of four and Metta and Mitchell's value of 4.683 (Table 9.1 and 9.8). It is also interesting to note that the ratio $\alpha/\beta=2.33$ is about 4% larger than the 2.25 factor usually used to ratio fat against carbohydrate in metabolisable energy calculations. However it should be remembered that in these regressions the resulting equations should not be extrapolated beyond the region of the diet space in which the experimental diets were located.

Figure 9.1 shows the regions of the two dimensional nutrition space in which the experiments of Metta and Mitchell and Carpenter and Clegg were performed. The region occupied by the diet points of Carpenter and Clegg is very small and crowded into the corner of the diet space near the pure NFE point $\kappa=1$ and separating NFE into starch, sugar, and residue did not improve the disposition of the points in the four dimensional nutrition space. This is a very poor situation for determining a response surface. The disposition of the three points, +, used by Metta and Mitchell is much better, but it is apparent that they might have included the

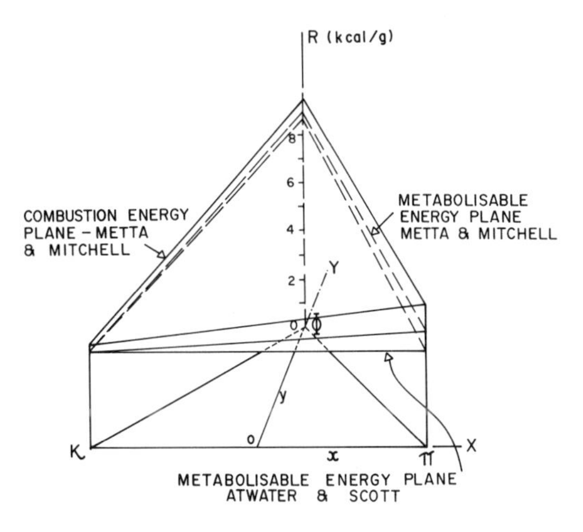


Fig. 9.6. Combustion and metabolisable energies as plane response surfaces over the nutrition space, $N=3$

four additional points, x, I suggest in Fig. 9.1 to cover a greater portion of the nutrition space. Diets with this kind of disposition cover a larger portion of the diet space offer the possibility of testing the response surface against a plane. Figure 9.6 illustrates how the combustion energy and metabolisable energy lines in Fig. 9.3 are the lines of intersection of the energy planes in Fig. 9.6 and a plane parallel to the nutrition space projected onto the nutrition space as contour lines. The implication of Fig. 9.6 is that the use of the Atwater values of 4, 4, 9 kcal/g of nutrient are not too far in error and are useful until more and better experiments are made.

## 9.5 Dietary Energy and Growth Parameters

Dietary energy is an energetic input, via food consumed, to an animal which continuously divides the building blocks of the food between growth and no growth.

We saw in Chap. 9.4 how metabolisable energy is defined and measured as the difference between combustion energies of food, fecal matter and urine and how the difference may be distributed among the nutrients in the food. Because the words "specific" and "density" refer respectively to reciprocal units of mass and volume, I shall define the amount of nutrients in one gram of food as the specific nutrient content, $\eta$, of the food, and the amount of metabolisable energy in one gram of nutrients as the specific metabolisable energy, $\varepsilon$, of the nutrients. Therefore the specific metabolisable energy $E_m$ of a food is the product $\varepsilon\eta$. If the bulk density (gm/ml) of the food is denoted by $\delta$, then the metabolisable energy density (kcal/ml) of the food is $\varepsilon\eta\delta$. Here it is seen how the various measures of the metabolisable energy of a diet can be changed by varying one or more of these factors.

If the bulk density, $\delta$, is held constant, the metabolisable energy density can be changed by changes in the specific metabolisable energy of the food ($\varepsilon\eta$) by varying $\varepsilon$ or $\eta$ or both. Changes of $\varepsilon$ arise from changes of the nutrient composition which shift the diet point around the nutrition space. Such shifts of the diet point confound the dietary energy effect on the response, $R$, of the animal, with the effects of varying the nutrients relative to each other. Keeping $\varepsilon$ constant and varying the specific nutrient content ($\eta$) varies $R$ at a particular point in the nutrition space. Changes of $\eta$ have been effected by adding nondigestible diluents to the ration, but some of the diluents also decrease the bulk density, $\delta$, of the ration, thus introducing a "gut fill" problem in analysing responses to changes in $\eta$. Holding $\varepsilon\eta$ constant, the bulk density ($\delta$) of diets can be increased by compacting the food into cakes, pellets, extruded rods, and so forth, thereby increasing the metabolisable energy density of the diet. Also the food can be ground and fluffed up to decrease $\delta$ and the metabolisable energy density of the diet.

Many experiments have been performed to uncover the effects on the animal's response of changes in one or more of the dietary energy factors, $\varepsilon$, $\eta$, and $\delta$, but there has been no effort to systematically explore how these factors affect the response surface. Practically all these researches are short term feeding trials and have expressed the responses in terms unsuitable for me to relate the growth parameters $A$, $(AB)$, $T_0$, and $t^*$ to the ways the rations have been manipulated.

Some information about the growth parameters can be obtained from ad libitum short term experiments provided they are designed with the ad libitum feeding and growth equations in mind. A short term experiment reflects the properties of a long term experiment, because a short segment of the trace of a growing animal reflects the properties of the trace. This can be seen by expanding

$$W-W_0=(A-W_0)\{1-\exp[-(AB)F/A]\},$$

and

$$F=C\{t-(1-D/C)t^*[1-\exp(-t/t^*)]\},$$

as infinite series using McLaurin's theorem, when $W_0$ and $D$ are very small compared to $A$ and $C$ respectively, as they are for day old chicks, or using Taylor's theorem when the animal has attained a significant fraction of its degree of maturity. However, in expanding these functions we can make use of the following formula,

$$1-\exp(-x)=x-x^2/2!+x^3/3!-x^4/4!\ldots,$$

with the general term $(-1)^{i+1}x^i/i!$, $i=1, 2, \ldots$, so that

$$1-\exp(-x)=\sum_{i=1}^{\infty}(-1)^{i+1}x^i/i!\,.$$

Using $a=1-W_0/A$ and $d=1-D/C$ and the preceding expansion with the proper $x$, expansions of $W$ in terms of $F$ and $F$ in terms of $t$ are respectively

$$W-W_0=a(AB)F\sum_{i=1}^{\infty}(-1)^{i+1}\,[(AB)F/A]^{i-1}/i!\,,$$

$$F=C(1-d)t-Cdt\sum_{i=2}^{\infty}(-1)^{i+1}(t/t^*)^{i-1}/i!\,.$$

These series show that $W$ and $F$ can be represented as polynomials of low degrees in $F$ and $t$ respectively if the time the experiment spans is such that $t/t^*$ is, say $<0.1$, and $(AB)F/A<0.1$. Since the terms of the series are further reduced by the factorials, taking $t/t^*=0.1$ and $(AB)F/A=0.1$ we need only the terms from $i=1$ to 4, because for the terms with $i>4$, $(t/t^*)^{i-1}/i!$ rapidly becomes much less than 0.001, therefore we get the polynomials

$$W-W_0=a(AB)F-[a(AB)^2/2A]F^2+[a(AB)^3/6A^2]F^3-[a(AB)^4/24A^3]F^4,$$

and

$$F=Dt+(dC/2t^*)t^2-(dC/6t^{*2})t^3+(dC/24t^{*3})t^4.$$

Here we see the influence of the growth and feeding parameters on these polynomials.

Dividing $W-W_0$ by $F$ to get gross growth efficiency and $F$ by $t$ to get gross food intake as measures of response, we have

$$(W-W_0)/F=a(AB)-[a(AB)^2/2A]F+[a(AB)^3/6A^2]F^2-[a(AB)^4/24A^3]F^3,$$

and

$$F/t=D+(dC/2t^*)t-(dC/6t^{*2})t^2+(dC/24t^{*3})t^3.$$

Thus what these responses measure depends on the age and weight of the animals when the experiment is begun as well as on its duration. For example, if the experimental treatments begin at 1 day or 1 week of age and have a duration of 1 or 2 weeks, then for chicks $W_0$ and $D$ are small compared to $A$ and $C$, $a$ and $d$ approximate unity, $(W-W_0)/F$ estimates $(AB)$ and $2F/t^2$ estimates the appetency factor $C/t^*$. But when the experiment is begun with older chicks or with weanling rats and has a duration of say 1 week, then $W_0$ and $D$ are not small compared to $A$ and $C$, $a$ and $d$ are less than unity, $(W-W_0)/F$ estimates $a(AB)$ and $F/t$ estimates $D$. The polynomials show that longer durations of experimental feeding progressively worsen the estimates of $a(AB)$ and $D$.

Generally animal scientists have been aware of the limitations of short term experiments but have been unable to do much about these limitations. They had no concept of the trace of animals and had no estimates of the growth parameters on which to base the design of their experiments and decide where along the trace an experiment should best begin and end. Furthermore their measures of response were not $W/F$, $F/t$, or $2F/t^2$, but feed efficiency $(F/W)$, growth rate $(W/t)$, food consumed per 100 g body weight etc., and seldom were the initial weight or the estimated initial food intake reported. Except for more recent experiments reported in the literature it has been difficult for me to estimate the effects of dietary energy on the growth parameters. The effects of dietary treatments on the growth parameters are best determined from long term experiments, but before designing such experiments it is useful to have prior estimates of the effects of the treatments from short term experiments. As we have seen in the $W/F$ and $F/t$ polynomials, fair estimates may be obtained of $(AB)$, $D$ or $C/t^*$ as the constant terms but not of $A$ and $t^*$. Before discussing a long term experiment of the effects on the growth parameters of changing the specific nutrient content, $\eta$, of a diet, I shall now discuss two short term experiments involving dietary energy effects on responses.

Sibbald et al. (1957) allotted two male and two female weanling rats to each of the diets in which substitutions of a non-nutrient (Alphacel, a purified cellulose) or protein or both were made gram for gram with sucrose. The object of the experi-

Table 9.9. Rations fed during a two week period. (Data from Sibbald et al. 1957)

| Diet | Alphacel (%) | Protein (%) | Sucrose (%) | Gross energy (kcal/g) | Gross nitrogen (mg/g) |
|---|---|---|---|---|---|
| 1a | 10 | 10.2 | 69.8 | 4.20 | 14.24 |
| 1b | – | 15.2 | 64.8 | 4.25 | 20.31 |
| 1c | – | 20.3 | 59.7 | 4.32 | 26.64 |
| 1d | – | 25.4 | 54.6 | 4.39 | 33.12 |
| Mean | | | | 4.29 | 23.61 |
| 2a | 20 | 10.2 | 59.8 | 4.12 | 13.63 |
| 2b | – | 15.2 | 54.8 | 4.20 | 21.22 |
| 2c | – | 20.3 | 49.7 | 4.29 | 27.16 |
| 2d | – | 25.4 | 44.6 | 4.35 | 34.00 |
| Mean | | | | 4.24 | 24.00 |
| 3a | 30 | 10.2 | 49.8 | 4.18 | 14.90 |
| 3b | – | 15.2 | 44.8 | 4.35 | 20.92 |
| 3c | – | 20.3 | 39.7 | 4.28 | 28.11 |
| 3d | – | 25.4 | 34.6 | 4.58 | 34.28 |
| Mean | | | | 4.35 | 24.57 |
| 4a | 40 | 10.2 | 39.8 | 4.26 | 14.18 |
| 4b | – | 15.2 | 34.8 | 4.26 | 20.72 |
| 4c | – | 20.3 | 29.7 | 4.33 | 26.80 |
| 4d | – | 25.4 | 24.6 | 4.48 | 33.49 |
| Mean | | | | 4.32 | 23.80 |

Each ration contained 5% Mazola oil, 4% salts and 1% vitamin mix. Alphacel is non-nutritive cellulose, Nutritional Biochemicals Corpn. Protein composed of 58.7 g casein, 68.8 lactalbumin, 0.5 g DL-methionine, 1.5 g L-histidine. HCL and 1.0 g DL-threonine

ment was fourfold, but I shall discuss here only the relations of digestible energy and nitrogen intake to food consumption. The diets were characterised by four levels of non-nutrient and four levels of protein within each level of non-nutrient (Table 9.9). The protein source was composed of casein and lactalbumin with some synthetic amino acids added at low levels. The diets were fed for two weeks with the first week as an acclimatisation period. During the second week the weight of the rats and the food consumed by them were measured, and faeces and urine were collected and analysed for energy and nitrogen. At the end of the experiment the animals were analysed for protein, fat and water, and the diets were analysed for energy and nitrogen.

Table 9.9 shows the gross energies and nitrogen levels of the diets and Table 9.10 shows the triangular coordinates of the data points in the nutrition space and the values of $\varepsilon$ calculated using Atwater's 4, 4, 9 factors, and $\beta=\varepsilon\eta$ where $\eta=1-(\%\ \text{Alphacel}+4\%\ \text{salts}+1\%\ \text{vitamins})/100$. Figure 9.7 shows the spread of the diet points in the nutrition space as a quadrilateral ABCD which is narrow in the $\Phi$ co-ordinate but long in the direction of increasing protein fraction. The variation of dietary energy is a combination of increases in nutrient energy and in-

Table 9.10. Triangular co-ordinates and specific metabolisable energies $\varepsilon$ of the nutrients and dietary energies $\beta$ of the diets in Table 9.9

| Diet | $\phi$ | $\Pi$ | $\kappa$ | $\varepsilon$ [a] | $\beta$ [b] |
|---|---|---|---|---|---|
| 1a | 0.059 | 0.120 | 0.821 | 4.29 | 3.65 |
| 1b | | 0.179 | 0.762 | | |
| 1c | | 0.239 | 0.702 | | |
| 1d | | 0.299 | 0.642 | | |
| 2a | 0.067 | 0.136 | 0.797 | 4.33 | 3.24 |
| 2b | | 0.203 | 0.730 | | |
| 2c | | 0.271 | 0.662 | | |
| 2d | | 0.339 | 0.594 | | |
| 3a | 0.077 | 0.157 | 0.766 | 4.38 | 2.85 |
| 3b | | 0.234 | 0.689 | | |
| 3c | | 0.312 | 0.611 | | |
| 3d | | 0.391 | 0.532 | | |
| 4a | 0.091 | 0.185 | 0.724 | 4.45 | 2.45 |
| 4b | | 0.276 | 0.633 | | |
| 4c | | 0.369 | 0.540 | | |
| 4d | | 0.462 | 0.447 | | |

[a] Calculated using Atwater values 4, 4, 9 kcal/g nutrient
[b] Calculated as $\beta=\varepsilon\eta$, where $\eta=1-$(% Alphacel+4% salts+1% vitamin mix)/100

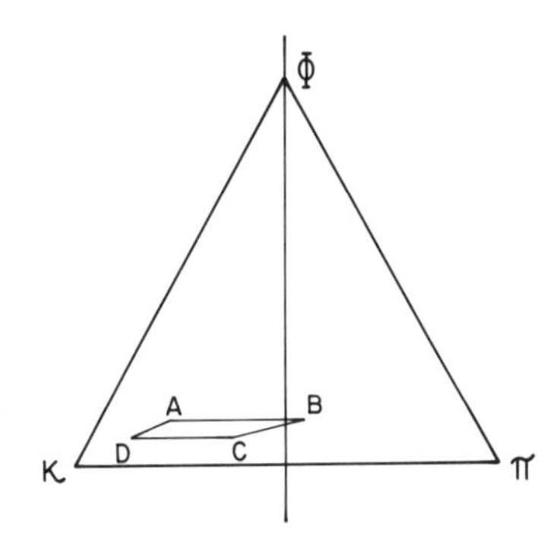


Fig. 9.7. The region ABCD of the nutrition space, $N=3$, covered by the diets of Sibbald et al. (1957)

creases in the non-nutrient fraction which overshadows the change in nutrient energy (Table 9.10). Here is seen about a 4% increase in $\varepsilon$ accompanied by a 35% increase in $\beta$.

The results of the experiment, with regard to my objectives, are shown in Table 9.11. Since Sibbald et al. felt it necessary to remove body weight effects from the food consumed, they expressed the food and the gross nitrogen consumed on the basis of 100 g body weight. The body weights are the arithmetical mean of the weights at the start and end of the experiment. The apparent digestible energy (ADE) was determined by the difference between gross energy intake and the total faecal energy. The apparent digestible nitrogen (ADN) was determined similarly.

Results of calculating correlations between food consumed per 100 g body weight, designated here by $F'$, and the other factors, gross nitrogen (N), ADE and

Table 9.11. Mean values of experimental responses to diets in Table 9.9. (Data by Sibbald et al. 1957)

| Diet | $W$, g[a] | $F'$ g[a] | Gross $N$ Consumption/100 g body wt, g | ADE of food, kcal/g | ADN of food, Mg/g |
|---|---|---|---|---|---|
| 1a | 76 | 84 | 1.205 | 3.74 | 12.99 |
| 1b | 83 | 84 | 1.723 | 3.80 | 18.88 |
| 1c | 87 | 85 | 2.256 | 3.84 | 24.83 |
| 1d | 85 | 88 | 2.914 | 3.91 | 30.56 |
| Mean | 82.8 | 85.3 | | 3.82 | |
| 2a | 79 | 100 | 1.362 | 3.27 | 11.92 |
| 2b | 86 | 93 | 1.968 | 3.36 | 19.11 |
| 2c | 96 | 98 | 2.657 | 3.42 | 24.00 |
| 2d | 87 | 92 | 3.124 | 3.54 | 31.46 |
| Mean | 87.0 | 95.8 | | 3.40 | |
| 3a | 81 | 106 | 1.579 | 3.00 | 12.86 |
| 3b | 86 | 110 | 2.282 | 3.18 | 18.64 |
| 3c | 82 | 104 | 2.911 | 3.17 | 25.42 |
| 3d | 85 | 105 | 3.576 | 3.40 | 30.66 |
| Mean | 83.5 | 106.3 | | 3.19 | |
| 4a | 78 | 120 | 1.702 | 2.79 | 11.89 |
| 4b | 79 | 116 | 2.413 | 2.75 | 17.77 |
| 4c | 77 | 124 | 3.342 | 2.82 | 22.59 |
| 4d | 83 | 112 | 3.761 | 2.83 | 29.50 |
| Mean | 79.3 | 118 | | 2.80 | |

[a] $W=(W_1+W_0)/2$ and $F'$ = grams of food consumed/100 g body weight

ADN, (Table 9.11) led the authors to the conclusion that the rats ate proportionately more of the lower energy foods, i.e., weanling rats eat to satisfy their energy requirements but not to satisfy protein needs. These conclusions can be supported by a quick search of Table 9.11 for apparent relations between $F'$, gross N, ADE, and ADN of the food. The response $F'$ within sets of diets appears to have no drift attributable to the comparatively large systematic increases in gross nitrogen or apparent digestible nitrogen of the food or the low systematic increases in the apparent digestibility of the energy of the food. But there is an inverse relation between the mean $F'$ and mean ADE down the sets of diets as seen by the 38% increase of $F'$ with a 27% decrease of ADE and the 33% decrease of $\beta$ in Table 9.10. Using $Y=KX$ [where $Y$ = food consumed per 100 g body weight, $X$ = reciprocal of ADE (kcal per 100 g) and K = constant] as a regression equation, Sibbald et al. found $K=33{,}188 \pm 2{,}484$ from all the individual rat data. They recognised the fact that K as a constant should be more accurately determined by $K=F'$ (ADE) and found the mean value of $K=33{,}064$, not significantly different from the previous value.

Their response $F'$ = grams of food consumed per 100 g body weight is actually food consumed per 100 g body weight per week, because the duration of the experiment was one week, so $F'$ estimates $D'=D/100$ g body weight. They have shown

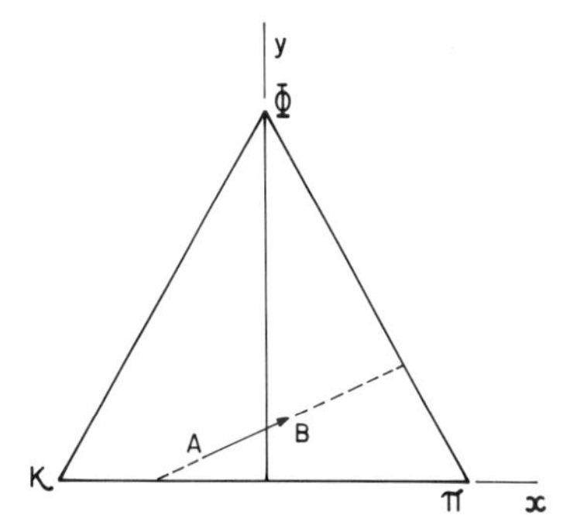


Fig. 9.8. Line AB in the nutrition space, $N=3$, along which Carew and Hill (1964) isocalorically exchanged carbohydrates for fat

indirectly that the initial food intake, $D$, per 100 g body weight is inversely proportional to the apparent digestible energy, ADE, or the specific metabolizable energy, $\beta$, of the diet, but is uninfluenced by changes in dietary protein level. It is too great an extrapolation to suggest that these findings apply to $C$ and $t^*$. However, in some long term experiments on rats, reported later in this section and in the next, the findings of Sibbald et al. from the feeding of weanling rats will be shown to apply to the mature feed intake $C$ and to the whole course of feeding from immaturity to maturity. Unfortunately Sibbald et al. did not report the initial weights $W_0$ and weight gained $\Delta W$ of the rats on the various treatments so that $a(AB)$ could be estimated to show its relation to dietary energy and protein. In this experiment $\beta$ was reduced from 3.65 kcal by adding a non-nutrient to the diet, whereas in the experiment discussed next, $\beta$ was increased from 3.36 kcal/g by adding corn oil (Carew and Hill 1964).

The diets used by Carew and Hill were a low fat basal (Table 9.12) and the basal plus 20% corn oil added isocalorically as a substitute for the glucose in the basal, using metabolisable energies of 3.64 and 8.8 kcal/g for glucose and corn oil respectively. Figure 9.8 illustrates how this manoeuvre shifted the diet point, A, of the basal to point B of the basal plus 20% corn oil. All the points along line AB satisfy

Table 9.12. Composition (%) of the basal diet used by Carew and Hill (1964)

| Ingredients | % | % |
|---|---|---|
| Glucose | | 51.8 |
| Constant meal | | |
| Soybean meal (50% protein) | 16.0 | |
| Crude casein | 11.0 | |
| Gelatin | 2.5 | |
| Fish meal, menhaden | 4.0 | |
| Brewer's yeast, dried | 2.5 | |
| Whey, dried | 2.0 | |
| Fish solubles, condensed | 2.0 | |
| Corn oil | 2.5 | |
| Chromium oxide mix (30% $Cr_2O_3$) | 1.0 | |
| Mineral and vitamin mixtures | 4.7 | 48.2 |
| | | 100.0 |

Table 9.13. Results from the Carew and Hill experiment number 1 using ad libitum food intakes only

| | $\beta$ (kcal/g) | $\Delta F/2$ (g/wk) | $\Delta W$ (g) | $\Delta W/\Delta F$ |
|---|---|---|---|---|
| Basal | 3.36 | 223 | 246 | 0.553 |
| Basal + 20% corn oil | 4.25 | 170 | 231 | 0.679 |

the equations $3.64\ \kappa + 8.8\ \phi = \text{constant}$, which expresses the isocaloric replacement of carbohydrate with fat. The energies, $\beta$, of the two rations are respectively 3.36 and 4.25 kcal/g for the basal and basal plus 20% corn oil.

The diets were fed for two weeks to chicks of two weeks of age. At two weeks of age the chicks were weighed and selected to give a constant average value of $W_0$ between replicates and treatments. Of their data only the two-week ad libitum food intakes, and weight gains on the basal, and basal plus 20% corn oil diets are of interest here (Table 9.13).

The data from Carew and Hill makes it possible to view $F/2$ as an estimate of initial food intake $D$ and $\Delta W/\Delta F$ as an estimate of $a(AB)$, which is the growth efficiency factor, $(AB)$, biased by $a = 1 - W_0/A$. Although these experiments used only two diets, namely, the basal and the basal plus 20% corn oil, the data in Table 9.13 can give some ideas of the effect of changing $\beta$, keeping $\eta$ constant, on $D$ and $a(AB)$. If $D$ behaves as Sibbald et al. found on rats it may be expected that $D\beta$ is the same for both diets. Since $D\beta$ for the basal and the basal plus 20% corn oil differ only by 3.3%, we may say the expectation is fulfilled. The possible effect on $a(AB)$ appears to be a proportional increase with increased $\beta$, since $a(AB)/\beta$ is 0.165 for the basal and 0.159 for the basal plus 20% corn oil. If a is unaffected by $\beta$, it appears $(AB)$ may be proportionally related to dietary energy.

This short term experiment by Carew and Hill was the first I found which suggested a possible effect of dietary energy on one of the feeding and growth parameters, namely $(AB)$. Here it became apparent to me that long term experiments would be needed to explicitly find how changes in nutrition would change the course of the trace through the effects of nutrition on the parameters.

These short term experiments on chicks and weanling rats show that the effects on $D$ and $a(AB)$ of changing dietary energy $\beta = \varepsilon\eta$ may be indifferent to changes in $\beta$ by changing $\varepsilon$ or $\eta$ or both.

I met the need of studying the effect of dietary energy on the course of the trace by designing a 20 week ad libitum feeding and growth experiment on four groups of individually caged highly inbred Sprague-Dawley male albino rats each (Parks 1970a). The dietary energy, $\beta$, was changed between the groups by varying $\eta$ only on adding various amounts of a stable finely divided indigestable powder which did not change the bulk density, $\delta$, of the ration.

The basal ration was a laboratory rat chow produced as a meal by Allied Mills Inc. the composition of which is shown in Table 9.14 with $\beta = 3.37$ kcal/g and $\varepsilon = 4.18$ kcal/g of nutrients. The basal was designated as 100% relative specific energy, RSE, from which three diets with RSE's of 91.42, 83.04, and 74.56% were produced by blending the basal with precipitated silicic acid. Each of these diets were

Table 9.14. Proximate analysis, triangular co-ordinates ($\Pi$, $\phi$, $\kappa$) and the specific metabolisable energy of Wayne Lab-Blox meal

| Constituent | % |
|---|---|
| Crude protein | 24.65 |
| Fat | 2.80 |
| N.F.E. | 53.26 |
| Crude fibre | 2.70 |
| Moisture | 10.15 |
| Ash | 6.34 |
| Specific metabolisable energy ($\varepsilon\eta$) | 3.37 kcal/g[a] |
| $\Pi$ | 0.305 |
| $\phi$ | 0.035 |
| $\kappa$ | 0.660 |
| $\varepsilon$ | 4.18 kcal/g nutrients[a] |

[a] Calculated using the Atwater values of 4, 9, 4

Table 9.15. Effect of dietary energy on the growth parameters of male Sprague Dawley albino rats (Dan Rolfsmeyer Company Strain)

| RSE % | $A$ (kg) | $(AB)$ | $C$ (kg/wks) | $t^*$ (wk) |
|---|---|---|---|---|
| 100 | 0.431 | 0.390 | 0.175 | 2.45 |
| 91.52 | 0.434 | 0.370 | 0.194 | 2.58 |
| 83.04 | 0.391 | 0.330 | 0.208 | 2.17 |
| 74.56 | 0.383 | 0.294 | 0.236 | 2.86 |

ad libitum fed to each of the four groups of rats. It should be emphasised that the position of the 100% RSE ration in the nutrition space was unchanged by the dilutions so that effects, if any, on the growth and feeding parameters would be due to changes of $\beta$ only.

Data of individual weekly food consumption and body weight from weaning were collected. The population means of weight and weekly food intake are in Table A-11.

The mean data of each group of rats were analysed as Class I data by the method discussed in Chap. 4 for values of $A$, $(AB)$, $C$, and $t^*$ (Table 9.15). Here it is seen that $(AB)$ and $C$ monotonically decrease and increase respectively as the RSE of the diets decreases. There is no particular evidence of a systematic effect on $A$ and $t^*$. This is not to say that $A$ and $t^*$ would remain unaffected if the RSE were reduced to values lower than 74%. The trends of $(AB)$ and $C$ might also be affected at lower RSE than in this experiment.

The short term experiments of Sibbald et al. and Carew and Hill together implied that $D$ is inversely proportional to $\beta$, and the experimental of Carew and Hill also implied that $(AB)$ is directly proportional to $\beta$. If the implications of these

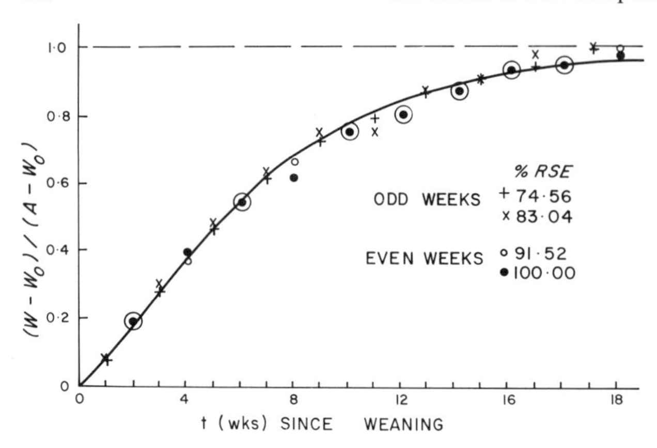


Fig. 9.9. The equal growth in time of rats fed a diet diluted by up to about 75% of its original energy content

short term experiments are extended to the results of this long term experiment we can expect

$$C \times \mathrm{RSE} = \text{constant}$$

and

$$(AB)/\mathrm{RSE} = \text{constant}.$$

From high to low RSE in Table 9.15 the products of C × RSE are 17.5, 17.8, 17.6, and 17.6 with a mean value of 17.6, and the ratios $(AB)$/RSE are 0.00390, 0.00404, 0.00397, and 0.00394 with a mean value of 0.00396. The very near constancy of these sets of numbers implies that, within a range of zero to 25.44% dilution of food,

$$C = 0.175 \times 3.37/\beta = 0.590/\beta \text{ kg/wk},$$

and

$$(AB) = 0.390\,\beta/3.37 = 0.116\,\beta.$$

The ad libitum growth and feeding equations (without the initial values of body weight, $W_0$ and food intake, $D$) become

$$W = A[1 - \exp(-0.116\,\beta F/A] \text{ and} \tag{9.23}$$

$$dF/dt = (0.590/\beta)[1 - \exp(-t/t^*)]. \tag{9.24}$$

The cumulative food consumed [integral of DE (9.24)] is

$$F = (0.590/\beta)\{t - t^*[1 - \exp(-t/t^*)]\}, \tag{9.25}$$

which when substituted in Eq. (9.23) shows that the course of growth in time is independent of the dietary energy [Eq. (9.26)],

$$W = A\langle 1 - \exp[-(0.0688/A)\{t - t^*[1 - \exp(-t/t^*)]\}]\rangle. \tag{9.26}$$

Figure 9.9 shows that the rats grew equally in time by virtue of adjusting their food consumption according to Eq. (9.25) so that their energy consumptions were the same. The solid curve is computed by Eq. (9.26). Figure 9.10 shows the upward

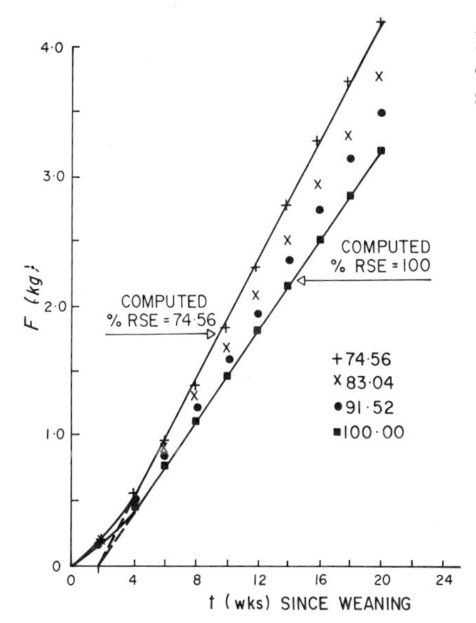


Fig. 9.10. The spread, inversely to the energy content of diets, of the cumulative food consumed by rats versus time

spread of cumulative food consumed as the RSE of the diets were decreased. Equation (9.25) shows that if the data of cumulative food consumed were multiplied by the appropriate $\beta$ all the resulting data values would lie on a single curve.

In Chap. 5 I introduced the concept of the $Z$ function defined as

$$Z(t) = (AB)F(t)/A.$$

This function was used to normalise the cumulative food consumed in discussing the biotraces of the animals. Since $(AB)$ is proportional and $F$ is inversely proportional to dietary energy, $\beta$, it can be concluded that the $Z$ function is independent of dietary energy, at least within the limits of the range of $\beta$ used in this experiment. This finding suggests that within presently unknown limits the biotraces of animals are independent of dietary energy.

Short term experiments on other animals have led nutritionists to conclude that "animals eat for energy" and occasionally adding "but not for protein" (on pigs, Owen and Ridgman 1967; on rats, Harte et al. 1948; chicks, Carew et al. 1964). The experiment discussed here shows that, under laboratory conditions, rats do in fact eat for energy throughout their postnatal growth from weaning to at least early adulthood. The role dietary protein plays in growth is the subject of discussion of the next section.

The phrase "animals eat for energy" is usually applied to homeotherms, but Rozin and Mayer (1961) have shown its applicability to goldfish and a report from the Research Branch of the Canadian Department of Agriculture (1967) observed that grasshoppers increase their food intake as the nutrient content of their food decreases. These observations show that poikilotherms may also have this property in common with homeotherms. However none of these statements can be considered more than an isolated observation until long term experiments on poikilotherms have been used to establish Eqs. (9.23)–(9.26) for these families of animals.

## 9.6 Dietary Protein and the Growth Parameters

Before designing a long term experiment on rats to study the effects of varying the levels of dietary protein on growth and feeding, I searched the literature for short term and long term data which might be analysed for effects of dietary protein level on one or more of the growth parameters. I found some short term data on rats and chickens, and one long term experiment on chickens which clearly show the effects of the level of dietary protein in the diet on the growth efficiency parameter, $(AB)$. It has been shown that $\Delta W/\Delta F$ in short term experiments, estimates $a(AB)$ where $a=1-W_0/A$. In the following discussions of short term experiments I shall consider $A$ insensitive to variation of dietary protein level and source so that $a$ is a constant of proportionality which can be considered absorbed in the estimates of $(AB)$ from data.

The short term experiments by Hegsted and Chang (1965 a, b) and Hegsted and Worcester (1968) were suggestive of the variation which might be expected of $(AB)$ when the dietary protein is varied in level and source. Their experiments were 3 weeks feedings of weanling rats so $(AB)$ could be estimated from $\Delta W/\Delta F$. The protein sources were lactalbumin, casein, soya, and wheat gluten. The diet consisted of 8% fat, 4% salt mixture, 2% cod liver oil, 1% vitamin mix and proteins plus cornstarch to make up the balance, thus $P+C=85$, where $P=\%$ protein and $C=\%$ cornstarch. Adjusting $P$ and $C$ in this fashion gives isocaloric diets of 4.1 kcal/g using the Atwater values of 4, 4, 9.

The data in Table 1 of the Hegsted and Chang paper were used to estimate $(AB)$ for each level, $P$, and source of protein (Table 9.16). Figure 9.11 shows the graphs of this data and it is seen that for each protein source the disposition of the data points suggests that the equation of diminishing increments as a function of $P$ may apply, [Eq. (9.27)].

$$(AB)=a\{1-\exp[-b(P-p_0)]\}. \tag{9.27}$$

Here $P$ is the percent dietary protein, $a$ is the maximum value of $(AB)$ attainable by increasing the level of protein; $b=\ln 2/P_{1/2}$, where $P_{1/2}=P-p_0$ is the amount of protein for which $(AB)=a/2$, and $p_0$ is the displacement of the curve along the percent protein axis where $P=p_0$ yields $(AB)=0$. The growth efficiency factor becomes negative when $P$ is less than $p_0$.

Equation (9.27) was used as the nonlinear least squares regression equation for determining the parameters $a, b$, and $p_0$ from all the $(AB)$ versus $P$ data discussed in this section. In the following discussion of other short and long term experiments showing the effect of $P$ on $(AB)$ the experiments will be briefly presented and the results illustrated graphically. The value of the parameters, the rsd of fit of Eq. (9.27) and the metabolisable energies, $\beta$, of each experiment are presented together in Table 9.22 for convenience in the general discussion of the effects of varying dietary protein level and type on $(AB)$ for rats and chickens.

Table 9.22 displays the values of the parameters, $a$, b, and $p_0$ and the rsd of fit for each source of protein. The rsds indicate that the equation of diminishing increments satisfactorily describes the data. The calculated curves in Fig. 9.11 are solid and illustrate the fit of Eq. (9.27) to the data.

Table 9.16. Growth efficiency, grain/feed, for rats grown for 21 days after weaning, as related to percent dietary protein. (Calculated from data in Table 1 of Hegsted and Chang 1965)

| Protein source | $P$ | $a(AB) \simeq \Delta W/\Delta F$ |
|---|---|---|
| Lactalbumin | 2.21 | −0.058 |
| | 3.69 | 0.110 |
| | 5.16 | 0.203 |
| | 7.38 | 0.307 |
| | 9.59 | 0.349 |
| | 11.80 | 0.429 |
| | 14.75 | 0.407 |
| Casein | 3.47 | −0.031 |
| | 6.08 | 0.109 |
| | 8.69 | 0.244 |
| | 12.16 | 0.361 |
| | 15.64 | 0.469 |
| | 19.11 | 0.517 |
| | 26.06 | 0.571 |
| Soya | 8.51 | 0.112 |
| | 11.92 | 0.180 |
| | 15.32 | 0.236 |
| | 18.73 | 0.264 |
| | 25.54 | 0.338 |
| | 34.05 | 0.359 |
| | 42.56 | 0.379 |
| Wheat gluten | 11.38 | 0.025 |
| | 14.63 | 0.055 |
| | 17.88 | 0.103 |
| | 21.33 | 0.156 |
| | 24.38 | 0.161 |
| | 32.50 | 0.223 |
| | 40.63 | 0.278 |

Velus et al. (1971) reported a short term experiment (8 to 21 days of age) on chicks using rations in which the source of nitrogen was a crystalline amino acid mixture neutralised with bicarbonate of soda and the source of carbohydrate was cornstarch (Table 9.17). The amino acid mix was adjusted up from zero replacing cornstarch on a gram for gram basis in 100 g of diet. In the following discussion the amino acid mix will be denoted by AA.

Analysis of the AA showed that the percent dietary protein, $P$, is the product of 0.7541 and the percent of AA in the ration. The data gathered were weight gain, food consumed, body protein, and energy stored over the 13 day period for nine diet compositions with $P$ ranging from zero to 23.3% protein. From these data the growth efficiency factor, $(AB)$; the protein retention efficiency, PRE; and the thermochemical efficiency of growth, TCE, were calculated respectively from the ratios of weight gain and food consumed, the ratios of 100 times the protein retained and the protein consumed, and the ratios of body energy stored and the metabolizable energy consumed. The values of these measures of the nutritive effectiveness of the diets as functions of $P$ are listed in Table 9.18 and plotted in Fig. 9.12.

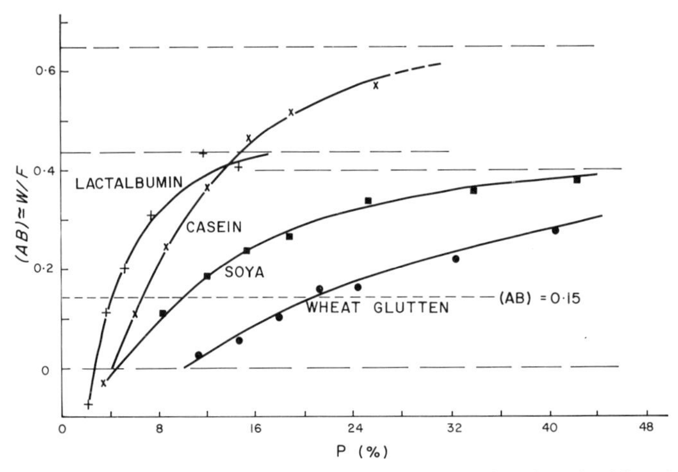


Fig. 9.11. Graphs of $(AB)$ versus percent dietary protein, where $(AB)$ is estimated (Table 9.15) for several different proteins, from the data of Hegsted and Chang (1965). All data for each protein suggests that the law of diminishing returns applies. The *dashed horizontal line*, $(AB)=0.15$, shows the level of each protein source from lactalbumin to wheat gluten which yield the same value of the growth efficiency

Table 9.17. Composition of basal diet and crystalline amino acid mixture used by Velu et al. (1971)

| Basal diet | % | Amino acid mix | g |
|---|---|---|---|
| Cornstarch [a] | 75.43 | L-Arginine.HCL | 1.21 |
| Amino acid mix (AA) [a] | – | L-Histidine.HCL.$H_2O$ | 0.41 |
| Corn oil | 15.00 | L-Lysine.HCL | 1.19 |
| Salt mix | 5.37 | L-Tyrosine | 0.45 |
| Cellulose | 3.00 | L-Trutophan | 0.15 |
| Sodium bicarbonate | 1.00 | L-Phenyalanine | 0.50 |
| Choline chloride | 0.20 | DL-Methionine | 0.35 |
| Vitamins etc. | + | L-Cystine | 0.35 |
| | 100.00 | L-Threonine | 0.65 |
| | | L-Leucine | 1.20 |
| | | L-Isoleucine | 0.60 |
| | | L-Valine | 0.82 |
| | | Glycine | 1.20 |
| | | L-Proline | 0.20 |
| | | L-Glutamic acid | 10.00 |
| | | | 19.28 [b] |

[a] AA + cornstach = 75.43
[b] The diet containing 19.28 amino acid mix and 56.15 cornstarch was calculated to have $\beta = 4.1$ kcal/g and 14.54% protein ($P = 0.7541 \times \mathrm{AA}\%$)

The figure shows that the growth efficiency factor $(AB)$ continues to rise with diminishing increments, with increase of $P$ after PRE and TCE have levelled out and begun to drop. Velus et al. explain this from the way the carcase fat to carcase protein ratio varies as the dietary protein level is increased. Whatever the reason, the disposition of the $(AB)$ data points in Fig. 9.12 indicates that Eq. (9.27) may ap-

Table 9.18. The variations of the growth efficiency factor (*AB*), the protein retention efficiency (*PRE*) and thermochemical efficiency (*TCE*) of chicks as functions of percent dietary protein *P*. (Calculated from data published by Velu et al. 1971)

| *P* | (*AB*) | *PRE* | *TCE* |
|---|---|---|---|
| 0 | −0.184 | $-\infty$ | 0.077 |
| 2.70 | 0.080 | −0.053 | 0.239 |
| 5.82 | 0.309 | 0.515 | 0.336 |
| 8.72 | 0.443 | 0.633 | 0.409 |
| 11.6 | 0.521 | 0.662 | 0.413 |
| 14.5 | 0.643 | 0.688 | 0.460 |
| 17.4 | 0.712 | 0.669 | 0.454 |
| 20.4 | 0.725 | 0.613 | 0.412 |
| 23.3 | 0.763 | 0.577 | 0.414 |

ply. The values of $a$, $b$, and $P_0$ found for this case are shown in Table 9.22. The residual standard deviation, rsd, and the calculated curve, shown solid in Fig. 9.12, show that the equation of diminishing increments adequately describes the data of (*AB*) versus *P*.

Hammond et al. (1938) reported a long term experiment with male chicks in which *P*, in isocaloric diets, was varied from 13% to 25%, in 2%-steps by proportionate mixing of a high and a low protein diet of equal dietary energy. The diets were composed of natural foodstuffs and fed ad libitum. Proximate analyses of each ingredient were used to establish the proportion required to give the desired *P*. The birds were reared from hatching to 52 weeks of age.

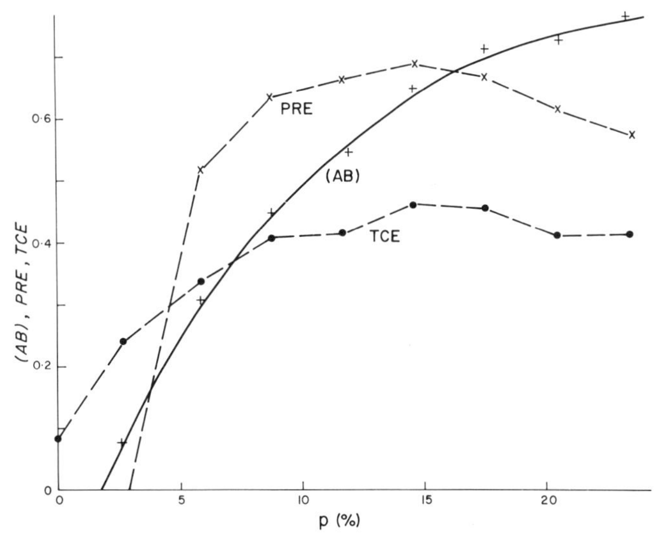


Fig. 9.12. (*AB*), the protein retention efficiency (*PRE*), and the thermochemical efficiency (*TCE*) versus percent dietary protein equivalent to amino acid mix (*AA*) level in the diet. (*AB*, *PRE*, and *TCE* estimated from the data by Velu et al. 1971.) The data for (*AB*) suggests the law of diminishing returns

Table 9.19. Effect of level of dietary protein on growth parameters $A$ and $(AB)$ for male chickens. (Data Hammond et al. 1938)

| $P$ | $A$, kg | $(AB)$ |
|---|---|---|
| 25 | 3.27 | 0.41 |
| 23 | 3.30 | 0.39 |
| 21 | 3.27 | 0.38 |
| 19 | 3.20 | 0.38 |
| 17 | 3.20 | 0.36 |
| 15 | 3.26 | 0.34 |
| 13 | 3.50 | 0.24 |

Since these authors presented their data for live weight and cumulative food consumed by age, I recalculated the feeding and growth functions by the nonlinear regression methods discussed in Chap. 4. Table 9.19 shows the estimates of $A$ and $(AB)$ versus $P$. Here $(AB)$ increases with increasing $P$, but $A$ appears to be independent of $P$. Figure 9.13 of $(AB)$ versus $P$ is similar to the curves in Figs. 9.11 and 9.12 and Eq. (9.27) was expected to apply. The values of the parameters $a, b$, and $p_0$ and the rsd of fit are shown in Table 9.22. Here it is important to see that, in a long term experiment on chickens, the law of diminishing increments applies to $(AB)$ as a function of $P$ and that the estimation of $(AB)$ from short term data does give information about $(AB)$ in the long term.

Study of the ad libitum feeding data of Hammond et al. showed that $C, D$, and $t^*$ as well as $A$ are independent of the dietary protein level within the limits of variation of $P$ in this experiment. This experiment was the model of my long term experiment on rats briefly described next (Parks 1970b).

Four groups of 25 male Sprague-Dawley albino rats were fed for 14 weeks from weaning on four isocaloric diets of different protein levels. The diet ingredients selected were lactalbumin, dextrose, corn oil, and silicic acid with Lab-Blox rat "chow" as the basal diet to which the other ingredients were added to form the experimental diets. The ingredients were purchased in single lots, analysed and mixed according to my specifications for four diets of different levels of $P$. An

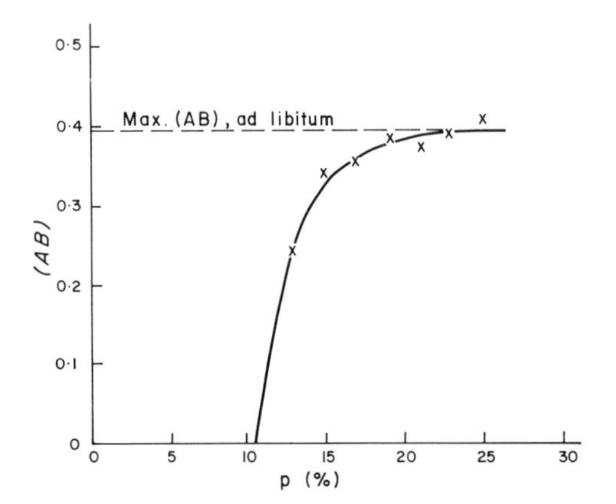


Fig. 9.13. Variation of efficiency factor, $(AB)$, versus percent dietary protein. (Data by Hammond et al. 1938)

Table 9.20. Proximate analyses (%), metabolisable energies and triangular co-ordinates of experimental diets

| | Diet | | | |
|---|---|---|---|---|
| | 1 | 2 | 3 | 4 |
| Protein, $P$ | 36.15 | 25.20 | 19.63 | 14.05 |
| NFE, $C$ | 41.01 | 51.44 | 57.01 | 63.40 |
| Fat | 3.63 | 4.18 | 3.80 | 3.88 |
| Ash + fibre | 9.28 | 9.30 | 9.18 | 8.87 |
| Moisture | 9.93 | 9.88 | 10.38 | 9.80 |
| $M.E.$, kcal/g | 3.41 | 3.44 | 3.41 | 3.45 |
| $\Pi$ | 44.75 | 31.18 | 24.40 | 17.28 |
| $\kappa$ | 50.76 | 63.67 | 70.88 | 77.95 |
| $\phi$ | 4.49 | 5.15 | 4.72 | 4.77 |

Analyses by the Analytical Laboratories of the Merchant's Exchange of St. Louis, Missouri

Table 9.21. Growth parameters obtained by nonlinear regression of weekly data of live weight and food consumed by Younger Laboratory strain of male Sprague-Dawley albino rats

| $P$, % Protein<br>Parameter | 36.15 | 2.520 | 19.63 | 14.05 |
|---|---|---|---|---|
| $A$, (kg) | 0.431 | 0.450 | 0.445 | 0.437 |
| $(AB)$ | 0.491 | 0.478 | 0.462 | 0.390 |
| $C$, (kg/wk) | 0.153 | 0.153 | 0.156 | 0.154 |
| $D$, (kg/wk) | 0.057 | 0.054 | 0.066 | 0.059 |
| $t^*$, (wk) | 1.6 | 1.5 | 1.59 | 1.5 |

Values are means ± SD

antioxidant was added to prevent the corn oil from becoming rancid. Proximate analyses of the finished diets showed that the metabolisable energies were about 3.43 kcal/g with fat levels at about 3.85%. The percent protein, $P$, was, percent for percent, changed with the percent carbohydrate, $C$, so that $P+C \simeq 76.97\%$ (Table 9.20). The proximate analyses and the triangular coordinates of the diets numbered one to four are in Table 9.20. Diet two is 100% Lab-Blox meal.

Data collection was as in the experiment varying dietary energy only, and the data are listed in Appendix Tables A-12. Nonlinear regression analysis of the data was by the methods of Chap. 4 for Class I data. The values of the growth parameters are shown in Table 9.21. Here it can be seen that parameters $A, C, D$, and $t^*$ appear independent of $P$, but $(AB)$ systematically varies with $P$.

The crosses (×) in Fig. 9.14 are the experimental values of $(AB)$ plotted versus $P$. Equation (9.27) was expected to fit these data and the values of $a, b$, and $P_0$ and the rsd of fit are shown in Table 9.22. Admittedly, fitting a three parameter function to four data points is poor practice but Fig. 9.14 shows that the equation of diminishing increments also adequately describes the data points but with much

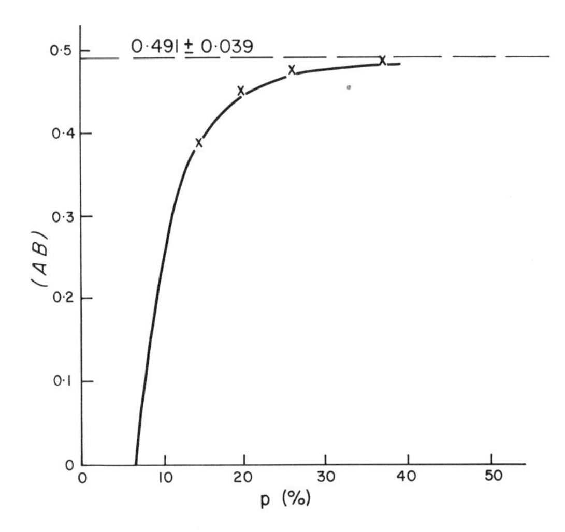


Fig. 9.14. Graph of $(AB)$ from long-term feeding and growth of rats, versus percent dietary level of protein $P$, showing again that the law of diminishing returns may be applicable

Table 9.22. Results of fitting the law of diminishing returns to growth efficiency factor $(AB)$ (isocaloric diets, rats and chicks)

| Protein | $a$ | $b\,(1/P)$ | $p_0$ | rsd[a] | $\beta$[b] |
|---|---|---|---|---|---|
| | | Rats | | | |
| Lactalbumin | 0.437 | 0.260 | 2.7 | 0.020 | 4.3 |
| Casein | 0.648 | 0.103 | 4.1 | 0.014 | 4.3 |
| Soy | 0.398 | 0.082 | 4.5 | 0.008 | 4.3 |
| Wheat gluten | 0.400 | 0.038 | 10.0 | 0.012 | 4.3 |
| Lab-Blox and lactal | 0.491 | 0.200 | 6.0 | 0.037 | 3.44 |
| | | Chickens | | | |
| Corn and casein | 0.395 | 0.420 | 10.7 | 0.011 | 3.49 |
| Amino acid mix | 0.856 | 0.106 | 1.86 | 0.018 | 4.1 |

[a] rsd is residual standard deviation of fit of Eq. (9.27)
[b] $\beta$ = metabolisable energy of diets (kcal/g)

less confidence, with only one degree of freedom for error. However the data points do not contradict Figs. 9.11–9.13.

Figure 9.15 shows how decreasing the dietary protein level $P$ causes retardation in the growth of the rats. Hammond et al. showed the same effect on long term growth of chickens. My experiment on rats and the experiment of Hammond et al. on chickens confirm the deductions from the short term experiments of Hegsted and Chang and of Velus et al. Figure 9.16 is a graph of weight versus time since weaning of some rats fed isocaloric diets of different dietary protein levels (Mayer and Vitale 1967). This figure and Fig. 9.15 show clearly the retardation or delay of the rats in getting to some specified weights when fed diets of lower protein levels and show the asymptotic rise of the growth curve as a whole toward a curve beyond which growth cannot be improved by adjusting $P$ above about 25%. Figure 9.17 shows that cumulative food consumed by the rats versus time was unaffected by dietary protein level, all four groups were alike in energy consumption because $C$ and $t^*$ are independent of dietary protein level at least within the range of protein

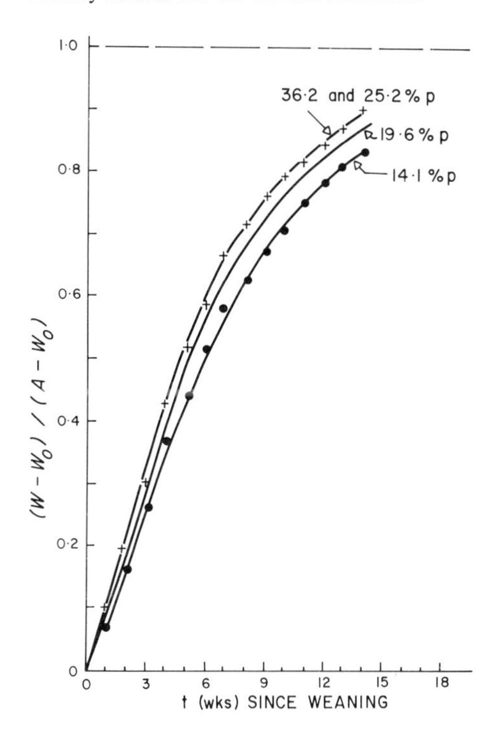


Fig. 9.15. Graphs of the fraction of total gain of rats versus time and percent dietary protein level $P$, showing how low protein diets delay growth and how the growth curve for a 36% protein diet is approached asymptotically as $P$ is increased

levels tested. The long term experiments on rats and chickens give support to the statement that animals eat for energy but not for protein.

The question now arises as to what happens when the animals eat the same amount of energy but are delayed in growth. Does the extra energy leave the animal via increased maintenance or is it retained as energy stored making the animal fatter? The effect is probably some combination of both maintenance and fattening. The experiment of Velus et al. gave some data on carcase analysis of the 21 day old chicks which show that carcase fat, as percent of fresh carcase weight, increased from 10.9% to 21.3% as $P$ was reduced from 23.3% to 8.72%. It is not known

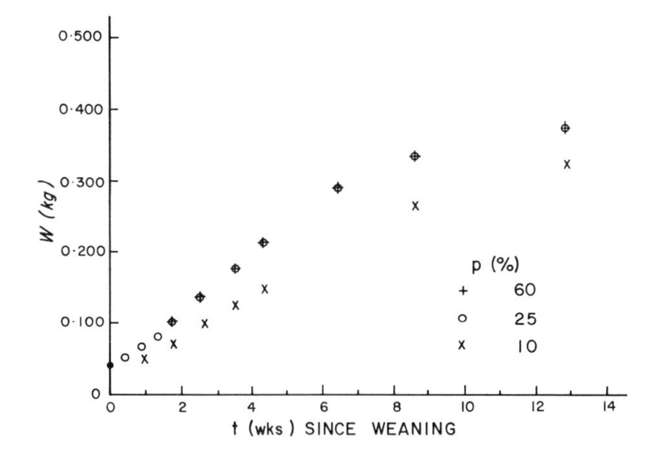


Fig. 9.16. Growth curves of rats showing the same properties of delay of growth by lower levels of dietary protein. (Data by Mayer and Vitale 1957)

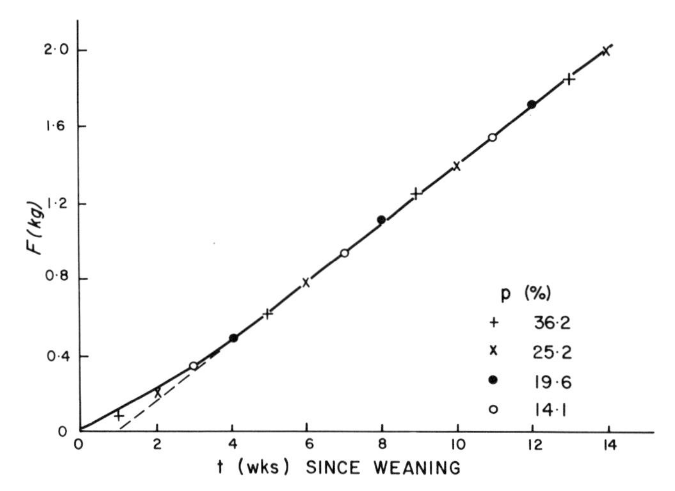


Fig. 9.17. Cumulative food consumed by rats versus time and percent dietary protein level $P$, showing that in the long-term these rats eat for energy

whether this fattening effect persists in the long term. It is regrettable that the long term experiments by Hammond et al. on chicks and by me on rats are incomplete in not providing experimental estimates of carcase composition of the animals as they grew on the various diets. The question is re-examined in Chap. 11.

Recalling the discussion of the equation $dW/dt=(dW/dF)/(dF/dt)$ in Chap. 1.5, we have here experimental evidence of a treatment, namely varying dietary energy, which leaves growth, $dW/dt$, invariant because of the direct and inverse effects which this treatment has on the growth efficiency, $dW/dF$, and food intake, $dF/dt$, respectively. Here is an example of a treatment which may be misjudged if $dW/dt$ only is regarded as the response. There may be other treatments that have this kind of effect on the growth process.

Also there is here experimental evidence that the experimental effects of varying dietary protein level on growth is primarily through the effect of the treatment on the growth efficiency only; the food intake being independent of this type of treatment. These remarks should give pause to nutritionists who design dietary treatment experiments using the growth of the animal as the primary response. I shall consider this matter in more detail later.

The experiments I have thus far discussed show that of the feeding and growth parameters, only the growth efficiency factor, $(AB)$, and the mature food intake, $C$, are generating response surfaces over the nutrition space. Equation (9.28) is $(AB)$ as a function of $\beta$ and $P$, and Eq. (9.29) expresses $C$ as a simple inverse function of $\beta$.

$$(AB)=a'\beta\{1-\exp[-b(P-p_0)]\} \tag{9.28}$$

$$C=c'/\beta. \tag{9.29}$$

We already know how to get $\beta$ and the triangular coordinates, $\Pi$, $\phi$, and $\kappa$ in terms of $x$ and $y$ for various diets so that Figs. 9.18 and 9.19 respectively illustrate the variation of $(AB)$ and $C$ over the nutrition space.

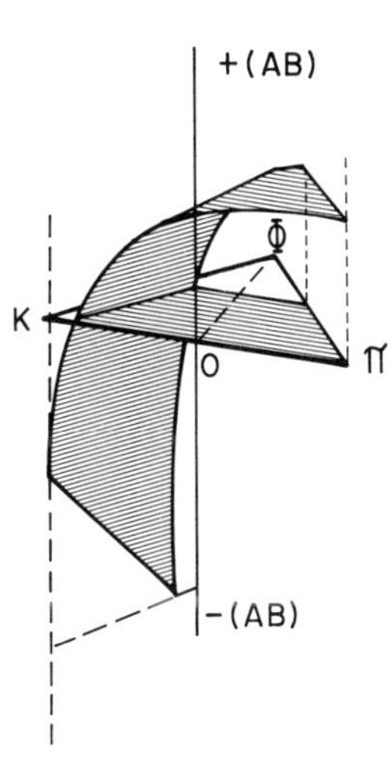


Fig. 9.18. The $(AB)$ response surface over a portion of the nutrition space, $N=3$, showing the basic property of the law of diminishing returns in the direction of increasing $\Pi$

The surface in Fig. 9.18 rises from some value less than zero at the $\kappa\phi$ side of the nutrition space in diminishing increments towards a maximum at the $\Pi$ side, and rises at a lower constant rate in the direction of increasing $\phi$. The surface is composed of an infinity of parallel response curves which are the equations of diminishing increments when $\phi$ is held constant at any value between 0 and 1 and $\Pi$ varies from 0 to 1. When $\Pi<\Pi_0$, corresponding to $P<p_0$, the factor $\{1-\exp[-b(P-p_0)]\}$ becomes negative, thus making $(AB)<0$ and the animal will lose weight as it eats. When $P=p_0$, $(AB)=0$, meaning the animal cannot grow on the food it consumes, and therefore $p_0$ is the percent protein of a maintenance diet.

Because $C$ is independent of $P$, the response surface in Fig. 9.19 is an infinity of parallel straight lines which are horizontal to the nutrition space when $\phi$ is held constant. The surface curves downward from a high of $C'/\beta_0$ at $\phi=0$, to a low of $C'/\beta_1$ at $\phi=1$, because $\beta$ is directly related to $\phi$. These surfaces are hypothetical since Eqs. (9.28) and (9.29) were established in two different experiments, one with only $\eta$ varying and the other with only $P$ varying. I know of no long term experiments in which $\beta$ and $P$ are varied together in an experimental design in which the response surfaces of the growth parameters are more accurately revealed over more extensive regions of the nutrition space.

## 9.7 Growth Promoting Ability of Proteins

Since $(AB)$ appears to be a stable quantity over a wide range of species and the effect of dietary protein level is represented by the equation of diminishing returns [Eq. (9.27)], it becomes feasible to define and compare the value of proteins for animal growth which is dependent on the proteins only and independent of the level of protein in the diet. In the context of the results of the experiments discussed in the preceding section, the reasoning leading to a definition of the value of a protein for growth is as follows.

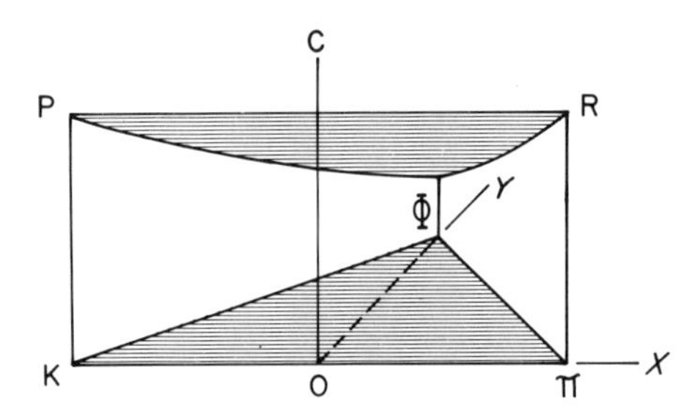


Fig. 9.19. Response surface of mature food intake, $C$, as a function of location of the diet points in the nutrition space, $N=3$, showing the downcurving of the surface due to the inverse relation of $C$ to $\phi$

The amount of food $F$, in units of mature live weight $A$, required by an animal to increase its weight from $W$ to $(W+A)/2$ is

$$F/A=\ln 2/(AB),$$

and from Eq. (9.27), the percent, $P$, of a particular dietary protein required to make $(AB)=a/2$ is

$$P=\ln 2/b+P_0.$$

Therefore, the weight of the protein per unit of mature live weight which the animal must consume to increase its weight from $W$ to $(W+A)/2$ is

$$PF/100\,A=2\ln 2(\ln 2/b+p_0)/100\,a, \tag{9.30}$$

a function only of the parameters of the equations of diminishing increments applicable to the effect of the particular dietary protein level on $(AB)$. Equation 9.30 shows that the lower the protein consumption required to halve $(W+A)$, the better the protein is for growth, and that better proteins will have larger values of $a$ and $b$, the smaller values of $p_0$. Because $F$ is inversely proportional to dietary energy, $\beta$, $PF/100\,A$ must be corrected to some chosen dietary energy, $\beta_r$. The amino acid mix in Table 9.22 is seen to be the best growth promoter in the list so its dietary energy was chosen and the corrected values of $(PF/100\,\mathrm{A})\beta/\beta_r$ are shown in Table 9.23. Here it seen that the amino acid diet requires the least and wheat gluten the greatest, consumption to halve $(W+A)$. In the following discussion it will be assumed that Eq. (9.30) has been multiplied by $\beta/\beta_r$ as a correction factor.

Table 9.23. The dietary energy corrected weights of the proteins per unit of mature weight in Table 9.22 which an animal must consume to increase its weight from $W$ to $(W+A)/2$. The reference dietary energy, $\beta_r$, is 4.1 kcal/g of diet

| | $(PF/100\,A)\,\beta/\beta_r$ |
|---|---|
| Amino acid mix (Illinois) | 0.136 |
| Lactalbumin | 0.178 |
| Lab-Blox plus Lactal | 0.224 |
| Casein | 0.234 |
| Corn plus casein | 0.369 |
| Soya | 0.473 |
| Wheat gluten | 1.02 |

Table 9.24. Absolute and relative growth promoting abilities, GPA (x) and RGPA (x) respectively, of the proteins listed in Table 9.23

| Protein x | GPA (x) | RGPA (x) |
|---|---|---|
| Amino acid mix (Illinois)[a] | 7.33 | 1.000 |
| Lactalbumin | 5.62 | 0.764 |
| Lab-Blox plus lactalbumin | 4.46 | 0.608 |
| Casein | 4.12 | 0.561 |
| Corn plus casein | 2.71 | 0.370 |
| Soya | 2.11 | 0.288 |
| Wheat gluten | 0.98 | 0.133 |

[a] Reference protein r

This measure of required protein consumption provides a means of defining growth ability, GPA(x), of a protein named x as the reciprocal of $(PF/100\,A)$, namely

$$\mathrm{GPA(x)} = 100\,a/2\ln 2(\ln 2/b + P_0). \tag{9.31}$$

This is reasonable because the larger the GPA(x) the better the protein is for growth. If there were some protein named r generally accepted as a standard, the relative growth promoting ability, RGPA(x), of protein x, may be defined as the ratio

$$\mathrm{RGPA(x)} = \mathrm{GPA(x)}/\mathrm{GPA(r)}, \tag{9.32}$$

where r denotes the name of the reference protein. The value of GPA(x) and RGPA(x), using the amino acid mix as the reference, are collected in Table 9.24, where it is seen that the RGPA's of lactalbumin, casein, soy, and wheat gluten are in the same order and have the same relative values that Hegsted and Chang assigned them. However, the RGPA's of these proteins were calculated from all the data points (Table 9.16) whereas Hegsted and Chang used only the initial portions of the data which appeared to lie along straight lines. Most methods of evaluating proteins depend on data lying along straight line responses for animals in negative nitrogen balance (Thomas 1909, later modified by Mitchell 1924, and with Carman 1926, Osborne et al. 1919).

The RGPA's are not closely related to the amino acid compositions of the proteins, because the GPA's reflect the properties of the protein in the diet and are influenced by the digestibilities of the proteins in combination with other components of the diet. The various amino acid mixes, representing proteins, are intriguing experimental tools, the digestibilities of which could be 100% for all such "proteins." Furthermore, the amino acid vectors of the natural proteins are bunched together in the amino acid hyperspace leaving a large portion of the hypervolume empty of protein vectors (Chap. 8). This empty region could be used to study proteins which may not occur naturally, but which could be fabricated from crystalline amino acids. The use of RGPA(x) in such experiments with growing animals could be quite useful in conjunction with the hypervector of x relative to the hypervector of r because digestibility would pose no problems.

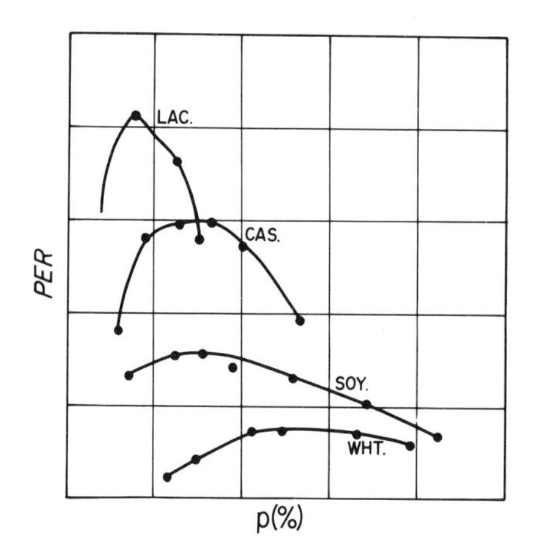


Fig. 9.20. Protein efficiency ratio, *PER*, versus percent dietary protein, showing the curves passing through maxima characteristic of *PER* as a measure of the growth promotion of protein. (Data by Hegstead and Chang 1965a, b)

The growth efficiency factor ($AB$) follows the law of diminishing returns as a function of the dietary protein level, at least for the proteins I have studied, and this property leads to some interesting relations to other methods of evaluating proteins. For example, in an attempt to express the biological value of proteins for growing animals, Osborne et al. (1919) suggested weight gain per gram of protein consumed, $\Delta W/P\Delta F$, as a "proper" measure of the efficacy of a protein for growth. They found this ratio (later called the protein efficiency ratio, $\mathrm{PER} = 100\,\Delta W/P\Delta F$) passes through a maximum as the level of the dietary protein was increased in isocaloric diets. Figure 9.20 shows this property of the four proteins studied by Hegsted et al. See Fig. 9.11 for the way ($AB$) varies with $P$. They proposed ranking proteins for growth promotion by the maximum values of the PER, a proposal later stressed by Mitchell. Later this proposal was relaxed with the acceptance of the values of the PER, at dietary protein level, $P$, of 10%, as the rank indicator. Time and expense of determining the maxima of PER for various proteins were arguments for standardising at the 10% level.

Why the PER passes through a maximum with subsequent decrease was poorly understood as a biological phenomenon. Block and Mitchell (1946) proposed that "The upward and then downward trend of the PER as the level of dietary protein is progressively increased seems to be merely a resultant of two opposing tendencies: first, a tendency for the ratio to increase as the proportion of dietary protein used for growth increases with increasing intake; and, second, a tendency for metabolic wastage of protein to increase at higher levels of intake, shows clearly in nitrogen metabolism studies."

There is a nonbiological explanation for the maximum of the PER which is related to the equation of diminishing increments as expressed by $\Delta W/\Delta F = (AB)$ varying with the dietary protein level, $P$, namely Eq. (9.27) (Figs. 9.11–9.14).

By definition

$$\mathrm{PER} = 100\,\Delta W/P\Delta F,$$

where $P$ is the percent of dietary protein, but $\Delta W/\Delta F = (AB)$ so that

$$\mathrm{PER} = 100\,(AB)/P.$$

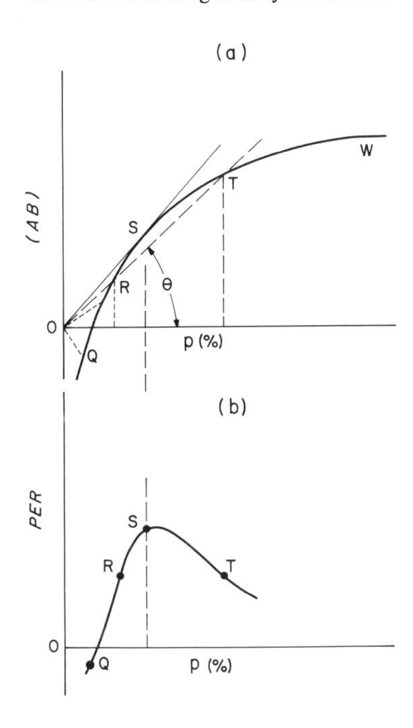


Fig. 9.21 (a, b). The relation of *PER* to (*AB*) as a function of dietary protein. (a) Geometric construction showing how *PER* is related to the tangent angle $\theta$. (b) The curve of *PER* calculated from the law of diminishing return for (*AB*) as point $\phi$ moves along it. The maximum point S corresponds to the slope of the line OS tangent to (*AB*). Their figure shows the maximum in *PER* is arithmetic rather than biological

Curve QSW in Fig. 9.21 a is the graph of (*AB*) versus *P*. As *P* is increased point Q moves along the curve passing sequentially through points R, S, and T. Simultaneously the line connecting Q to the origin O sweeps out angle $\theta$ from negative values through zero to a positive maximum where the line becomes tangent to the curve at point S and diminishing as Q moves towards T and beyond. Since $(AB)/P$ is $\tan\theta$, then

$$\mathrm{PER} = 100\tan\theta = 100\,a\{1-\exp[-b(P-P_0)]\}/P \tag{9.33}$$

and the PER must rise to a maximum and subsequently diminish (Fig. 9.21 b) as *P* increases. Points R and T have the same PER at two different values of *P* because line OR extended intersects curve QSW at T so that R and T have a common value of $\theta$. Here we see that the "biological explanation" of the behaviour of PER is really an artifact of the calculation of PER.

The growth efficiency factor (*AB*) is also related to two other biological measures of the effects of dietary protein level on the growth of immature animals. These measures are the protein retention efficiency, PRE = body protein gain/protein consumed, and the thermochemical efficiency, TCE = body energy stored/energy consumed. Since the protein fraction of the body weight is nearly a constant denoted by $\gamma$, the PRE is given by

$$\mathrm{PRE} = \gamma \Delta W/P\Delta F,$$
$$= \gamma(AB)/P = \gamma\tan\theta,$$

and it is seen that the PRE could have the property of PER, namely that of passing through a maximum value as *P* is increased. Figure 9.12 shows this property in the PRE calculated from the data of Velus et al. in experiments with chicks fed diets with graded dietary levels of an amino acid mix.

The body energy gain in short term experiments can be estimated as being the product of a weight average body energy per unit body weight

$$\gamma = (\gamma_0 W_0 + \gamma_1 W_1)/(W_0 + W_1)$$

and the gain $\Delta W$ so that TCE is estimated by

$$\begin{aligned} \text{TCE} &= \lambda \Delta W / \beta \Delta F, \\ &= (\lambda/\beta) \Delta W / \Delta F = (\lambda/\beta)(AB), \end{aligned}$$

so that TCE versus $P$ should share the equation of diminishing increments with $(AB)$ as a function of $P$. Again Fig. 9.12 shows evidence of this sharing. From these remarks it can be concluded that most, if not all, of the conventional measures of efficacy of dietary proteins in promoting growth contain the properties of $(AB)$ as a function of the dietary level $P$ of protein x. Since the GPA(x) is dependent only on $a, b$, and $p_0$, which are independent of dietary protein level within the ranges of $P$ studied, the GPA(x) is a good measure of the growth promoting ability of protein x.

## 9.8 Effect of Single Amino Acids on the Growth Parameters

The effect of the level of dietary protein on the growth parameters $A$, $(AB)$, $C$, and $t^*$ was discussed in the previous section. The growth efficiency factor $(AB)$ was shown to follow the equation of diminishing increments, whereas $A$, $C$, and $t^*$ were unaffected over the range of protein levels used. Also a measure of growth promoting ability, GPA, based on the effect of level of dietary protein on $(AB)$, was suggested with the possibility of its relation to biological value of proteins which is thought to be related to the level of the limiting essential amino acid of the protein. Literature search revealed no papers describing either short term or long term experiments with data that could be used to study the effect on $(AB)$ of adding an amino acid to a protein known to be deficient in that amino acid. However the papers read revealed effects on relative growth rate or some other assumed biological measure due to adding a single amino acid or combination of two amino acids. A discussion by Almquist et al. (1942) of the effect of adding methionine to soybean based diets was typical. The only readily useable data available to me were from a large interlaboratory experiment designed in 1965 by the animal scientists of the Monsanto Company to compare a synthetic analogue of methionine with L-methionine as diet supplements to improve the growth of chicks. Among those participating were Professors G.F. Combs of the University of Maryland, E.J. Day of Mississippi State University, and D.E. Greene of the University of Arkansas (Parks 1971).

For the experiment to be a success, the growth of chicks on the basal plus supplement had to be as sensitive as practicable to methionine supplementation. Grau and Almquist (1946) reported on a basal diet which, when supplemented with methionine, gave improved growth of chicks as a function of the level of supplementation. Combs, following the ideas of Grau and Almquist, designed a special basal

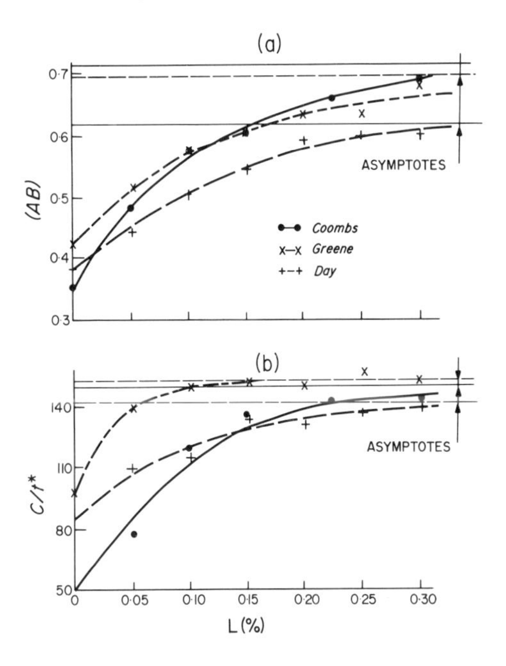


Fig. 9.22(a, b). The results of an interlaboratory experiment on the effect of L-methionine as an additive to a soya-basal diet. (a) Growth efficiency factor $(AB)$ as the law of diminishing returns versus level $L$ of the additive. (b) Appetency $C/t^*$, of animals for the diet, as the law of diminishing returns

diet designated as CP-24. CP-24 is soybean based and supports relatively poor growth because soy protein is deficient in methionine.

The experimental diets were CP-24 plus levels of L-methionine in steps of 0.05% of the basal diet, up to and including 0.30%. During the first 7 days from hatching a suitable pre-experimental ration was fed, followed by the experimental diets for 21 days. The number of replicates were 5 or 10 chicks each for each experimental diet. The data were the three week mean weight gains and food consumed per bird in each replicate. The experiments were short term starting with week old chicks and lasting 3 weeks compared to about 40 weeks to grow a chick to maturity. Since $D$ is small compared to $C$ and $W_0$ is small compared to $A$, $(AB)$, and $C/t^*$ are estimated by $\Delta W/\Delta F$ and $2\Delta F/\Delta t^2$, respectively with little bias by the factors $1-D/C$ and $1-W_0/A$. Here $2\Delta F/\Delta t^2$ is the ad libitum feeding acceleration during the 3 week period.

Figure 9.22 graphically shows the systematic differences between the results of the participators. The data of each participator appear to define curves which rise with diminishing increments toward limiting asymptotic values defined by the horizontal straight lines. The appearance of these data led me to fit each set of data points with the equation of diminishing increments

$$y=a'[1-\exp(-b'L)]+c', \qquad (9.34)$$

to obtain values of $a'$, $b'$, and $c'$ (Table 9.25). Here $y$ represents either $(AB)$ or $C/t^*$.

The curves in Fig. 9.22 are computed using Eq. (9.34) and the values found for $a', b'$, and $c'$. Here it is seen that the equation adequately describes each set of data with the possible exception of the $\Delta F/4.5$ data of Combs.

The effects of adding methionine to CP-24 being describable by the equation of diminishing increments means that the approach used in defining the growth

Table 9.25. Growth efficiencies, feeding accelerations, and results of fitting Eq. (9.34) to the experimental data

| | Combs | | Day | | Greene | |
|---|---|---|---|---|---|---|
| | $\Delta W/\Delta F$ $(AB)$ | $\Delta F/4.5$ $C/t^*$ g/wk$^2$ | $\Delta W/\Delta F$ $(AB)$ | $\Delta F/4.5$ $C/t^*$ g/wk$^2$ | $\Delta W/\Delta F$ $(AB)$ | $\Delta F/4.5$ $C/t^*$ g/wk$^2$ |
| $a'$ | 0.361 | 107 | 0.286 | 57 | 0.271 | 58 |
| $b'$ (1/% $L$) | 9.00 | 10 | 6.14 | 10 | 8.0 | 29 |
| $c'$ | 0.352 | 47 | 0.378 | 86 | 0.423 | 95 |
| rsd | 0.00768 | 9.30 | 0.0107 | 3.99 | 0.0137 | 2.94 |

promoting ability, GPA(x), of proteins may be used to define the efficacy, EFA(x), of a food additive x in promoting the utilisation of a particular dietary protein by reducing the amount of food an animal needs to consume to attain some specified weight. The approach to defining EFA(x) begins with $F/A = \ln 2/(AB)$, where $F/A$ is the food, in units of mature weight $A$, required for the animal to grow from weight $W$ to $(W+A)/2$. Let $\Phi(½)$ denote the above value of $F/A$ for the basal diet, the $(AB)$ of which is $c'$, so that

$$\Phi(½) = \ln 2/c'.$$

On adding $L(½) = \ln 2/b'$ amount of x to the basal, $(AB)$ becomes $(a'+2c')/2$ thus reducing $\Phi(½)$ to $\Phi'(½)$ given by

$$\Phi'(½) = 2\ln 2/(a'+2c'). \qquad (9.35)$$

Equation (9.35) is the efficacy of x for improving the utilisation of the basal diet if the effect of increasing $(AB)$ or $C/t^*$ by adding x to the diet is describable by the equation of diminishing increments. Because $c'$ appears in the denominator of Eq. (9.35) and $c'$ is the $(AB)$ for a particular level, $P$, of the protein in the basal diet, the EFA(x) becomes a more sensitive test of the efficacy of the additive as $P$ is adjusted downwards towards $p_0$ thus making $c'$ small but greater than zero.

Since parameters $a', b'$, and $c'$ are independent of the level of the additive they are characteristic of the additive, x, so that EFA(x) is not dependent on any particular level in the diet, which is an advantage over the other definitions of additive effectiveness, such as gain per unit of additive consumed. The reader will see in Fig. 9.23 that gain per unit of additive consumed has exactly the same disadvantage for comparing efficacies of additives that protein efficiency ratio, PER, has for comparing the growth promoting abilities of the proteins. Figure 9.23 shows that the gain per unit additive, x, consumed is the slope of the chord OQ of the curve of $(AB) = \Delta W/\Delta F$ versus the level $L$ of the additive or $\Delta W/L\Delta F = \tan\theta$.

It is important to note that each addition of the L-methionine to the soya protein in basal diet CP-24 produced a new protein which was more effective for efficiency of growth. Recalling from Chapter 8 the description of the amino acid composition of a protein as a vector and the exercise of changing the direction of the wheat gluten vector by adding lysine to get improvement by making it more like beef muscle protein, the effect of adding methionine to soya protein becomes

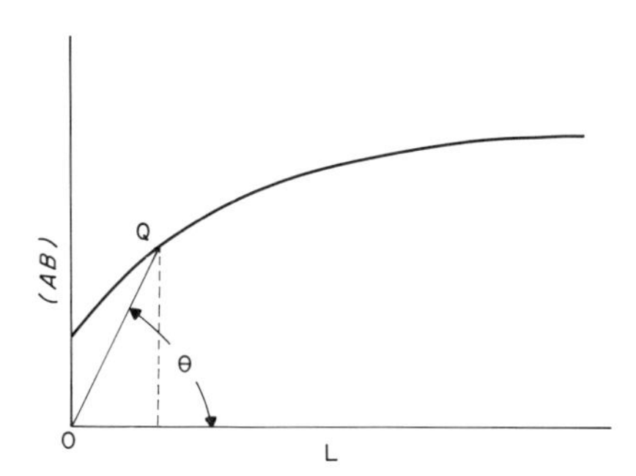


Fig. 9.23. Geometrical construction showing the possible arithmetic artifact introduced by expressing the effectiveness of a food additive as the response divided by the amount of additive consumed

clearer. Figure 8.5, Chap. 8, implies that there is operating a diminishing returns effect, because the first additions of lysine cause the wheat gluten vector to rotate more rapidly toward the beef muscle vector than subsequent additions of lysine and still higher additions cause a slower rotation away from the beef muscle vector with loss of improvement. This leads me to expect that loss of improvement of basal CP-24 plus added L-methionine may have occurred if the additions of L-methionine had been made much higher than 0.3%.

The definition of EFA(x) depends on whether $(AB)$ follows Eq. (9.34) as a function of the level of x added to the diet. The behaviour of L-methionine in basal CP-24 is one example only. The ranges of levels of the additives discussed in the nutritional literature are all too low to adequately test adherence of the data to Eq. (9.33). More extensive experimental work is needed to take advantage of the possibilities of the definitions of GPA(y) and EFA(x) in relating the biological usefulness of proteins to their amino acid compositions.

## 9.9 Requirement

The concept of the required level of some nutrient in the diet for maximum response of the animal probably dates back to the law of the minimum announced by Liebig (ca. 1855). At first this law was interpreted as meaning a constant increment of response for each unit increase of the nutrient known to be at a minimum level in the diet.

From 1900 to about 1920 experiments by Mitscherlich (1909) and work by Spillman and Lang (1924) showed that the linear increase of response to increasing levels of the minimum nutrient should be replaced by the equation of diminishing increments. These workers also discussed the possible economic value of this equation in relating the cost of the level of the nutrient to economic gain from the response. More interestingly, Spillman and Lang also gave evidence that nutrients other than the one at the minimum level of the ration showed the property of increasing the response according to the equation of diminishing increments as their levels were increased. The implication of this evidence was that all the nutrients of a ration had potentials of greater or lesser degree for improving responses to the diets.

Some time in the intervening years between then and now someone pointed out that Liebig's law of the minimum could be retained up to a break point of the response at an increased level of the minimum beyond which any increase gave no increase in response. This became the broken line equation of the response versus the level of the dietary constituent declared to be deficient. The level at which the slope of the response line discontinuously fell to zero was declared the requirement for that nutrient. Statisticians pointed out the difficulty of distinguishing between the broken line function and the equation of diminishing increments in describing the response data (Carpenter 1971). The broken line function became the standard procedure for determining requirements of dietary nutrients for what came to be known as optimum growth. Statistical methods were developed for determining the break point and its confidence level (Robbins et al. 1979). Here was a definite statement of how much return could be had from a costly dietary additive when the response is growth rate or weight gain over some fixed period of time.

When I joined the animal science research group of the Monsanto Company during the early 1960's I became acquainted with the broken line method of comparing the requirements of chicks for various synthetic sources of dietary methionine. During my studies of the feeding and growth of animals, which I have reported in the first few chapters of this book, I became intuitively dissatisfied with the broken line technique, even though it was highly useful to nutritionists in developing least cost diets using linear programming methods.

My criticisms led the group to design the experiment involving Combs, Day, and Greene which I have discussed in the preceeding section. There I explored how the additive modified the growth rate through its modification of both the growth efficiency and the food intake of the chicks for the rations through modification of the growth efficiency factor, $(AB)$, and the appetency, $C/t^*$. The equation for diminishing increments was highly useful (Parks 1971).

Lately there has been renewed interest in the use of functions other than the broken line equation for description of response to dietary factors. Robbins et al. (1979) have used a modified form of the logistic function to study weight gains of chicks over the gain on the basal versus levels of L-histidine added as percentages of the diet to the basal. They also used a form of the equation of diminishing increments to study the weight gains of chicks versus graded levels of biotin (ug/kg of diet) added to the basal diet. In addition they fitted the broken line function to the same data for dinging the requirements of the chicks for L-histidine and biotin. They assumed that the level of L-histidine, which gave 95% of the asymptotic response from the logistic function, was the requirement for L-histidine. They also assumed that the level of biotin, which gave a response of 95% of the difference between the asymptote of the equation of diminishing increments and the gain on the basal, was the requirement for biotin. They found these values for the requirements were nearly the same as those found by the broken line equation.

Gibney et al. (1979) discuss the use of another form of the equation of diminishing increments, which they name the Mitscherlich equation, in the study of ten sets of data used by the Agricultural Research Council, U.K., to derive the lysine requirement of the growing chick. Eight of the ten sets of data gave satisfactory fits to the data. The response considered was growth rate as a function of the added lysine.

They point out the difficulties experienced using the broken line method of determining that rather vaguely defined biological quantity designated requirement, and suggest the use of the equation of diminishing increments to determine the level of lysine required for a desired rate of gain. This is a turning towards the economic use of the equation of diminishing increments, as a function of the cost of the added dietary constituents, discussed by Mitscherlich (1909) and Spillman and Lang (1925) sixty to seventy years ago.

I am glad to see this drift away from the broken line concept despite its popularity because I think it is economically misleading in its reference to the level of some dietary additive for what has been habitually called optimum growth in the literature. I have already shown how growth rate is related to the product of the growth efficiency and the food intake. If growth is optimised by an additive, it may well be solely due to an increased food intake which could result in a profit loss due to increased consumption of the costly additive.

I have pointed out in Chap. 2.3 how these researchers and others who use nonlinear regression equations to describe experimental data can bias their results by using forms of the equations in which the characteristic parameters are confounded with the initial condition parameters. The most suitable nonlinear regression equations are those in which the characteristic parameters are not confounded with the initial condition parameters. These types of regression equations are those in which the characteristic parameters are clearly distinguished from the initial condition parameters.

The term requirement has been used in so many biological contexts that it is difficult to know just what the term means. For example there are requirements of protein for growth, protein for maintenance, individual amino acids for growth, and for maintenance etc. almost ad infinitum.

If the concept of requirement has not become a fixation in the science of nutrition there may be a wide open field for research on the relationship of such quantities like GPA(y) of protein, y, and EFA(x) of dietary additive, x, to the biochemistry of metabolism. Tentative steps in this direction have been taken by nutritionists as reported in Newer Methods of Nutritional Biochemistry, ed. A.A. Albanese (1963). Professor D.M. Walker and his students in the Department of Animal Husbandry, University of Sydney (1964, 1975a, b, 1976a, b) have been using blood plasma analysis in their studies. There are, no doubt, many other scientists involved in this kind of work with the newer tools available for study of blood chemistry in relation to diet composition.

A more proper use of the term "requirement" is in relation to dietary levels of vitamins, cofactors, and other micronutrients including minerals. If such dietary factors are below or above certain levels, nutritional pathologies develop in man and animal. These levels should be met, independently of the nutritive content of the diet for continued health, well being, and growth. The reader should refer to F.A.O. Bulletin Number 16 (1955) for an informative discussion of the various definitions of requirement.

## References

Albanese AA (ed) (1963) Newer methods of nutritional biochemistry with applications and interpretations. Academic Press, London New York

Almquist HJ, Mecchi E, Kratzer FH, Grau CR (1942) Soybean protein as a source of amino acids for the chick. J Nutr 24(4):385–392

Atwater WO (1899) 12th Annual Report No 69, Conn Agric Exp Stn
Atwater WO, Bryant AP (1903) Bull 28 US Dep Agric
Block RJ, Mitchell HH (1946) The correlation of the amino acid composition of proteins with their nutritive value. Nutr Abstr Rev 16:249–278
Carew LB Jr, Hill FW (1964) Effect of corn oil on metabolic efficiency of energy utilization of chicks. J Nutr 83(4):293–299
Carew LB Jr, Hopkins DT, Nesheim MC (1964) Influence of amount and type of fat on metabolic efficiency of energy utilization by the chick. J Nutr 83(4):300–306
Carpenter KJ (1971) Problems in formulating simple recommended allowances of Amino acid for animals and man. Proc Nutr Soc 30:73–83
Carpenter KJ, Clegg KM (1956) The metabolisable energy of poultry feeding stuffs in relation to their chemical composition. J Sci Food Agric 7:45–51
Chami DB, Vohra P, Kratzer FW (1980) Evaluation of a method for determination of true metabolizable energy of feed ingredients. Poult Sci 59:569–571
Donaldson WE, Combs GF, Romoser GL (1956) Studies on energy levels in poultry rations. I. Effect of calorie protein ratio of the ration on growth, nutrient utilisation and body composition of chicks. Poult Sci 35:1100–1105
Fox BA, Cameron AG (1968) A chemical approach to food and nutrition. Univ London, London
Gibney MJ, Dunne A, Kinsella IA (1977) The use of the Mitscherlik equation to describe amino acid response curves in the growing chick. Nutr Rep Int 20:501–510
Grau CR, Almquist HJ (1946) The utilisation of the sulfur amino acids by the chick. J Nutr 26:630–640
Hafez ESE, Dyer IA (1969) Animal growth and nutrition. Lea and Febiger, Philadelphia
Hammond JC, Hendricks WA, Titus HW (1938) Effect of protein in the diet on growth and feed utilisation of male chickens. J Agric Res 56:791–810
Haresign W, Lewis D (eds) (1980) Recent advances in animal nutrition 1979. Butterworths, London
Harte RA, Travers JH, Sarlich P (1948) Voluntary caloric intake of the growing rat. J Nutr 36(6):667–669
Hegsted DM, Chang Y (1965a) Protein utilisation in growing rats I. Relative growth index as a bioassay procedure. J Nutr 85:159–168
Hegsted DM, Chang Y (1965b) Protein utilisation in growing rats at different levels of intake. J Nutr 87:19–25
Hegsted DM, Worcester J (1968) Determination of the relative nutritive value of proteins: factors affecting precision and validity. J Agric Food Chem 16:190–195
Klain CJ, Greene DE, Scott HM, Johnson BC (1960) The protein requirement of the growing chick determined with amino acid mixtures. J Nutr 71:209–212
Leveille G, Shapiro R, Fisher H (1960) Amino acid requirements for maintenance in the adult rooster. IV. The requirements for methionine, cystine, phenylalinine, Tryosine, and Tryptophan, the adequacy of the determined requirements. J Nutr 72:8–15
Liebig JV (1855) Die Grundsätze der Agrikulturchemie. Vieweg, Braunschweig
Mayer J, Vitale JJ (1957) Thermochemical efficiency of rats. Am J Physiol 189:9–42
McCollum EV (1957) A history of nutrition: the sequence of ideas in nutrition investigations. Houghton Mifflin, Boston
McDonald P, Edwards RA, Greenhalgh JFD (1973) Animal nutrition, 2nd edn. Longman, New York
Metta VC, Mitchell HH (1954) Determination of the metabolizable energy of organic nutrients for the rats. J Nutr 52(4):601–611
Mitchell HH (1924) A method of determining the biological value of protein. J Biol Chem 58:873–903
Mitchell HH (1962) Comparitive nutrition of man and domestic animals. Academic Press, London New York
Mitchell HH, Beadles JR (1930) The paired feeding method in nutrition experiments and its application to the problem of cystine deficiencies in food proteins. J Nutr 2:225–243
Mitchell HH, Carman GG (1926) The composition of the gains in weight and the utilisation of food energy in growing rats. Am J Physiol 76:389–410
Mitscherlick EA (1909) The law of the minimum and the law of diminishing return from the soil. Landwirtsch Jahrb 38:522–537
Morrison FB (1954) Feeds and feeding. A handbook for the student and stockman. Morrison, Ithacal
Osborne TB, Mendel LB, Ferry EL (1919) A method of expressing numerically the growth promoting value of proteins. J Biol Chem 37:223

Owen JB, Ridgman WJ (1967) The effect of dietary energy content on the voluntary intake of pigs. Anim Prod 9:107–113

Parks JR (1970a) Growth curves and the physiology of growth. II. Effects of dietary energy. Am J Physiol 219:837–839

Parks JR (1970b) Growth curves and the physiology of growth. III. Effects of dietary protein. Am J Physiol 219:840–843

Parks JR (1971) Growth curves and the physiology of growth. IV. Effects of dietary methionine. Am J Physiol 221:1845–1848

Parks JR (1973) Diet space and response surfaces. J Theor Biol 42:349–358

Phillips DD, Walker DM (1978) Milk replacers containing isolated soybean protein for preruminant lambs: influence of experimental design on estimates of requirements for supplementing methionine. Aust J Agric Res 29:1031–1042

Robbins KR, Norton HW, Baker DH (1979) Estimation of nutrient requirements from growth data. J Nutr 109(10):7170–7114

Rozin P, Mayer J (1961) Regulation of food intake in the goldfish. Am J Physiol 201:968–974

Scott ML, Nesheim MC, Young RJ (1969) Nutrition of the chicken. Scott and Associates, Ithaca

Sibbald IR, Bowland JP, Robblee AA, Berg RT (1957) Apparent digestible energy and nitrogen in the food of the weanling rat. J Nutr 61:71–85

Spillman WJ, Lang E (1924) The law of diminishing increment. World, Yonkers

Thomas K (1909) Arch Anat Physiol 219

Velu JG, Baker DH, Scott HM (1971) Protein and energy utilisation by chicks fed graded levels of a balanced mixture of crystalline amino acids. J Nutr 101:1249–1256

Walker DM, Faichney GJ (1964) Nitrogen balance studies with the milk fed lamb. I. Endogenous urinary nitrogen, metabolic faecal nitrogen and basal heat production. Br J Nutr 18:187–200

Walker DM, Kirk RD (1975a) The utilisation by preruminant lambs of milk replacers containing isolated soya bean protein. Aust J Agric Res 26:1025–1035

Walker DM, Kirk RD (1975b) The utilisation by preruminant lambs of isolated soya bean protein in low protein milk replacers. Aust J Agric Res 26:1037–1052

Walker DM, Kirk RD (1976a) Plasma urea nitrogen as an indicator of protein quality. I. Factors affecting the concentration of urea in the blood of the preruminant lamb. Aust J Agric Res 27:109–116

Walker DM, Kirk RD (1976b) Plasma urea nitrogen as an indicator of protein quality. II. Relationships between plasma urea nitrogen, various urinary nitrogen constituents and protein quality. Aust J Agric Res 27:117–127

# Chapter 10 The Growth Parameters and the Genetics of Growth and Feeding

Some time ago Brody (1945, Table 16.1) suggested that his growth parameters $A, k$, and $t^*$ are quantities geneticists could select on if they were seeking alterations of the growth curve of a breed of animal. He knew that factors such as nutrition, environment and management could affect these parameters, but under controlled conditions he thought there was sufficient genetic variability of these parameters to be the basis for selection experiments continuously to change the course of growth.

In the past the subject matter of quantitative genetics has been those characters fixed at some chosen age or over some chosen interval of age and measurable on individuals but showing continuous differences between individuals within a population. Examples of such characters are litter sizes, birth weights, back fat thickness, production of eggs, milk, wool, and so on. There has also been selection for increased body weights and feed efficiencies at some fixed age, weight gain or feed efficiency between two ages with measurements of food consumption, body composition and so forth for studies of the changes of these characters correlated with the character directly selected upon (Pym and Nicholls 1979; Pym and Solvyns 1979). Most of these studies have been short term and contain no reference to the animal as a growing system for inferring correlated changes in growth and feeding parameters.

## 10.1 Involvement of Growth Functions

The quantitative geneticist meets a challenge in changing his attention from the kinds of characters just mentioned to the types of characters developed by Brody and others to describe the dynamic growth of an animal from infancy to young adulthood (Chap. 1). Characters such as these are considered fixed over the age span from birth or weaning to young adulthood for individuals but are variable between the individuals of a population.

The characters alluded to are not directly measurable at instants of age or over short age spans and can only be estimated using techniques of nonlinear regression to apply growth functions thought applicable by the experimenter to his growth and or feeding data gathered in long term experiments (Chaps. 4 and 7). The problems of using growth functions are many. A major one is the meaning of the various parameters in the chosen function. The least baffling of these parameters is the asymptotic weight designated as the mature weight, $A$, an example of which is in Brody's equation of diminishing increments of $W$ as a function of age. Other parameters, such as Brody's $k$ and $t^*$, pose real problems of interpretation because their meanings depend solely on the formulation of the mathematical function.

Most animal scientists are still struggling with the mathematics of growth functions and the criteria for choosing the most suitable. Little headway is being made because of the multitude of growth functions that have been proposed over the years (Chap. 1; De Boer and Martin 1978, Sect. 4).

There has been a steady increase in the appreciation of the usefulness of growth curves in experimental quantitative genetics (Eisen et al. 1969; Bakker 1974; Eisen 1974). Lately there have been reports of quantitative genetic experimental work on bending the growth curves of animals by selecting for high and low weights at two different ages (on mice, McCarthy 1971, 1974; McCarthy and Bakker 1979; on poultry, Ricard 1975). I direct the reader's attention to the extensive sets of references listed by McCarthy and Ricard as background material for their work.

### 10.1.1 An Experiment by Timon and Eisen on Mice

Timon and Eisen (1970) selected on a breed of mice for increase of one character only, namely the gain in weight from weaning at 21 days of age to 42 days of age. They carried this experiment of the males and females for nine generations while maintaining their base population by random selection. This experiment is uniquely suited to our purpose in that long term *ad libitum* food intake data, $q^*(t)$, was also taken with the weights, $W(t)$, over the 21 day experimental period (Fig. 4.7, Chap. 4). Timon and Eisen presented their ninth generation $(T, q^*, W)$ data for males and females graphically from which I developed tables of Class V data (Tables A-17 to A-20) suitable for use with my feeding and growth functions, Eqs. (10.1) and (10.2), as nonlinear regression equations

$$q^*(t) = (A/T_0 - D)[1 - \exp(-t/t^*)] + D \tag{10.1}$$

$$W(t) = (A - W_0)[1 - \exp\langle -[(AB)/T_0] \{t - t^*(1 - DT_0/A)[1 - \exp(-t/t^*)]\}\rangle] + W_0 . \tag{10.2}$$

The results of these analyses are shown in Table 10.1. Because of the inherent inaccuracies in Class V data, the values of the feeding and growth parameters and their percent changes from the parameters of the base population should be considered in terms of orders of magnitude only. The percent change of each parameter from that of the base population is a measure of the correlated response of that parameter to selection on a single character, namely the increase of weight gain of the mice over the 21 days following weaning at 21 days of age.

Here was the first opportunity I had to explore the possibility that my growth and feeding parameters, $A$, $(AB)$, $T_0$, and $t^*$ have the properties of being genetic characters as Brody suggested. Let $X$ denote the single character Timon and Eisen selected on, and let $P$ denote any one of the growth and feeding parameters associated with the selection on $X$. In general the correlated response, $CR_P$, of character $P$ is

$$CR_P = i h_X r_{XP} h_P \sigma_{p^P} . \tag{10.3}$$

Here $i$ is the intensity of selection on $X$; $h_X^2$ is the heritability of $X$, $r_{XP}$ is the genetic correlation between characters $X$ and $P$; $h_P^2$ is the heritability of $P$; and $\sigma_{p^P}$ is the phenotypic standard deviation of $P$ (Falconer 1960).

Table 10.1. The growth parameters of ninth generation male and female mice compared with those of the base population. (Data by Timon and Eisen 1970). Class V data (see Tables A-17 to A-20 and Chap. 4)

| Parameter | Male | | | Female | | |
|---|---|---|---|---|---|---|
| | Ninth gen. | Base pop. | %[a] Change | Ninth. gen. | Base pop. | %[a] Change |
| $A$, kg | 0.0376 | 0.0334 | 12.6 | 0.0316 | 0.0284 | 11.3 |
| $(AB)$ | 0.428 | 0.410 | 4.39 | 0.408 | 0.302 | 35.1 |
| $T_0$, wks | 0.819 | 0.763 | 7.34 | 0.740 | 0.707 | 4.67 |
| $t^*$, wk | 0.806 | 1.11 | −27.4 | 0.781 | 0.701 | 11.4 |

[a] % Chg = 100 (ninth gen. − base pop.)/base pop.

Timon and Eisen found non-zero values for $h_X^2$ for both males and females. It therefore follows from Eq. (10.3) and the percent changes in each parameter $P$ in Table 10.1, that $h_P^2$, $r_{XP}$, and $\sigma_{P^P}$ cannot be zero. Even though the magnitudes of the percent changes of $P$ for the males and females may be doubtful, the signs of these changes are not. Overall the signs show the $CR_P$ are in the direction favouring the increase of weight selected for over the 21 days following weaning. In the case of the females the increase of $t^*$ means that the ninth generation females had reduced appetites (Chap. 7.1). However the reduced appetite was more than offset by the increase in the growth efficiency factor $(AB)$. The increases in $T_0$ mean that the ninth generation animals permitted less of the live weight equivalent of the food intake to go to no growth. In the case of the ninth generation males the decrease of $t^*$ means their appetites were increased quite drastically. These detailed remarks based on Table 10.1 are supported by the general conclusion of Timon and Eisen that their selection character and procedure produced increased weights and appetites, especially in the case of the males.

### 10.1.2 An Experiment by Ricard on Chickens

From a base population, $X-88$, of chickens Ricard (1975) produced four strains, namely $X-44$, $X-22$, $X-11$, and $X-33$, with remarkably different growth curves after selecting for high and low weights at 8 weeks and 36 weeks of age over thirteen generations (Fig. 10.1). I obtained this graph from Ricard in 1973 which, for lack of data, I used to create five tables of Class V data (Table A-49).

The dispositions of these growth curves relative to each other indicated to me they might be members of a two parameter family of curves described by

$$W(t) = (A - W_0)\langle 1 - \exp\{-\alpha[t/t^* - 1 + \exp(-t/t^*)]\}\rangle + W_0 \tag{10.4}$$

as a two parameter function if $\alpha$ is set to unity leaving $A$ and $t^*$ as the two parameters given the initial weight, $W_0$.

Equation (10.4) is the same as Eq. (10.2) except $D$ is negligible compared to $A/T_0$ for day old chicks making $(1 - DT_0/A)$ unity, and $t^*$ has been factored out of the exponent to change $(AB)/T_0$ to $\alpha = (AB)t^*/T_0$. Setting $\alpha$ to unity is justified since chickens are among the few animals for which the ad libitum feeding and growth discriminant, $\alpha$, is very nearly unity (Tables 5.1 and 5.2, Chap. 5). The re-

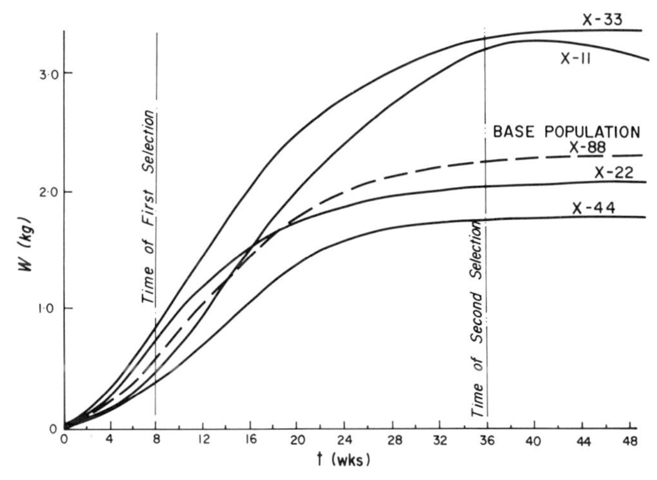


Fig. 10.1. The growth curves of four populations of female chickens derived by Ricard from his base population. (Published with Ricard's permission)

Table 10.2. The growth parameters $A$ (mature live weight) and $t^*$ (the internal resistance to the growth of appetite) for the four populations of chickens obtained by selection by Ricard compared to the parameters of his base population

| Type | Selected wts. | | $A$ kg | $t^*$ wks |
|---|---|---|---|---|
| | 8 wks | 36 wks | | |
| Base pop. X-88 | – | – | 2.162 | 8.106 |
| X–44 | Low | Low | 1.927 | 10.236 |
| X–22 | High | Low | 2.039 | 6.881 |
| X–11 | Low | High | 3.766 | 13.648 |
| X–33 | High | High | 3.339 | 8.731 |

sults of applying Eq. (10.4) to the data, with the initial weights, $W_0$, set to 0.039 kg for the five curves are listed in Table 10.2. The values of the figures of merit, *FM*, of fit of Eq. (10.4) to the data were between 5 and 12 indicating the equation more than adequately described the data.

Figure 10.2 shows the five growth curves obtained by Ricard are in fact members of a two parameter family of curves. Here are plotted the fractions of maturity, $u = W/A$, versus the normalized ages, $T = t/t^*$, for each set of data read from the growth curves in Fig. 10.1. The way in which these normalized curves cluster together show they are in the same family of curves. The minor divergences of $u$ from unity, both above and below, shown in the approaches to maturity, for large normalized ages, $T$, are to be expected from the variability of the data.

It is readily seen that Ricard's selection procedure caused correlated changes in $A$ and $t^*$. However, a real difficulty arises in considering these correlated changes from the point of view of Eq. (10.3) because Ricard did not select on a single trait but on a vaguely defined combination of the two traits, namely the weights at 8 weeks of age and 36 weeks of age. We can side step this difficulty by taking a

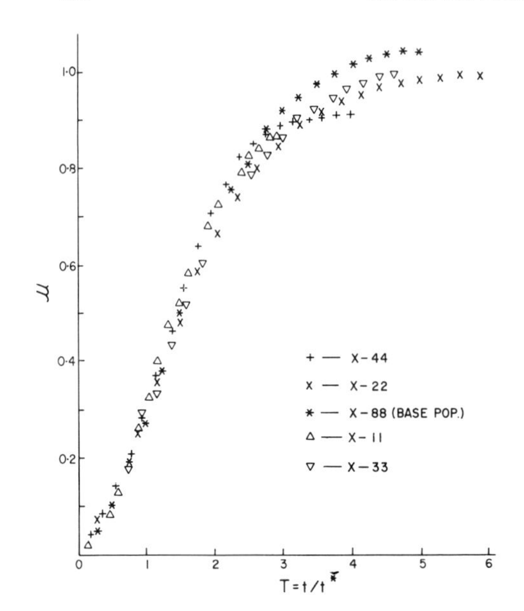


Fig. 10.2. Graph of fraction of maturity $u$ versus normalised age $T$, indicating Ricard's $(t, W)$ data are from a two parameter family of growth curves

simple approach. Let LL designate the selection for low 8 week weights and low 36 week weights, and also let HH represent the selection for high 8 week weights and high 36 week weights. This is about as close as we can get to single trait selection in thise case.

Exploring the correlated responses of $A$ and $t^*$ in the LL and HH lines as percent changes from $A$ and $t^*$ of the base population, $X-88$, we see from Table 10.2 that $CR_A$ for LL is $-19.9\%$ and for HH it is 54.4%. Similarly $CR_{t*}$ for LL is 26.3% and HH it is $-7.2\%$. Based on this simple analysis we can quite validly say from Eq. (10.3) that $h_A^2$, $\sigma_{pA}$, $h_{t*}^2$and $\sigma_{p_{t*}}$ are all non-zero. However the meanings of the genetic correlations $r_{\mathrm{LL}A}$, $r_{\mathrm{HH}A}$, $r_{\mathrm{LL}_{t*}}$ and $r_{\mathrm{HH}_{t*}}$ are not clear and inferring their signs from the $CR$ may not be useful in this simple interpretation of Ricard's selection procedure.

In the experiments of Timon and Eisen on mice and Ricard on chickens I have used my feeding and growth equations to show that the parameters $A$, $(AB)$, $T_0$, and $t^*$ are likely to have the quantitative characteristics expected of genetic traits postulated by Brody. The next subsection on the involvement of growth curves in selection experiments discusses the use of the Gompertz equation as the growth function for study of its parameters as quantitative genetic traits.

### 10.1.3 An Experiment by McCarthy and Bakker on Mice

This experiment is a large step in the direction in which I hope experimental work will move in studies of modifying the trace of an animal by an appropriate selection procedure. McCarthy and Bakker (1979) report the effects of three different selection procedures on the parameters of the Gompertz growth function, namely the mature weight, $A$, and $k$, usually designated as the maturing rate.

$$W(t) = A\exp[-b\exp(-kt)]. \qquad (10.5)$$

Table 10.3. Phenotypic correlations (above diagonal), genetic correlations (below diagonal) and heritabilities (diagonal) of body weight traits and the growth curve parameters, $A$ and $k$. [Table 4 from McCarthy and Bakker (1979)]

| | $W_{03}$ | $W_{05}$ | $W_{10}$ | $W_{21}$ | $A$ | $k$ |
|---|---|---|---|---|---|---|
| $W_{03}$ | 1.40[a] | 0.67[a] | 0.50[a] | 0.46[a] | 0.50[a] | 0.15[a] |
| $W_{05}$ | 0.74 (0.05) | 0.84[a] | 0.74[a] | 0.59[a] | 0.56[a] | 0.20[a] |
| $W_{10}$ | 0.57 (0.07) | 0.85 (0.03) | 0.90[a] | 0.81[a] | 0.81[a] | 0.06[n.s.] |
| $W_{21}$ | 0.57 (0.07) | 0.79 (0.05) | 0.94 (0.02) | 0.81[a] | 0.95[a] | −0.30[a] |
| $A$ | 0.63 (0.06) | 0.78 (0.05) | 0.93 (0.02) | 0.99 (0.00) | 0.87[a] | 0.37[a] |
| $k$ | −0.40 (0.10) | −0.08 (0.12) | 0.17 (0.11) | − 0.41 (0.11) | − 0.49 (0.11) | 0.46[a] |

s.e. of genetic correlations given in parentheses [a]$p < 0.01$ respectively

The parameter, $b$, was ignored as having no particular physiological significance. The possibility of error here will be examined later.

The three types of selection were; (1) single trait selections over 15 generations of high and low weights at 5 weeks of age, and high and low weights at 10 weeks of age; (2) independent culling over 22 generations for high weights at age 5 weeks and low weights at 10 weeks of age, and low weights at 5 weeks of age and high weights at age 10 weeks; (3) selection over 14 generations for high and low weights at age 5 weeks while restricting the 10 weeks weights to a narrow range, and high and low 10 week weights restricting the 5 week weights to a narrow range. They maintained unselected populations for each of the selection techniques.

In all the lines, litter size was kept to not more than 12 animals each. All were weaned at 3 weeks of age and not more than 3 of each sex were sampled for growth study. They were weighed weekly from 3–11 weeks of age and weighed bi-weekly to 21 weeks of age. Equation (10.5) was applied to these data for nonlinear regression to estimate A and k for the individual animals in each selection line. Using analysis of variance methods (Harvey 1960), they calculated heritabilities and the phenotypic and genetic correlations, of $A$ and $k$ along with the weights at several chosen ages (Table 10.3).

Their general conclusions regarding the growth parameters from the entire experiment, according to the type of selection in the above list, were; (1) single trait selections led to large changes of $A$ and small changes in $k$ from the base population with the differences between the high 5 week and high 10 week lines, and the low 5 week and low 10 week lines, being minimal among themselves; (2) independent culling selection led to minimal differences in $A$ and large differences in $k$ from the base population; (3) alternate selection for high 5 week and 10 week weights with changes of weights at 10 and 5 weeks of age respectively restricted led to large differences of $A$ and $k$ from the base population.

Turning back to the results of their statistical analysis over all the lines to obtain heritabilities, phenotypic and genetic correlations for $A$ and $k$ and weights at 3, 5, 10, and 21 weeks, some interesting ideas appear (McCarthy and Bakker 1979, Table 4, p. 62). Table 10.3 is their Table 4 containing the heritabilities on the diagonal, the phenotypic correlations between the traits above the diagonal, and the genetic correlations below the diagonal. The part of this table of most interest to us is the $2 \times 2$ table in the lower right hand corner.

Here we see that the mature weight $A$ has the very high heritability of 0.87, the maturing rate $k$ a medium heritability of 0.46 with medium phenotypic and genetic correlations of 0.37 and $-0.49$ respectively. Which means that selection on $A$ as a single trait is much easier than using $k$ as the trait, and that selecting on either will have a negative influence on the other. Equation (10.5) and Table 10.3 suggest that the heritabilities, and the phenotypic and genetic correlations of $W(t)$ may be estimated for any chosen set of times, $t$, when the heritabilities, and correlations of the parameters $A$ and $k$, and the growth equation are known. The estimates of $h^2[W(t)]$ and $\mathrm{cor}[W(t_i), W(t_j)]$, genetic and phenotypic, could be found with the usual quantitative genetic and statistical techniques if the probability distribution functions of $A$ and $k$ were known to be normal. This suggestion means that the portion of the matrix in Table 10.3 having to do with the statistical properties of $W_{03}$, $W_{05}$, $W_{10}$, and $W_{21}$ may be redundant information even though this information may be highly useful in other contexts.

McCarthy and Bakker might have found Eq. (10.6) a more biologically sound function than the Gompertz function for their work.

$$W(t) = [A - W(3)]\{1 - \exp[-k(t-3)]\} + W(3). \tag{10.6}$$

Here $A$ is the mature weight; $k$ is Brody's parameter; $W(3)$ is the initial weights of the mice; and $t$ is the age of the mice in weeks.

Also Eq. (10.6) clearly separates the characteristic parameters $A$ and $k$ from the initial condition $W(3)$, which the Gompertz function as stated in Eq. (10.5) does not do. Contrary to the statement by McCarthy and Bakker that $b$ is merely a scaling factor without particular physiological significance, I showed in Chap. 2.3 that $b$ is a function of the i.c., $W(3)$, and the characteristic parameter, $A$, namely $b = \ln[A/W(3)]$. The error in using Eq. (10.5) as it stands and ignoring $b$ may have caused underestimation of $A$. Figure 2 of McCarthy and Bakker (1979, p. 62), which contains graphs of percent maturity $= 100\ W/A$ versus age for the three types of selection, shows all the curves exceeding 100% at ages between 18 and 22 weeks. If the values of $A$ were correctly estimated it would be expected by chance alone that some of these curves would be below 100% as well as above in that age range (Fig. 10.2).

It is regrettable that both Ricard (1975) and McCarthy and Bakker (1979) did not measure food intake so that the effects of their selection techniques on the parameters $(AB)$, $T_0$, and $t^*$ could be explored.

The next subsection will briefly discuss some recent work on mice in which not only $(t, W)$ data but also $(t, f_i)$ Class I data were taken simultaneously on all lines so that all the feeding and growth parameters could be estimated.

### 10.1.4 An Experiment by Andrew Parratt

As part of an initial phase of a comprehensive project involving studies of alternative growth models, individual weights and food intake were recorded between 21 and 84 days of age at 3-day intervals on 572 male and female mice. These were progeny of 35 sires and 91 dams split over two generations of a three generation experiment.

The feeding and growth parameters $A$, $T_0$, $t^*$, and $(AB)$ were determined by nonlinear regression on the weights and feed intake data of individual mice using Eqs. (10.7) and (10.8).

$$f_i=(A/T_0)\{3+t^*(1-DT_0/A)[1-\exp(3/t^*)]\exp[-(i-3)/t^*]\} \qquad (10.7)$$

$$W_i=(A-W_0)[1-\exp\langle-[(AB)/T_0]\{3i-t^*(1-DT_0/A)[1-\exp(-3i/t^*)]\}\rangle]+W_0\,. \qquad (10.8)$$

Here $A$ is the mature weight in grams; $(AB)$ = growth efficiency factor, $T_0$ the internal resistance to the live weight equivalent of the food intake going to no growth in units of days $t^*$ is the internal resistance of the animal to increase its appetite in units of days; and $i$ is the index number of the 3 day interval at which $f_i$ and $W_i$ were measured (Chap. 4). Table 10.4 shows estimates of the heritabilities, genetic and phenotypic correlations of the growth parameters, A, $(AB)$, $T_0$, and $t^*$ based on paternal half sib analysis using Harvey's LSML 76 mixed model least squares analysis (1978).

Although A. Parratt is not presently proceeding with selection experiments to utilize the parameter estimates (Table 10.4) to alter the shapes of growth curves it is apparent that significant direct and correlated responses could be achieved with a properly designed selection experiment.

A defect in the present methods of extracting values of the characteristic parameters by fitting functions that are intrinsically nonlinear in the parameters to

Table 10.4. Genetic (below diagonal), phenotypic (above diagonal) correlations, and heritabilities (diagonal) of growth curve parameters, $A$, $T_0$, $t^*$, and $(AB)$. These results are from a presently unfinished PhD thesis by Andrew Parratt and are published with his and Prof. J.S.F. Barker's permission

| | $A$ | $T_0$ | $t^*$ | $(AB)$ |
|---|---|---|---|---|
| $A$ | 0.594<br>(0.10) | 0.965 | 0.499 | 0.202 |
| $T_0$ | 0.996<br>(0.07) | 0.663<br>(0.10) | 0.199 | − 0.646 |
| $t^*$ | 0.249<br>(0.12) | 0.220<br>(0.08) | 0.371<br>(0.08) | 0.155<br>(0.12) |
| $(AB)$ | − 0.903<br>(0.09) | 0.946<br>(0.09) | − 0.577<br>(0.12) | 0.528<br>(0.10) |

s.e. of genetic correlations and heritabilities given in parentheses experimental data, is the assumption that the parameter values are normally distributed about the true value (Draper and Smith 1966). This assumption is not known to be true or false. A. Parratt is presently doing simulation studies using Eqs. (10.7) and (10.8) in an effort to estimate the p.d.f.'s of the feeding and growth parameters, $A$, $(AB)$, $T_0$, and $t^*$, and test these p.d.f.'s against the assumed normal p.d.f. used presently in nonlinear regression software.

A fundamental idea contained in Parratt's work may be expressed as follows: knowledge of the genetic and phenotypic properties of the feeding and growth parameters (Table 10.4) of a genotype of animal permits theoretical comparisons of the efficience of various selection procedures to achieve desirable modifications of the feeding and growth curves of the genotype. The work of Gunsett et al. (1981) is relevant to the above idea in that they have experimentally shown that selecting mice over four generations for increased gain on a fixed quantity of food consumed (their FF treatment) or for decreased food consumption for a fixed weight gain (their FG treatment) is about 30% more effective than selecting for increased weight at 56 days post weaning. Gunsett et al. used my early work (Parks 1970a–c) as a basis for planning their experiment on both sexes of a strain of mice, and reduced the data to evaluate means and standard errors of the parameters A and B in the weight, $W$, versus cumulative food consumed, $F$, function [Eq. (2.2), Chap. 2] using the methods discussed in Chap. 4. This work showing (AB) invariant over the FF and FG treatments lends support to my second prediction from the theory (Chap. 7), namely that (AB) is independent of feeding treatment whether ad libitum or not.

I should like to make a conjecture here to the effect that the above approach to bending the growth curve of any breed of animal will yield results not only useful to animal scientists in agriculture but also in a broad sense to other bioscientists. This is not to say that the present methods of short term genetic selection for desirable characters at slaughter will be or should be regarded as less useful tools than they have been in the past (Pym and Nicholls 1979, Pym and Solvyns 1979).

## 10.2 An Economic Problem and the Feeding and Growth Parameters

The economic problem considered here involves the growout facility for an intensive meat animal production system. Young animals of known weights are put into the growout facility where they are fed to a known mean slaughter weight over a known period of time. The mean slaughter weight per animal is fixed at $\overline{W}$ and the mean growout time, $\overline{t}$, is sufficiently flexible, through nutritional and management techniques, for the growout facility to meet varying market demands. Here I am assuming that the facility is operated at full capacity at all times and the time and clean up cost between turn arounds are all constant. Consequently the yearly cash flow of the facility depends on the cash flow of each growout period, namely the income from the slaughter house less the cost of feed. Two problems arise at this point.

### 10.2.1 The First Problem and Its Solution

Suppose the manager of the growout facility, with the help of his technical personnel, has become aware of the possibilities of modifying the trace of his genotype of animals through modification of their growth and feeding parameters $A$, $(AB)$, $T_0$, and $t^*$. With this knowledge he would reason that he could increase his cash flow if the mean trace of his animals could be so modified that the value obtained by reduction of the mean turn around time would offset the cost of the increased food consumption to reach the mean slaughter weight.

Now let us suppose this idea is attractive enough that he sets up a long term project to determine $A$, $(AB)$, $T_0$, and $t^*$ for the animals he presently uses. He already knows the value, $v$, per unit reduction of turn around time, the cost, $c$, per unit of feed, the present mean turn around time, $\bar{t}$, mean food consumption per animal, $\bar{F}$, and the fixed mean slaughter weight, $\bar{W}$, per animal. With this information at hand he sees that he wants to know how the time of his present animals should be so changed that

$$v\Delta t - c\Delta F > 0. \tag{10.9}$$

Here $\Delta t$ is the desired reduction of the mean turn around time, $\bar{t}$, and $\Delta F$ is the change of the mean food consumed, $\bar{F}$, depending on the shift of the trace of the animals. The second problem is how to shift the trace to insure inequality Eq. (10.9) under the constraint that the change, $\Delta W$, in the mean slaughter weight is zero.

### 10.2.2 The Second Problem and Its Solution

Here I shall use functional notation to avoid writing the growth and feeding equations of the trace in full.

The forms of the righthand sides of Eqs. (10.10) and (10.11) are already known respectively from Eqs. (4.3) and (4.2) in which the mature food intake $C$ is set to its equivalent, namely $A/T_0$ (Chap. 4).

$$W = W[A, (AB), F] \tag{10.10}$$

$$F = F(A, T_0, t^*, t). \tag{10.11}$$

A slight change in the trace can be achieved by slight changes of the parameters, namely $\Delta A$, $\Delta(AB)$, $\Delta T_0$, and $\Delta t^*$. The change in the trace will also lead to slight changes, $\Delta t$, $\Delta F$, and $\Delta W$ of the mean turn around time, $\bar{t}$, the mean food consumed, $\bar{F}$, and the mean slaughter weight, $\bar{W}$, respectively. However only those changes in the trace are sought for which $\Delta W = 0$ and $\Delta t < 0$.

If the changes of all these quantities are small enough the total changes of Eqs. (10.10) and (10.11) can be written in terms of the partial derivatives of the right hand sides with respect to each of the quantities suffering change while considering the others not changed. In other words the total changes of $W$ and $F$ are respectively

$$\Delta W = (\partial W/\partial A)\Delta A + [\partial W/\partial(AB)]\Delta(AB) + (\partial W/\partial F)\Delta F = 0, \tag{10.12}$$

$$\Delta F = (\partial F/\partial A)\Delta A + (\partial F/\partial T_0)\Delta T_0 + (\partial F/\partial t^*)\Delta t^* - (\partial F/\partial t)\Delta t. \tag{10.13}$$

Table 10.5. List of the partial derivatives of $W$ and $F$ with respects to each variable of which they are functions

| Partial derivative | Its evaluation |
|---|---|
| $\partial W/\partial A$ | $1-[1+(AB)\bar{F}/A]\exp[-(AB)\bar{F}/A]$ |
| $\partial W/\partial(AB)$ | $\bar{F}\exp[-(AB)\bar{F}/A]$ |
| $\partial W/\partial F$ | $(AB)\exp[-(AB)\bar{F}/A]$ |
| $\partial F/\partial A$ | $(1/T_0)\{\bar{t}-t^*[1-\exp(-\bar{t}/t^*)]\}$ |
| $\partial F/\partial T_0$ | $-(A/T_0^2)\{\bar{t}-t^*[1-\exp(-\bar{t}/t^*)]\}$ |
| $\partial F/\partial t^*$ | $-(A/T_0)[1-(1+\bar{t}/t^*)\exp(-\bar{t}/t^*)]$ |
| $\partial F/\partial t$ | $(A/T_0)[1-\exp(\bar{t}/t^*)]$ |

Table 10.6. List of the economic pressures $\alpha$, $\beta$, $\gamma$, and $\delta$ as functions of the partial derivatives in Table (10.6)

| Economic pressure | Its evaluation |
|---|---|
| $\alpha$ | $v[(\partial W/\partial F)(\partial F/\partial A)/\partial A]/[(\partial F/\partial t)(\partial W/\partial F)]$ $+c(\partial W/\partial A)/(\partial W/\partial F)$ |
| $\beta$ | $\partial W/\partial(AB)(v/(\partial F/\partial t)+c)/(\partial W/\partial F)$ |
| $\gamma$ | $v(\partial F/\partial T_0)/(\partial F/\partial t)$ |
| $\delta$ | $v(\partial F/\partial t^*)/(\partial F/\partial t)$ |

The negative sign of the last term in Eq. (10.13) means $\Delta t$ signifies a reduction of $t$ as desired in this problem. It is understood that the partial derivatives are evaluated using the known characteristic parameters of the present breed of animal and the present values of $\bar{t}$ and $\bar{F}$. Table 10.5 is a listing of the evaluations of the partial derivatives in the terms already mentioned.

Solving Eq. (10.12) for $\Delta F$ and Eq. (10.13) for $\Delta t$ and substituting them in inequality Eq. (10.9) an expression is found which after algebraic reduction can be written as

$$\alpha\Delta A+\beta\Delta(AB)+\gamma\Delta T_0+\delta\Delta t^*>0. \tag{10.14}$$

Here the co-efficients $\alpha$, $\beta$, $\gamma$, and $\delta$ are algebraic combinations of the partial derivatives and the economic parameters $v$ and $c$. Because the algebraic expressions for these co-efficients are cumbersome they are separately listed in Table (10.6). Here they are seen to involve the economic parameters $v$ and $c$ as well as the partial derivatives listed in Table 10.5, which means these co-efficients can be regarded as economic pressures for shifting the present trace of the animal towards a trace required for a more profitable growout operation.

It is interesting and important to note in Table 10.6 that the economic pressures to change $A$ and $(AB)$, namely $\alpha$ and $\beta$, are linear combinations of $v$ and $c$, whereas the pressures to change $T_0$ and $t^*$, namely $\gamma$ and $\delta$ are functions of $v$ only. The numerical signs of these pressures are important because a negative sign of a particular pressure would mean pressure to decrease the parameter associated with that pressure in the inequality Eq. (10.14). In order to insure this inequality all changes in the parameters must be made in accordance with the signs of the pressures when they are computed for a particular growout operation. If an economic pressure is

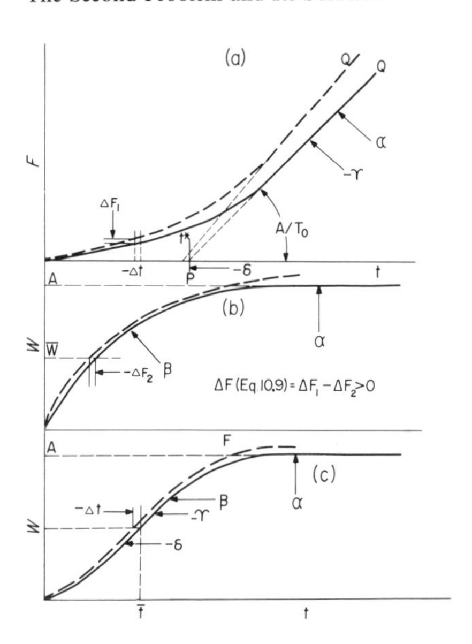


Fig. 10.3. An illustration of how the economic pressures tend to shift an animal's Trace. Parts (a–c) are the projection of the Trace on the $(t, F)$ $(F, W)$, and $(t, W)$ planes respectively

negative, selection pressure should be exerted to reduce the particular parameter; if it is positive, selection pressure should be exerted to increase the parameter. Tables 10.5 and 10.6 are bases for some general remarks about the economic pressures and their signs.

Table 10.5 tells us that in the general case all the partial derivatives involved in the calculation of the economic pressures $\alpha$ and $\beta$ are always positive. Because $v$ and $c$ are positive, Table 10.6 tells us what the economic pressures will be to increase $A$ and $(AB)$ respectively by amounts $\Delta A$ and $\Delta(AB)$ in any growout equation. Table 10.5 also tells us that the partial derivatives $\partial F/\partial T_0$ and $\partial F/\partial t^*$ are always negative while $\partial F/\partial t$ is always positive. Therefore by Table 10.6, the economic pressures $\gamma$ and $\delta$ are always negative and selection pressure should be to decrease $T_0$ and $t^*$ by amounts $\Delta T_0$ and $\Delta t^*$ respectively. If selection on the growth and feeding parameters can be accomplished, it appears prudent to select for a breed of animal with a higher mature weight, a higher efficiency factor and lower Taylor time constant, and a lower resistance to appetite increase dependent on the economic selection pressures (Table 10.6) involving the value, $v$, per unit time saved and the cost, $c$, per unit of feed to get the animals to market weight.

Figure 10.3 illustrates the meaning of $\alpha$, $\beta$, $\gamma$, and $\delta$ as forces, exerted in the direction of the arrows, to shift the trace of the present animal, denoted by the solid curves in the $(t, F)$, $(F, W)$, and $(t, W)$ planes towards the trace of a more economical animal in the growout operation denoted by the dashed curves. A positive $\alpha$ combined with a negative $\gamma$ tends to rotate the asymptotic mature food intake to an increased food intake position and a negative $\delta$ tends to shift $t^*$ to lower values (Fig. 10.3a), all of which tends to increase $\overline{F}$ by amount $\Delta F$, which in effect tends to reduce $\overline{t}$ as economically desirable. However a positive $\alpha$ combined with a positive $\beta$ will tend to shift the $(F, W)$ curve so that $\overline{F}$ is reduced by $\Delta F_2$ (Fig. 10.3b), which tends to counter the effects of the pressures $\alpha$, $-\gamma$, and $-\delta$ in producing the economically desirable reduction, $\Delta t$, of the mean growout time $\overline{t}$ (Fig. 10.3c).

G.M. Tallis (1968) reported a very interesting procedure for constructing a selection index to shift a growth curve gradually towards another declared an optimum growth curve according to some criteria. His primary assumption was that growth in the time domain is the realisation of a stochastic process with $W(t)$ as the expected growth function. I showed in Chap. 3 that this assumption is hardly tenable. It would be interesting to know how his deduction of a selection index would have to be modified if his primary assumption was discarded and his problem became the development of a selection index on $A$, $(AB)$, $T_0$, and $t^*$ to shift the animals trace in the $(t, F, W)$ space so that inequality Eq. (10.9) would be insured (Fig. 10.3).

The implication of all the chapters so far is that nutrition and genetics can be joined with the theory of feeding and growth to give intensive systems of meat animal production a technological base.

## References

Bakker H (1974) Effects of selection for relative growth rate and body weight of mice on rate, composition and efficiency of growth. Meded Landbouwhogesch Wageningen 8:1–94

Brody S (1974) Bioenergetics and growth. First published: Reinhold, New York (Reprinted Hafner Press, New York, 1974)

DeBoer H, Martin J (eds) (1978) Growth and development in cattle, Sect 4. Martinus, Hague

Draper NR, Smith H (1966) Applied regression analysis. Wiley, New York

Eisen EJ (1974) The laboratory mouse as a mammalian model for the genetics of growth. Proc World Congr Genet Appl Anim Prod Madrid 1:467–492

Eisen EJ, Lang BJ, Legates JE (1969) Comparison of growth functions within and between lines of mice selected for large and small body weight. Theor Appl Genet 39:251–260

Falconer DS (1960) Introduction to quantitative genetics. Oliver and Boyd, Edinburgh

Gunsett FC, Baik DH, Rutledge JJ, Hauser ER (1981) Selection for feed conversion on efficiency and growth in mice. J Anim Sci 52:1280–1285

Harvey WR (1960) Least squares analysis of data. US Dep Agric Agric Res Serv 20–80

McCarthy JC (1971) Effects of different methods of selections for weight on the growth curve of mice. 10th Congr Int Zootech Versailles, Thème VII

McCarthy JC (1974) Insights into genetic variation in the deposition of fat from selection experiments in mice. 1er Congr Mond Genet Appl Prod Anim, Madrid 3:529–531

McCarthy JC, Bakker H (1979) The effects of selection for different combinations of weights at two ages on the growth curve of mice. Theor Appl Genet 55:57–64

Parks JR (1970a) Growth curves and the physiology of growth. I. Animals. Am J Physiol 219:833–836

Parks JR (1970b) Growth curves and the physiology of growth. II. Effects of dietary energy. Am J Physiol 219:837–839

Parks JR (1970c) Growth curves and the physiology of growth. III. Effects of dietary protein. Am J Physiol 219:840–843

Pym RAE, Nichols PJ (1979) Selection for food conversion in broilers: direct and correlated responses to selection for body weight gain, food consumption and food conversion ratio. Br Poult Sci 20:73–86

Pym RAE, Solvyns AJ (1979) Selection for food conversion in briolers: body composition of birds selected for increased body weight gain, food consumption and food conversion rate. Br Poult Sci 20:87–97

Ricard FH (1975) Essai de selection sur la forme de la courbe de croisance chez le poulet. Ann Genet Select Anim 7:427–443

Richards FJ (1959) A flexible growth function for empirical use. J Exp Bot 10:290–300

Tallis GM (1968) Selection for an optimum growth curve. Biometrics 24:169–177

Timon VM, Eisen EJ (1970) Comparisons of ad libitum and restricted feeding of mice selected and unselected for postweaning gain. I: Growth, food consumption and feed efficiency. Genetics 64:41–57

# Chapter 11 Energy, Feeding, and Growth

Up to now I have been dealing with the animal as a black box with input of food and output of live weight and have noted some regularities between the input and output. No doubt these regularities are due to the unique ways living systems use the free energy and the body building materials in the food for growth and maintenance. It is well known open living and nonliving chemical systems are so constructed that the material and energy balances are interconnected to maintain them as stable, steady state, consuming, and producing systems.

All the presently known biological facts and theories of the in vivo metabolic exergonic reactions driving endergonic cycles for growth and maintenance do not lead me to the why of my differential equations of feeding and growth. They also do not lead to more fundamental meanings of the coefficients of the equations, namely the mature weight, A; the growth efficiency factor ($AB$); the Taylor time constant, $T_0$; and Brody's time constant, $t^*$ (Kleiber 1961, Goodwin 1963, Trincher 1965, Milsum 1966, Morowitz 1968).

## 11.1 The Power Balance Equation

From the point of view of the energy balance of a growing, or mature, animal, I am still forced to consider it a black box with energetic input, energy stored, and energy output. This simple statement implicitly contains all the possible ways the chemical energy and body building materials in the food are used by the animal to build and maintain its body structure (pre-ordained at conception) and all the possible ways unusable chemical energy and heat generated within the body are rejected to the environment. This statement says nothing about the kind of energy being considered.

Nutritionists and animal scientists in general have gone to great lengths to find ways of expressing energy input, energy stored and energy output so that the simple statement, of energy balance, made above can have more scientific meaning [the Introduction and Chaps. 1 and 2 of Brody (1945), Chaps. 6 and 7 of Kleiber (1961), McDonald et al. (1973) 2nd ed., among many texts on animal nutrition]. From a scientific point of view it would be best if the simple energy balance statement for an animal referred to the Gibbs free energy input less the free energy of the solids, liquids, and gases eliminated, the Gibbs free energy stored, and the heat output, all measured or expressed in kilocalories per unit mass under standard conditions of temperature and pressure.

The heat output of an animal generally is due to the inefficient use of Gibbs free energy in driving those endergonic reactions needed to maintain the animal in a steady energetic state above equilibrium. To this heat generation there must be added heat due to the degradation of Gibbs free energy to heat in driving electrical currents against the electrical resistances in the nervous system and to the degradation to heat of the free energy driving all the voluntary and involuntary mechanical muscular work against internal and external viscous frictional resistances. This total heat output is lost to the environment by radiation, conduction and convection through the body surface, and by evaporation of water from the lungs and skin. The heat loss is regulated so that the core temperature of the animal has minimum excursions about a steady state mean value if the animal is a homeotherm. These simple ideas based on Gibbs free energy passing through the animal can lead to many fundamental theoretical and experimental studies of growth and feeding.

However I shall substitute for this approach to my simple energy balance statement the usual approach using combustion energy instead of Gibbs free energy. Here a small inaccuracy is encountered in not correcting the combustion energy for entropy changes between reactants and products (Klotz 1957, Scott 1965). The equation I shall use to express my simple statement for an animal is:

$$\beta dF/dt = d(\Gamma W)/dt + M(t), \tag{11.1}$$

where $\beta$ is the metabolizable energy of the food or is the net energy if $\beta$ must be corrected for the heat increment; $dF/dt$ is the food intake and $\beta dF/dt$ is the energy intake; $\Gamma$ is the specific combustion energy of the body weight, $W$, and $(\Gamma W)$ is the whole body combustion energy considered here as stored energy and therefore $d(\Gamma W)/dt$ is the rate of storage of energy in the animal; $M(t)$ is the rate of dissipation of the internal heat, generated by the active processes mentioned above, to the environment.

Equation (11.1) is a power balance equation which must be met during the animal's existence as an ongoing entity. I have called Eq. (11.1) a power balance equation because the time rates of energy acquisition, use and dissipation in any manner are expressed in units of energy per unit of time. Until quite recently the units used were kilocalories per day or per week and so on. Now there seems to be some agreement that energy should be expressed in units of joules and the time rate of change of energy in units of joules per day and so on. However as of now the next logical step, namely to express joules per unit of time in terms of watts, has not been taken. Since all the terms of Eq. (11.1) are power terms, the equation is a power balance equation in whatever units power is expressed. In this Chap. I shall use the old units of power, namely kilocalories per unit of time.

For some purposes this equation is not specific enough; for example it does not contain a term for the rate of production of some product like eggs, milk, wool, or young. Neither does it contain terms for the rate of storage of thermal energy due to change of body temperature and rate of working according to some assigned schedule. The late H.L. Lucas Jr. (1971, private communication) suggested the following generalisation of Eq. (11.1).

$$\beta dF/dt = [1 + \eta(\Gamma^*)] d(\Gamma W)/dt + \mathrm{K}(\Gamma) g(W). \tag{11.2}$$

The functional term $\eta(\Gamma^*)$ is the heat of reaction extracted from the feed to produce a unit of stored energy and is a function of $\Gamma^*$ which is the specific combustion heat of a unit change of body composition. Hence the term $\eta(\Gamma^*)d(\Gamma W)/dt$ is the power extracted from the feed to effect the change in body stores; under ad libitum feeding $\Gamma^*$ and $d(\Gamma W)/dt$ are functions of dietary and environmental conditions. Under controlled feeding they are functions of the feeding rate, but the types of functions are unknown. The term $K(\Gamma)g(W)$ is the power extracted from the food for maintenance and is equivalent to $M(t)$ in Eq. (11.1). The factor $g(W)$ is that function of body weight (sometimes $g(W)=W^a$, $0<a\leqq 1$), to which maintenance is proportional at a standard value of the specific whole body combustion energy, $\Gamma$, under standard environmental conditions such as in the measurement of basal metabolism; $K(\Gamma)$ then is a function of environmental conditions and $\Gamma$, but not a function of a diet or of whether feeding is ad libitum or controlled (Brody 1945, Kleiber 1961, Blaxter 1968).

The unmodified power balance equation is useful in some general situations, examples of which are (1) if $\beta dF/dt < M(t)$, then $d(\Gamma W)/dt < 0$, and the animal will metabolize some of its own tissue at a rate sufficient to re-establish balance; (2) if $\beta dF/dt > M(t)$, then $d(\Gamma W)/dt > 0$ and the animal will grow and or fatten at a rate sufficient to restore balance. If the animal is neither growing nor fattening $d(\Gamma W)/dt = 0$, then the power intake, $\beta dF/dt$, will balance the power loss, $M(t)$, to the environment. These remarks relate directly to Chap. 6.5, which discusses possible energetic reasons for the dynamics of the weight change of an animal depending on how it is fed.

When the phase point $(q, W)$ is in the controlled growth region of the GPP the energy intake is greater than maintenance so that the growth rate $dW/dt > 0$ and the animal will grow. When the phase point is in the partial starvation region the energy intake is less than maintenance, so that the growth rate $dW/dt < 0$ and the animal losses weight. And when the phase point is on or near the Taylor diagonal the energy intake balances maintenance so that the growth rate $dW/dt = 0$ and the animals weight will be stable. At the time the concept of the GPP was developed, body composition data were not associated with the growth and feeding data under study. As a consequence I do not know how the specific whole body combustion energy, $\Gamma$, changes from point to point in the GPP and with time.

For this reason the second prediction I made from the theory in Chap. 7 may be false.

The second prediction was that the growth efficiency factor, $(AB)$, and the Taylor time constant, $T_0$, are independent of the way the animal is permitted to feed, in other words $(AB)$ and $T_0$ are independent of where the phase point $(q, W)$ may be in the GPP. In Chap. 6.6, I have discussed the plan of an experiment, illustrated in Fig. 6.16 which could demonstrate that $(AB)=h$ and $T_0=h/g$ of a genotype of animal may not have the same values above and below the Taylor diagonal of the GPP.

In a later section of this chapter it will be shown how an animal fattens as it feeds ad libitum. If the reader will refer to Fig. 6.16 he will see that $\Gamma$ of the animal at point P on the ad libitum curve must be considerably less than $\Gamma$ at point R. Now, suppose the experimenter has monozygotic twins which he feeds along the ad libitum curve until he reaches point P where he holds one twin at constant feed intake, $q_0$, and continues to feed the other to point R, where he reduces the food intake, $q$, to $q_0$ thereby putting the animal at phase point, T. The animal at point T must lose weight to arrive at point Q while the animal at point P must gain weight to get to Q. Although I think the dynamics of the gain and loss of weight from points P and T respectively are solutions of the controlled feeding (DE 7.3) according to the first prediction (Chap. 7.1), it is quite possible for $(AB)$ and $T_0$ to be functions of $\Gamma$ so that $(AB)$, $T_0$, and Brody's $k=(AB)/T_0$ cannot be the same above and below the Taylor diagonal, thereby falsifying prediction two of the theory.

My ad libitum and controlled feeding and growth DE's and Eq. (11.1) appear to be independent, because there is nothing about them which inhibit over or undereating during ordinary circumstances or adjustment of eating pattern under extraordinary conditions (Chaps. 7.4, 7.4.3, and 7.6). Mayer (1955) and Mayer and Thomas (1967) suggest that under ordinary conditions appetite is under the control of the endocrine and central nervous systems which regulate appetite on a short term basis to meet the requirement of the energetic balance, Eq. (11.1), with longer term corrections of the errors of the short term regulations (see Chap. 7.4.3). This idea implies that my feeding and growth DE's and Eq. (11.1) are not independent but are interrelated by an unknown biological control system which seeks steady state feeding and growth patterns which satisfy Eq. (11.1) in the long term and which have minimum divergences from the animal's trace defined by the solutions of my DE's (Chap. 5.2).

The maintenance terms in Eqs. (11.1) and (11.2) lump together all the uses of the free energy of the food ingested to maintain all the steady state activities of the animal; namely blood flow, heart rate, acquisition, ingestion, and digestion of food, muscle tone, etc. (Morowitz 1968, pp. 112–117; Brody 1945, Chap. 3). All these activities originating from free energy of the food must appear as heat within the animal because they are limited by the friction due to the visco-elastic nature of the body materials. The energetic levels of these steady states can be heightened by extra demands made on the animal by the environment but, when the extra demands are removed a general drift back to the previous energetic levels will ensue. These changes of steady activity levels are small relative to the causative factors but important in the studies of the details of reaction to environmental stress. The multitude of steady activities sum to the state of homeostasis in the animal and the reactions of the animal to changes of its own internal state and/or the environment are internally controlled to maintain homeostasis within very narrow limits (Brody 1945, Chap. 10; Mayer 1955; Milsum 1966).

There is nothing about Eqs. (11.1) and (11.2) which prevent animals from over- or undereating relative to energy intake for maintenance only. These equations are power balances only and state that if an animal overeats compared to its maintenance requirement the rate of storage of energy will increase, i.e. either $\Gamma$ or $W$ or both will increase, and if the animal undereats then the energy stores will be drawn on to make up the deficit. Mayer and Thomas (1967) suggest that appetite (food intake) is under the control of the central nervous system which regulates intake on a short term (daily) basis with long term correction to the errors of the short term regulation. As a consequence the growth equations, ad libitum or controlled feeding, and the power balance equations such as Eqs. (11.1) and (11.2) are the results of, and therefore already express the totality of all the internal controls the animal possesses to maintain itself as an ongoing open system. The deviations of growth and energy utilisation from the equilibrium functions used in this book may be interpreted as local and temporary fluctuations of this steady state complex which may be regarded as the metabolic state of the animal.

## 11.2 The Specific Whole Body Combustion Energy, $\Gamma$

The specific combustion energy, $\Gamma$, is important in itself because of its relation to the composition of the animal and how it varies as the animal ages or is put under some environmental stress. Because the fat component has a large combustion energy of 9.5 kcal/g (about double that of the other body components) the total body combustion energy, $(\Gamma W)$, can change appreciably without any marked change in the live weight, $W$. Even though the specific combustion energy of an organism is an average, it is an important economic number because it is a measure of the fatness of the animal as a whole.

The problem of determining body composition of animals as functions of age or weight is still a research problem which I shall not discuss here. However I have found some composition data on several species which are serially related to age and live weight, and therefore useful in getting some idea about how $\Gamma$ varies as the animal matures. If the mature weight, $A$, of the species is known or can be reasonably estimated, the data of $\Gamma$ versus degree of maturity, $u$, can be graphically represented to give a visual display of $\Gamma$ as a function of $u$, namely $\Gamma(u)$ (Fig. 11.1).

Kleiber (1961, Table 4.4, p. 49) presents composition data for Haecker's steers related to body weight in the range 45.5–545 kg. The composition was expressed as percentages of water, ash, protein, and fat. Since the body energy is mostly stored in fat and protein, $\Gamma$ could be estimated by summing the percentages of each of these components multiplied by their combustion energies per gram and dividing by 100. I estimated the mature weights of these steers to be about 1,100 kg (Brody 1945, Table 16.1). Figure 11.1 shows the dispositions of the data points relative to each other and they appear to define a straight line for the function $\Gamma(u)$. This

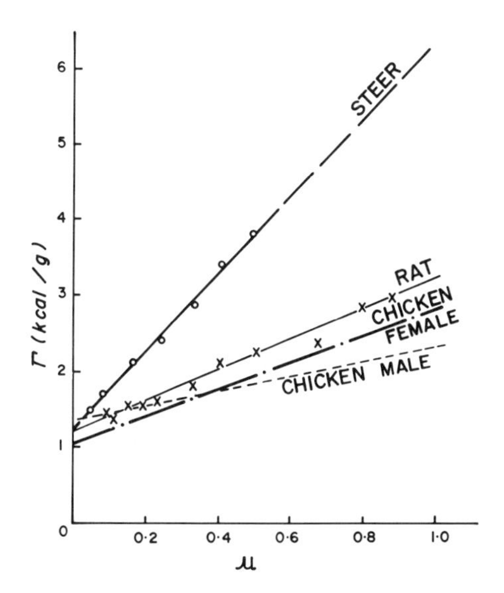


Fig. 11.1. The specific combustion energies, $\Gamma$, of steers, rats, and male and female chickens as straight line functions of the degree of maturity, $u$. The data points for the steers and rats are marked by 0 and X respectively. To reduce clutter the data points for the chickens are not shown

appeared to be a curious phenomenon which could be specific to steers or due to Haecker's (1920) use of dissection to isolate the body components. Later I found some data on the specific combustion energy, $\Gamma$, of some rats versus body weight and age since weaning (Mayer and Vitale 1957, Table 1, p. 40). Using my weight versus age equation, namely

$$W(t)=(A-W_0)[1-\exp\langle -k\{t-t^*(1-D/C)[1-\exp(-t/t^*)]\}\rangle]+W_0$$

as the nonlinear regression equation I found the mature weight, $A$, of these animals to be 422 g (Table 4.1, Chap. 4). Using this value for the mature weight I calculated the fraction of maturity, $u$, for each value of $\Gamma$ listed by Mayer and Vitale and plotted these data points on the same graph with the steer data. Again the data appeared to define a straight line function for $\Gamma(u)$ (Fig. 11.1). Since the $\Gamma$ data for these rats was obtained by chemical analysis of aliquot parts of the whole body finely ground and homogenized, it seems the straight line property of $\Gamma(u)$ may be independent of the method of measuring the whole body percentages of fat and protein, provided the method yields accurate estimates of the total fat in the body.

### 11.2.1 Cattle, Chickens, and Rats

I also found some carcase composition data versus empty body weight for male and female chickens from which I could calculate $\Gamma$ at each weight and estimate the mature weights from the weight versus age data to obtain the corresponding fractions of maturity (Mitchell et al. 1926, 1931). Graphs of these data also displayed the linear form of the function $\Gamma(u)$,

$$\Gamma(u)=a+bu. \qquad (11.3)$$

Linear regression of the steer, rat, and chicken $(u,\Gamma)$ data using Eq. (11.3) showed that it was a reasonable description of the data for each species. The regression lines are shown in Fig. 11.1. To reduce clutter in the figure the $(u,\Gamma)$ data points for the chickens are not shown.

The equations for these species are:

Steers, $\Gamma=1.25+6.15\,u$,

Rats, $\Gamma=1.21+1.97\,u$,

Chickens

Male $\Gamma=1.36+0.95\,u$,

Female $\Gamma=1.04+1.84\,u$.

I have not shown the statistical criteria of fit for Eq. (11.3) because I wish to emphasise the linear form of the $\Gamma$ versus u function and not give the reader the idea that I consider it as verified within the species let alone between the species. None of the data I have reported here were experiments designed to falsify Eq. (11.3). Experiments based on this equation to test its robustness as a descriptor of $(u,\Gamma)$ data would reveal the adequacy of fit and the precisions of the intercept, $a$, and slope $b$, between sexes within species and between species. Equation (11.3) is intriguing but needs some biological support.

### 11.2.2 Allometric Relation Between Fat and Body Weight

During discussion of Eq. (11.3) with some colleagues at the Agriculture Research Council's Animal Breeding Research Organisation (ABRO) at Edinburgh, a biological reason for the linearity of Eq. (11.3) was proposed. Louis Monteiro pointed out that if total body fat was allometrically related to the body weight with an exponent equal to two, then Eq. (11.3) must be linear in terms of the fraction of maturity, $u$. His reasoning proceeded as follows.

Suppose total body fat, $\phi$, in an animal of body weight, $W$, is given by the allometric equation

$$\phi = cW^2, \tag{11.4}$$

then the total body nonfat is

$$W - \phi = W - cW^2.$$

Letting $\alpha$ and $\gamma$ be the specific combustion energies of the nonfat and fat respectively, the values of $\Gamma$ at weight $W$, is

$$\Gamma = [\alpha(W - cW^2) + \gamma c W^2] \% W,$$

which reduces to

$$\Gamma = \alpha + (\gamma - \alpha) c W.$$

Since by definition of fraction of maturity, $W = Au$, this equations becomes

$$\gamma = \alpha + [(\gamma - \alpha) cA] u,$$

which is not only of the linear form of Eq. (11.3), but also gives meanings to the intercept, $a$, and the slope, $b$, in biological terms. Here it is seen that the intercept, $a$, is not only the total body specific combustion energy of the animal when its weight, or fraction of maturity, is zero as indicated by the mathematics of Eq. (11.3), but also is the specific combustion energy of the nonfat portion of the body weight of any weight, $W$. The slope, $b$, is the product of three factors, namely $\gamma - \alpha$, the difference between the specific combustion energy of the fat and nonfat, the factor $c$ in the allometric relation, Eq. (11.4), and the mature weight, $A$. These constant factors of $b$, therefore, show how the body composition and the mature weight of species and varieties within species affect the slope of the line but not its linearity. Furthermore the slope, $b$, of the line is related to the percentage of fat that may be expected in the mature body weight since by the postulated allometric relation, namely Eq. (11.4),

$$100\, cA = 100\, b/(\gamma - \alpha),$$

is the percentage of fat in the mature body weight, $A$.

In this way the empirical nature of the linearity of $\Gamma$ as a function of $u$, $\Gamma(u)$, is shown to be related to a biologically fundamental allometric relation of weight of total body fat, $\phi$, to the whole body weigth, $W$. This proviso is testable by searching the literature for those body composition experiments in which the allometric equation has been extensively used on the experimental results.

The search of the literature revealed many inconsistencies in the values of the exponent, $d$, in the allometric relation of the amount of body fat to the body weight, namely $\phi = cW^d$. For example Reid and several of his colleagues (1968) reported $d=1.89$ based on a range of empty body weights of sheep. J.M. Thompson did a study of serial data of whole body fat versus body weight on Merino sheep and found the value of $d=2.033$ (unpublished data from Butterfield 1978). Hilmi (1975) studied carcase composition versus body weight of two groups of Border Leicester and Merino wethers. One group was fed ad libitum over the growth range of 3–50 kg, and the other group was fed 60% of ad libitum over the same growth range. He found values of 1.91 and 1.96 respectively for $d$ in these groups of sheep. Whereas Tulloh (1963) found $d=1.540$ across several breeds of sheep; $d=1.899$ across breeds of cattle and $d=1.499$ for the Large White breed of swine. Much of this work was aimed at meat quality evaluation and in order to obtain a wide range of body weights, data across varieties within a species were used. Furthermore most of these studies used empty body weight for determination of $d$. The use of empty body weights instead of live weights may cause error in determination of $d$, but the error would be small compared to the error in the factor $c$.

The usefulness of Eq. (11.3) in Eq. (11.1) depends on establishing the specific whole body combustion energy as a function of the degree of maturity, $u$, for growing animals. The values of $\Gamma$ must be accurately determined from representative samples of the experimental animal population as it feeds and grows. I have applied the allometric equation to two such sets of data. One set is the composition data Kleiber (1961) gives for Haecker's steers and another set for a strain of Sprague Dawley albino rats from Table 1 of Mayer and Vitale (1957). The value of the exponent, $d$, for the steers is $1.814 \pm 0.061$, and for the albino rat, $d=1.652 \pm 0.061$. Even here Joy and Mayer (1968) cast doubts on the value I get for the albino rat by reporting in their Table 1 a value of $d=1.844 \pm 0.100$ for the same strain of rats. They also cast doubts on my Eq. (11.3) by showing in their Fig. 1 an allometric relation between the total combustable energy, $\Gamma W$, and the body weight, $W$, of these rats given by

$$\Gamma W = 0.0734\ W^{1.566}.$$

Here $W$ is what they designated as 4 h fasted body weight. This result contradicts Eq. (11.3) since dividing by $W$, the specific combustion energy is

$$\Gamma = 0.0734\ W^{0.566},$$

which bears no relation to Eq. (11.3) and which may represent a misuse of the concept behind allometry.

Here we see that the values of $d$ found in the literature vary between species from about 1.5 to about 2.00, the value predicted by Monteiro, but in a confusing manner since for the same species high and low values of $d$ have been reported. The only conclusive remark which can be made is that since $d$ is usually greater than unity the amount of total body fat increases exponentially as the body weight increases with age from infancy to adulthood. In other words there is no answer to a question such as "At what age does an animal begin to fatten?"

During the study of the literature on body composition and the use of the allometric equation it became apparent that many animal scientists consciously or subconsciously believe that the weights of all the growing components of an animal can be put into allometric relationships to the total body weight. I suspect this belief and think the animal scientists should be wary when acting on it in a study of experimental data. My suspicions are based on the following discussion. Let $w_i$ be the weight of the $i$th component of body weight $W$ and let it be assumed that $w_i$ is allometrically related to $W$. The total weight, $W$, is then given by

$$W = \sum_{i=1}^{n} w_i \tag{11.4a}$$

$$= \sum_{i=1}^{n} c_i W^{d_i}. \tag{11.4b}$$

Here $n$ is the number of components of which the animal is composed. In as much as the animal is growing, the growth rate, $dW/dt$, is the derivative of Eq. (11.4b) with respect to time, namely

$$dW/dt = \sum_{i=1}^{n} d_i c_i W^{d_i - 1} dW/dt.$$

Since the animal is growing $dW/dt$ is nonzero and $W^{di-1} = W^{d_i}/W$, the preceding equation can be reduced to

$$W = \sum_{i=1}^{n} d_i c_i W^{d_i}. \tag{11.4c}$$

By the assumption that the allometric equation, namely

$$w_i = c_i W^{d_i},$$

holds for each component, Eq. (11.4c) reduces to

$$W = \sum_{i=1}^{n} d_i W_i. \tag{11.4d}$$

Equation (11.4d) can be identical to Eq. (11.4a) only if all the $d_i$ are unity. However the equality expressed in Eq. (11.4d) can hold true if

$$\sum_{i=1}^{n} d_i = 1, \tag{11.4e}$$

signifying that the total body weight, $W$, is the sum of the arithmetically weighted component weights, $w_i$, where the $d_i$ are the arithmetic weights.

Experimental body composition studies using allometry on many species and breeds of animals have shown that not only are the $d_i$ not unity but also at least one of them, notably the fat component, is greater than unity. In this latter case at least one of the $d_i$ must be less than zero in order for Eq. (11.4e) to hold true, which violates the general notion that the $d_i$ must be greater than zero. These ideas cast strong doubt on the concept that all the component parts are allometrically related to the whole.

This is a mathematical argument which can be overridden by some compelling biological principle behind allometry of which I am presently unaware.

It appears that the question of how the specific combustion energy of an animal varies as it ages under normal and abnormal food management, is still an open subject for research. However I shall accept Eq. (11.3) as an accurate, if not a scientifically defensible, description of the linear character of $(u, \Gamma)$ data and use it in the following sections of this chapter.

### 11.2.3 The Relation of $\Gamma(u)$ to the Thermochemical Efficiency of Growth and Fattening

Mayer and Vitale (1957) used the phrase "thermochemical efficiency" to distinguish the gain of whole body combustion energy per unit of metabolizable energy consumed, $d(\Gamma W)/\beta dF$, from the more general concept of energetic efficiency defined as the useful work obtained per unit of free energy input. In the context of growth and fattening of animals the preceding definition of the thermochemical efficiency (TCE) is quite useful in study of how animals store combustion energy as they feed and grow with age. Brody (1945, p. 51) estimated what he called the gross efficiency of growth of Jersey and Holstein cattle and showed how it steadly dropped in value from about 0.35 to about 0.06 in the period from 1–21 months of age. Brody's gross efficiency is similar to TCE in that it was calculated as the ratio of the kilocalories per day of the weight gains and the kilocalories in the TDN consumed per day. It is interesting to note that Brody assigned a constant value of 909 kilocalories per pound as the specific combustion energy of the weight gains for both breeds of cattle. I think his concept of the steadily decreasing gross efficiency of growth has been accepted even though it definitely depended on the assumption of a constant specific combustion energy, $\Gamma$, for the weight gains independent of age. Equation (11.3) and Fig. 11.1 contradict Brody's assumption of a constant $\Gamma$ and should lead to something different from a steady decline of the TCE or gross efficiency from infancy to adulthood.

Since it is more likely that $\Gamma$ is not constant, the TCE of growth is given by

$$\mathrm{TCE} = (\Gamma dW/dF + W d\Gamma/dF)/\beta\,. \tag{11.5}$$

Noticing that $d\Gamma/dF$ can be written as $(d\Gamma/dW)(dW/dF)$ Eq. (11.5) becomes, on factoring the growth efficiency $dW/dF$,

$$\mathrm{TCE} = (\Gamma + W d\Gamma/dW)(dW/dF)/\beta,$$

which shows how a nonconstant $\Gamma$, i.e., $d\Gamma/dW \neq 0$, affects the way the animal stores combustion energy. The specific combustion energy, $\Gamma$, at any body weight is augmented by the increment of specific combustion energy per fraction of the increment of body weight at the weight, $d\Gamma/dW/W$. This formulation of TCE can be rewritten in a very interesting form for ad libitum fed animals by recalling from Chap. 2 the formula for the growth efficiency, $dW/dF$, and $\Gamma(u)$ from Eq. (11.3) and remembering $W = uA$ from the definition of fraction of maturity, $u$. The new form is

$$\mathrm{TCE} = (a + 2bu)(1 - u)(AB)/\beta.$$

If $b$ were zero as assumed by Brody, like his gross efficiency, the TCE would decrease steadily as $u$ approaches unity. In the cases we have studied, the steers, rats, and chickens, $b$ is greater than zero, so that the TCE is a product of an increasing factor and a decreasing one. At the end of Chap. 1 mention was made of the behaviour of a function composed of two such factors, with the conclusion that the function must pass through a maximum somewhere on the range of the independent variable, u, namely $0 < u \leqq 1$. Expanding this equation algebraically, but keeping the constant, $[(AB)/\beta]$, as a factor yields a quadratic function for TCE, namely

$$\mathrm{TCE} = [(AB)/\beta][a + (2b - a)u - 2bu^2]. \qquad (11.6)$$

Differentiating this with respect to $u$ and setting it to zero gives the value of $u$ where TCE has its maximum value, TCE(max). The values of TCE(max) and $u$ where the maximum occurs are respectively

$$\mathrm{TCE(max)} = [(AB)/\beta][a + (2b - a)^2/8b], \text{ and}$$

$$u = (2b - a)/4b.$$

The linear relation between $\Gamma$ and $u$, Eq. (11.3), and the quadratic relation between the thermochemical efficiency TCE, and $u$, Eq. (11.6) are new and have biological meaning through Monteiro's suggestion that the exponent of the allometric relation of total body fat versus body weight is exactly two, and my theory of ad libitum feeding and growth.

If these relations had been known to Mayer and Vitale in 1957, when they reported the results of their experiments measuring the body weights, food intake, and body composition versus age for calculating the TCE of growing rats, they may have arrived at different conclusions concerning the thermochemistry of feeding and growth on diets of constant energy and different dietary protein levels. The method they were forced to use was arithmetic analysis of their data by sequential segments over periods of time; a method which overemphasises minor variations in the data especially in computing ratios. Whereas the method proposed in this section subsumes the experimental variability of the body weight and feeding data into the growth and feeding parameters $A$, $(AB)$, $T_0$, and $t^*$, and the variability of the body composition data is subsumed into the parameters $a$ and $b$ of the specific body heat of combustion, $\Gamma$, versus the fraction of maturity, $u$, [Eq. (11.3)]. The experiments of Mayer and Vitale are examined in more detail in the next section.

Before leaving this Sect. I should point out an important property of Eq. (11.6). In Chaps. 9.5 and 9.6 is was shown that the growth efficiency factor, $(AB)$, follows the equation of diminishing increment as a function of dietary protein level, and is proportional to the dietary energy density. This means that the factor $[(AB)/\beta]$ in Eq. (11.6) is independent of the dietary energy but is a strong function of the dieatry protein level. In the neighbourhood of the 25% protein level, $[(AB)/\beta]$ is very near its asymptotic maximum value. Little if any increase in this factor can be expected for dietary protein levels greater than 25%, but for levels less than 25% the value of this factor falls very fast towards zero at a percent dietary protein greater than zero.

### 11.2.4 The Effects of Dietary Protein

The experiments of Mayer and Vitale on the Harvard strain of Sprague Dawley male albino rats were among the first in which body composition data were determined simultaneously with ad libitum growth and feeding data over the time period from weaning to 90 days after weaning. The food was of a purified type based on lactalbumin and dextrose with 12% oil and with a metabolisable energy of 4.44 kcal per gram. The animals were individually caged at 25 °C ambient temperature.

Three populations were grown on the same type of food with the dietary protein levels set to 60%, 25%, and 10% by adjusting the levels of lactalbumin and dextrose gram for gram so that their sum remained constant at 80.9%. All other ingredients were held at the same level so that the metabolisable energies of the three diets were constant at 4.44 kcal per gram.

At the times of weaning and thereafter when the mean weights, $W$, and food intakes, $dF/dt$, per animal in each population were measured, six animals with mean weight per animal equal to that of the population from which they were drawn were killed, frozen, homogenized, and chemically analysed for percent total body fat and protein. Using specific combustion energies of 9.5 and 5.7 kcal per gram for the fat and protein respectively, they calculated the specific whole body combustion energy, $\Gamma$, which leads to whole body combustion energy, $(\Gamma W)$, at each experimental time after weaning.

They reported in their Table 1 [t, $\beta dF/dt$, $W$, $\Gamma$, $(\Gamma W)$] data for the rats fed 25% protein diet for times t equal zero at weaning and 3, 6, 9, 12, 24, 30, 45, 60, and 90 days after weaning. The last column in their table contains their estimates of the thermochemical efficiency, TCE, of growth during the successive time periods of 0 to 3, 3 to 6, 6 to 9 days and so on. This table of data was suitable to examine from the point of view of the preceding section.

Dividing their value of daily caloric intake by 4.44, I obtained the food intake, $dF/dt$, in grams per day corresponding to the time, $t$. Using nonlinear regression I fitted to these data my theoretical function of food intake versus time, namely

$$dF/dt = (C-d)[1-\exp(-t/t^*)] + D,$$

with $D = 7.5$ grams per day at weaning, to get the values of the mature food intake $C = 18$ grams per day and the appetite factor $t^* = 11.2$ days. Using these values of $C, t^*$, and $D$ in my theoretical function relating cumulative food consumed, $F$, versus time, namely

$$F = C\{t - t^*(1-D/C)[1-\exp(-t/t^*)]\},$$

I could calculate the cumulative foods consumed at the experimental times after weaning and then use the equation of diminishing increments of weight, $W$, versus the calculated cumulative food consumed, $F$, namely

$$W = (A - W_0)\{1 - \exp[-(AB)F/A]\} + W_0,$$

where $W_0$ is the weight at weaning, to determine by nonlinear regression the mature weight, $A = 422$ g and the growth efficiency factor $(AB) = 0.618$. These values of $A$ and $(AB)$ are useful in exploring the uses of Eqs. (11.3) and (11.6) for this case of growth, feeding and fattening of the rats on the 25% protein diet.

Dividing the weights in their Table 1 by 422 gave the fractions of maturity, $u$, corresponding to their measured specific heat contents, $\Gamma$, for these rats. These $(u, \Gamma)$ data were shown plotted in Fig. 11.1 with the graphs for steers and chickens.

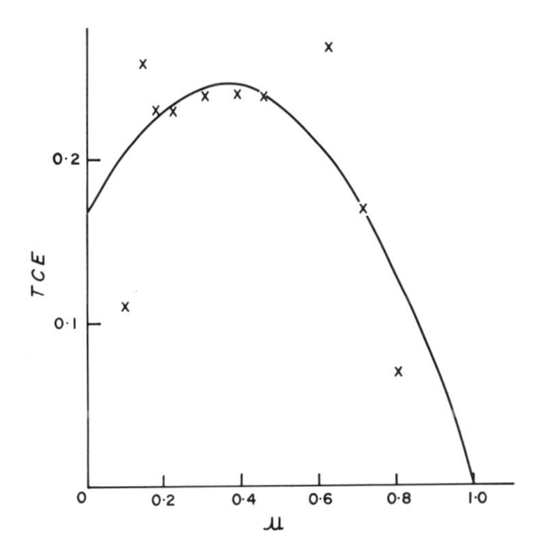


Fig. 11.2. The thermochemical efficiency, TCE, as a parabolic function of the degree of maturity, $u$, for the rats fed the 25% protein diet. The *solid curve* is computed and shows good fit to the data points X, from Table 1 of Mayer and Vitale (1957)

Here it is seen that the data appear to define a straight line the equation of which is

$$\Gamma = 1.21 + 2.15u.$$

With these values of 1.21 and 2.15 for $a$ and $b$ respectively in Eq. (11.3) and the values of 0.618 and 4.44 kilocalories per gram respectively for $(AB)$ and $\beta$, the parabolic equation [Eq. (11.6)] for the TCE versus u is

$$\mathrm{TCE} = 0.168 + 0.430u - 0.599u^2, \tag{11.7}$$

the maximum value of which is 0.246 occurring at the fraction of maturity of 0.36. It is of interest to see how this parabolic function compares with their estimates of the TCE. Figure 11.2 shows the solid curve calculated from Eq. (11.7) and the estimated TCE by Mayer and Vitale in their Table 1. The fit of the curve to the estimates is not too bad. However considering the way in which the estimates were calculated from the data over consecutive periods of time, I think the fit shown here is quite fortuitous.

The data and the calculated TCE's listed in their Table 2, for the populations ad libitum fed the 10% and 60% protein diets, are for the same overall time period of 0 to 90 days after weaning but taken over longer time intervals. The schedule of times for taking data were 0, 6, 12, 18, 24, 30, 60, and 90 days. In this table they report only the weight, percentage of body protein and TCE at the specified times after weaning. Here again I suspect their calculated values of the TCE, especially the sequences of the values for the population of rats fed the 60% protein diet. I can see no compelling reason for these values to be constant at $0.236 \pm 0.026$ over the 90 day period.

Knowing how the TCE were calculated and the metabolizeable energy, $\beta$, of the diets, if the food intakes were known it would be possible to estimate the specific whole body combustion energies, $\Gamma$, measured by Mayer and Vitale at the experimental times. It was shown in Chap. 9.6, that the feeding parameters $C$, $t^*$, and $D$ are constant for animals fed diets of constant energy density and varying only in the dietary protein level. It is therefore no mere assumption to use the values of

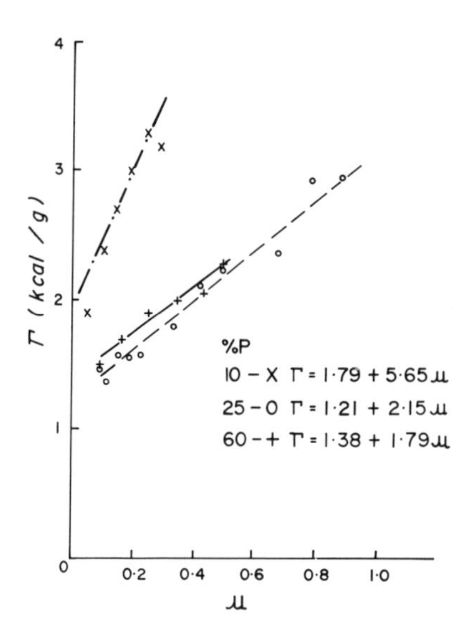


Fig. 11.3. The effects of dietary protein level on the slopes of the straight line relation between the specific combustion energies, $\Gamma$, of rats and their degrees of maturity, $u$. For a given value of $u$ the rats fed the diet with 10% protein are much fatter than the rats fed 25% and 60% protein diets

$C$, $t^*$, and $D$ calculated from the food intake data for the rats fed the 25% protein diet in Table 1 to find the mature weights, $A$, and growth efficiency factors $(AB)$, for these two populations by nonlinear regression using the equation of diminishing increments of live weight with the cumulative food consumed as previously discussed in this section for the population on the 25% protein diet. The energy consumed during any of the experimental periods could now be calculated as 4.44 $\Delta F$, where $\Delta F$ is the food consumed during the period, and multiplied by the tabular value of the TCE for that period to obtain the increment of whole body combustion energy, $\Delta(\Gamma W)$. Summing these increments sequentially, beginning with the 57 kcal whole body combustion energy for the rat at weaning, gave an estimate of the $(\Gamma W)$ at each age after weaning. Dividing these estimates of the whole body combustion energies by the weight data at that age gave a value for the specific combustion energies, $\Gamma$, for the populations of rats. Using the appropriate mature weight to obtain the fraction of maturity, $u$, the $(u, \Gamma)$ data for these populations of rats were plotted in Fig. 11.3 along with the data for the rats on the 25% protein diet.

This figure is very illuminating in that it graphically shows that the rats on the 10% protein diet fattened at a much higher rate than those on the 25% and 60% protein diets which fattened at about the same rate. When I reported the results of my experiments with male Sprague Dawley albino rats ad libitum fed diets of different dietary protein levels, I was very puzzled about what happened to the extra energy the rats on the low protein diets consumed to reach the same weights as the rats on the high protein diets (Parks 1970, Chap. 9.6). Figures 9.15 and 9.16 show how the growth of my rats and those of Mayer and Vitale on the low protein diets lagged behind the growth of those on the high protein diets: lags in time which increased as the animals aged and increased in weight towards their mature weights. Figure 9.17 shows how all my rats cumulatively consumed the same amount of food in the same time regardless of the dietary protein level. The high

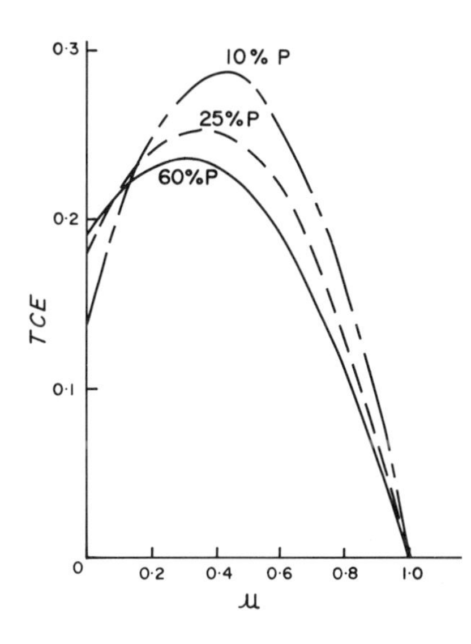


Fig. 11.4. The effects of dietary protein levels on the parabolic functions of the thermochemical efficiencies of rats versus their degrees of maturity, $u$. The peaks are higher and occur at higher values of $u$ as the percent protein is reduced

slope of the $\Gamma$ versus $u$ in line in Fig. 11.3 for the rats on the 10% protein diet indicates that the extra energy consumed was most likely being stored as fat.

Figure 11.4 shows the parabolic forms of the TCE versus $u$ for the varying protein diets. Here is seen the relative maxima and their position on the $u$ axis as the percent protein is reduced. There appears to be a general shift to higher maxima occurring at higher levels of $u$ as the dietary protein level is reduced. I am not at all sure that these parabolas are true expression of TCE versus $u$. The biological meanings of these types of curves depend on the TCE being proportional to a product of an increasing rate of body energy during growth and the decreasing decline of the growth efficiency. The best conclusion regarding the TCE of growth is a negative one, namely the TCE is not constant over the growth period as indicated by the calculated results of Mayer and Vitale, nor does it diminish as indicated by Brody.

The ideas discussed here indicate that the statement "The body composition of an animal of a given weight is independent of how it arrived at that weight" made by some animal scientists is most probably false. These ideas also indicate that the lean body weights of two animals of the same breed and weight may be quite different. I agree with Mayer and Vitale that body weight alone is a relatively poor index of the dynamics of growth, but I do not agree that thermochemical growth alone is any better index. I think my theory of growth and feeding plus a power balance equation, like Eq. (11.1), and an equation expressing the specific whole body combustion energy, $\Gamma$, as some function of body weight or fraction of maturity, like Eq. (11.3) (whether or not the function is a straight line) is a much more powerful approach to the dynamics of growth and fattening. I shall use this approach in Chaps. 11.3, 11.3.1, and 11.3.2 to discuss some experimental studies of the effects of work on feeding and growth.

It is regrettable that Wilson, Emmans, and I did not include body composition studies in experiment BG 54 we designed to put the theory of feeding and growth at risk (see Chap. 7). The way in which animals feed, grow, and fatten is still open for experimental research in long term experiments. Such experiments are extremely sparse in the literature.

Long term experiments in which body composition data are also taken could, in conjunction with my theory of feeding and growth, yield not only scientifically valuable information but also information of economic value to the meat industries.

### 11.2.5 The Effects of Ambient Temperature

The content of this section depends on finding whether or not the ad libitum feeding and growth differential equations are invariant under the conditions of changes of ambient temperature only and finding how ambient temperature affects the $\Gamma(u)$ function directly and the thermochemical efficiency, TCE, indirectly.

Joy and Mayer (1968) reported results of extensive measurements of ad libitum feeding rate, body weight, carcase composition, etc., of groups of Charles River variety of Sprague Dawley albino rats over a period of 64 days at ambient temperatures of 25° and 5 °C. The purpose of their work was to find how ambient temperature affected the partition of dietary energy intake between heat production, activity and accumulation of body energy. Fortunately the raw data were extant at this late date and Doctor Joy made them available to me (private communication, see Appendix Tables D-3 and D-4). I used the data first to show applicability of the feeding and growth equations and determined the growth parameters at each temperature, and second to approach the question of the relation of energy intake to the accumulation of body energy and the effect of ambient temperature on the relation (Parks 1971).

Initially, 125 rats were divided into two groups in which the average 4 h fasting live weight per rat was 215 g. There were 76 rats in the group to be held at 25° and 49 to be held at 5 °C. The raw data included 4 h fasting live weights, whole body heat contents and food intakes of the warm living and cold living rats at 0, 4, 8, 15, 22, 36, 50, and 64 days. At each experimental time seven rats randomly selected from the warm living group and five from the cold living group were weighed, killed, and frozen in liquid nitrogen. The homogenized carcases were chemically analysed for percent fat, percent protein, and percent of nonfat nonprotein. The specific whole body combustion energies, $\Gamma$, of each carcase was calculated by taking the combustion energies of fat, protein, and nonfat nonprotein respectively as 9.49, 5.70, and 4.32 kcal/g. The whole body combustion energies were the products of $\Gamma$ and the live weight, $W$. The food intakes were 7 day averages around each experimental time. The food was Purina rat chow with $\beta = 3.6$ kcal/g.

Figure 11.5 shows plots of the average live weight per rat versus time for both temperatures. Except during the first eight days at 5 °C, the data points of the cold and warm living groups appear to rise to maximum live weights as the animals age. The initial section ab of the growth curve of the cold living rats shows that the initial adaption of the rats was to lose weight at what appears to be a low but steady rate for eight days to point b followed by a positive growth rate which is lower than that of the warm living rats.

Figure 11.6 shows graphs of the average food intakes per rat versus time for both temperatures. The rats at 5 °C rapidly increased their food intakes, on intro-

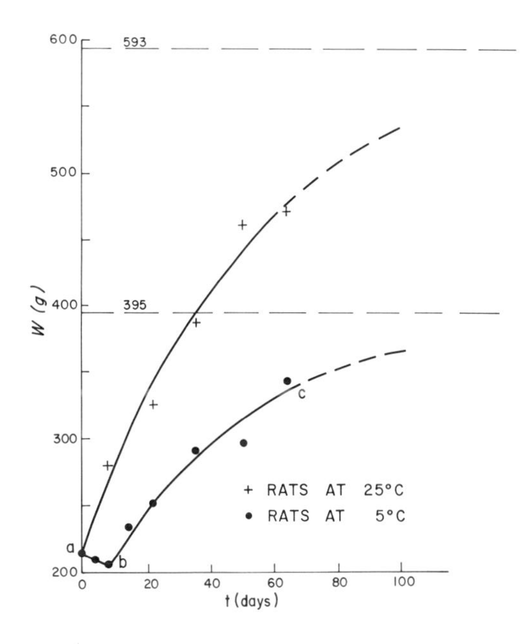


Fig. 11.5. The growth curves, $W(t)$, of two populations of rats of the same age and average live weight (data point *a*) versus the time, $t$, since one population was placed in a cold chamber at 5 °C ambient temperature and for the other simultaneously reared at 25 °C ambient. The *solid curves*, excepting the portion *ab* for the cold living rats, were calculated using the values of the feeding and growth parameters, $A$, $(AB)$, $T$, and $t^*$ obtained by nonlinear regression fitting the growth and feeding functions to the data points. (Data from Joy and Mayer 1968)

duction to the cold environment, from 19.8 g/day towards 39 g/day as a maximum. This curve shows the progressive increase of food intake capacity as they grew. The maximum appears to have been achieved in about 60 days. The average food intake of the 25 °C rats rose very slowly from 20.8 g/day initially to 23.5 g/day at 64 days, as is normal for animals already near their mature food intakes. For these

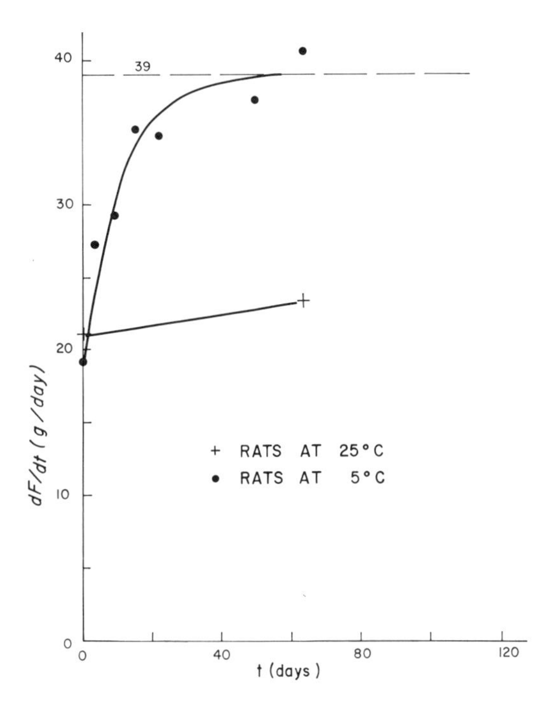


Fig. 11.6. Daily food intake, $dF/dt$, of the warm and cold living rats versus time, $t$, since one population was placed in the cold chamber. Here is seen the high acceleration of food intake (appetency) of the cold living rats and the near constancy of the food intake of the warm living rats because they are near their mature food intake during the period of this experiment. The solid curve for the cold living rats is calculated from the nonlinear regression values $A$, $T_0$, and $t^*$. (Data from Joy and Mayer 1968)

rats we can take the average food intake of 22.2 g/day as the mature food intake and the cumulative food consumed, therefore, is $F = 22.2t$.

The disposition of the data points for the cold living rats in Figs. 11.5 and 11.6 indicates that the ad libitum growth and feeding equations

$$W = (A - W_0)\{1 - \exp[-(AB)F/A]\} + W_0,$$

$$dF/dt = (C - D)[1 - \exp(-t/t^*)] + D,$$

are applicable for finding $A$, $(AB)$, $C$, and $t^*$ by the methods discussed in Chap. 4 for reducing class II data. It was expected that the data for the rats at 25° would conform to the growth equation with $F = 22t$ to find values for $A$ and $(AB)$. For these rats I assumed a value of $t^* = 14$ days (see values for rats in Table 4.1). Application of the above equations as nonlinear regression equations to the growth data of the 25 °C rats and the feeding and growth data of the 5 °C rats showed that reduction of ambient temperature from 25 °C–5 °C caused a 33% reduction of the mature weights, $A$, from 593 g–395 g; a 56% reduction of the growth efficiency factor, $(AB)$, from 0.50–0.22; and a 63% reduction of the Taylor time constant, $T_0$, from 26.6 days to 9.8 days. The value of the appetite resistance factor, $t^*$, of about 12 days for the cold living rats indicates they found the Purina rat chow no less appetising than it was to the warm living rats. Brody's maturing coefficient, $k$, was virtually the same, but the appetency, $C/t^*$, which is a measure of the urge to eat, was increased by 140%.

The Taylor time constant, $T_0 = A/C$, which is the time it takes for a mature animal to eat its own weight of food and is a measure of the energetic "leakiness" of the mature animal to the environment, has been reduced by 62%. Cold living has more than doubled the energy required to maintain the mature animal. It was surprising to find Brody's maturing coefficient, $k$, so little affected by the temperature reduction but this is because $k = (AB)/T_0$ and $(AB)$ was reduced by 56% as $T_0$ was reduced by 62%. This can mean that the animals in the cold adapted physiologically in such a manner, for example becoming hairier (Joy and Mayer 1968), that Brody's $k$ would be kept constant. However, this conclusion is based on just one experiment. I have been unable to find data similar to those of Joy and Mayer for rats or any other species. The increase of the appetency of the cold living rats by 140% was unexpected. Before studying these data I was unaware of and suprised at the possibility of animals being able to increase their appetency so greatly, especially in the low body weight range of their growth.

The effects of ambient temperature, shown here for animals adapted to the temperatures at which they are reared, and these effects on the growth parameters cannot be used to assess temperature effects on growth in circumstances of fluctuating ambient temperatures, the periods of which are too short for adaption by the animals to take place. It is quite possible that use can be made of the effects of ambient temperature on $(AB)$ and $k$ in assessing the effects of seasonal temperature variations on mature animals, because a period of a year is long enough for adaption to take place from season to season. However, to be more sure of the effects of ambient temperature on the growth parameters, experiments similar to Joy and Mayer should be carried out for at least two additional temperatures in the range 5–25 °C to establish existence of possible functional relationships.

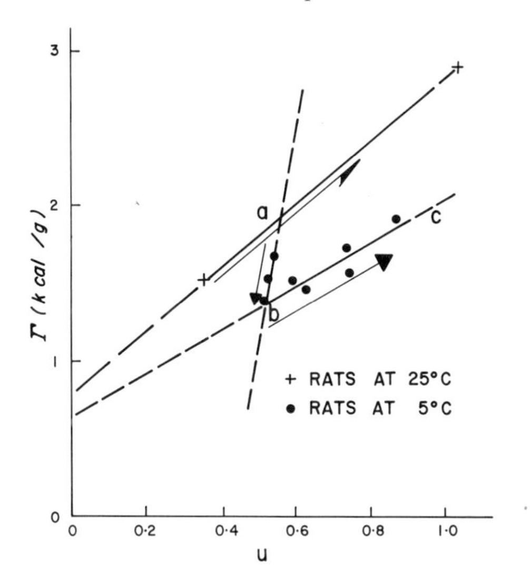


Fig. 11.7. The specific combustion energy, $\Gamma$, of the cold living rats versus their degree of maturity, $u$, compared to that for the warm living rats. The *arrows* indicate passage of time. In the beginning at point a the cold living rats lose fatness and weight until at point b the $\Gamma(u)$ as the straight line bc indicates these rats have become adapted to cold living and are gaining energy, but at a lower rate than the warm living rats. (Data from Joy and Mayer 1968)

With the mature weights, $A$, of the warm and cold living rats of 593 and 395 g respectively at hand, attention can be focused on the effects of ambient temperature on the specific whole body combustion energies, $\Gamma$, versus the fraction of maturity, and on the thermochemical efficiencies, $\mathrm{TCE}=d(\Gamma W)/\beta dF$, of these rats.

In the case of the warm living rats, the whole body combustion energy, $\Gamma W$, was divided by the live weight to give individual values of $\Gamma$ and the live weights were divided by $A=593$ g (Table D-3) to get 56 $(u,\Gamma)$ data points. These data points were treated by linear regression and found to define a straight line adequately, the formula of which is

$$\Gamma=0.80+2.03u,\ 0.28\leqq u<1.$$

It is interesting to note that this equation is not greatly different from the straight line equation found in Chap. 11.2.4 for the rats at 25 °C ambient temperature and fed a diet of 25% protein and 4.44 kcal/g metabolizeable energy.

The $\Gamma$ versus u equation for the warm living rats is shown in Fig. 11.7 for comparison with the lines for the cold living rats.

For each rat in the 5 °C group the total body combustion energy was divided by the live weight to obtain individual values of $\Gamma$. These values of $\Gamma$ were averaged over the five rats selected at each experimental time, and the average live weights were divided by $A=395$ g (Table D-4) to give eight $(u,\Gamma)$ data points shown plotted in Fig. 11.7. These points appear to define a broken straight line. Segment ab has a steep slope. The arrow showing passage of time indicates that the 215 g rats introduced into the cold reduced their specific whole body combustion energy, $\Gamma$, quite rapidly compared to the loss of fraction of maturity, $u$. Segment bc has a gradually rising slope with increasing u indicating subsequent recovery of growth and fattening after 8 days in the cold. Linear regression was used to find the equations of the two line segments (ab) and (bc) respectively,

$$\Gamma=-5.8+13.8u,\ 0.51\leqq u<0.54,$$

$$\Gamma=0.64+1.39u,\ 0.51\leqq u<1.$$

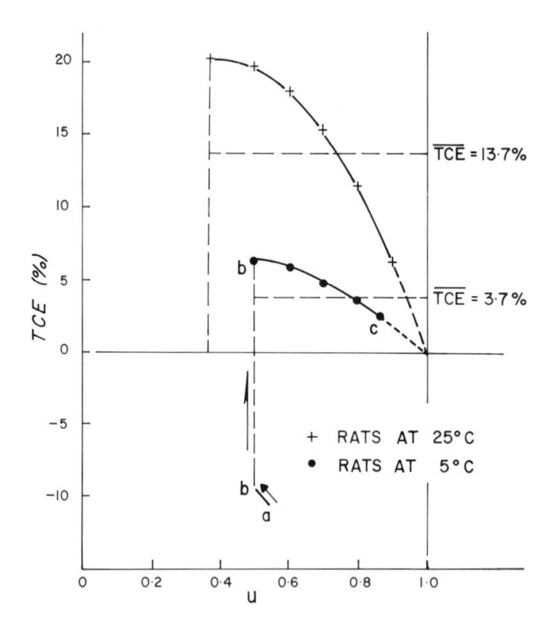


Fig. 11.8. Comparison of the parabolic arcs of the thermochemical efficiencies, TCE, of the warm and cold living rats versus degree of maturity, $u$, showing the jump of the TCE($u$) of the cold living rats at point b from the negative thermochemical efficiency region to the positive region after the first 8 days in the cold chamber. Here is seen the strong effect of ambient temperature on the thermochemical efficiency of growth

The steep drop in $\Gamma$ compared to the 6% reduction of $u$ on suddenly subjecting these rats to 5 °C ambient probably indicates these rats had sufficient disposable carcase fat in their initial body weight of 215 g for self consumption until the energy intake could be increased in excess of the increased maintenance for the cold living rats to begin to grow and fatten. The lower slope of the $\Gamma(u)$ line, namely 1.39, compared to the slope 2.03 for the warm living rats, means the cold rats were laying down carcase fat much more slowly so that the percentage of carcase fat at maturity will be considerably lower for the cold living rats.

Figure 11.8 shows the parabolic forms of TCE($u$), expressed here as percentages, found for the warm living and cold living rats. Because ($AB$) is a factor in the TCE equation and the 20 °C reduction of ambient temperature reduced ($AB$) by 56%, the cold living rats achieved a much lower thermochemical efficiency of growth. The maximum values of the TCE were reduced from 20.2% for the warm living rats to 6.1% for the cold living ones. In the period 0–8 days the weights of the cold living rats dropped, as did their specific combustion energies, $\Gamma$, (line segment ab in Figs. 11.5 and 11.7), because, in this period, energy intake had to be augmented by reduction of stored energy in order to maintain power balance. In this period, $\Gamma(u) = -5.8 + 13.8u$ and the growth efficiency $dW/dF = -0.042$. Using these values in $\mathrm{TCE} = 100(\Gamma + u d\Gamma/du) dW/dF$ gives the percent of TCE as

$$\mathrm{TCE} = 6.7 - 32u, \ 0.54 \geqq u \geqq 0.51,$$

which is shown as line segment ab in the negative TCE region of Fig. 11.8. The line, sloping up with the decreasing $u$, shows that these rats tended to increase their TCE as their live weights decreased until point b was reached, at which the TCE appears to jump discontinuously from −9.6%–6.1%. Although this discontinuity is exaggerated, the growth and fattening of these rats does show a marked turnabout in the neighbourhood of point b (Figs. 11.5 and 11.7). The TCE of the 25 °C and 5 °C rats both drop parabolically from highs of 20.2% and 6.1%, respectively, towards 0 as the fraction of maturity approaches unity. The curves show extrapolation to (0, 1), though this may be unwarranted from the range of the data in the experiment by Joy and Mayer. Because TCE decreases as $(AB)$ decreases, there may be a temperature at which the animal may grow but not fatten.

With greater demands to maintain body temperature plus other demands (maintenance needs independent of temperature, genetic urge to grow and fatten, etc.), the animal will compensate, in so far as its digestive capacity permits, by increasing $dF/dt$. There is, however, an upper bound on the compensation by increased $dF/dt$. Thus, if ambient temperature is continuously lowered below some critical point, fat accumulation and eventually non fat body growth will be depressed. If the animal has fat stores and the temperature is sufficiently low, the stores will be used to supplement the food consumed in order to accomplish non fat body growth. This will be reflected in decreased $\Gamma$ even though $W$ may increase. If the temperature is even lower, the use of fat stores can be rapid enough so that $W$ will decrease and a strong decrease of $\Gamma$ will be observed for a time. If the temperature is low enough so that maintenance needs exceed energy consumption, then, after fat stores are used, non fat body will decrease. If the temperature is very low but energy consumption exceeds maintenance needs, then, after an initial period of adjustment when fat stores are consumed, $dW/dt$ will be low compared to that above the critical temperature and $\Gamma$ will remain relatively low, although it may increase as $W$ increases, because, relative to the urge to grow, the urge to fatten increases with age. At very low temperatures not only will growth be slowed, but the animal may be permanently stunted. These tentative conclusions about $dF/dt$, $W$, and $\Gamma$ in relation to ambient temperature, based on very general biological facts and considerations, are well illustrated by comparing the data for animals in low and high ambient temperatures in Figs. 11.5 and 11.8. These remarks need to be falsified by experiment.

Joy and Mayer (1968) have shown that a large fraction of the input power goes to the extra mechanical energy demanded to move the body about in chronic cold living. This power is properly included in the term $M(t)$ [Eq. (11.1)] because it is demanded by maintenance activity. Imposed work routines should be added to Eq. (11.1) because they have effects on the specific combustion energy, $\Gamma$, and energy intake, $\beta dF/dt$. Equation (11.1) is not complete, even as modified, for a non-productive animal, but it is useful for gaining insight into the performance of the animal under various conditions. Various modifications of Eq. (11.1) have been found useful. Blaxter (1968) has used a form to study maintenance feeding of sheep and Morowitz (1968, Chap. 6) has used another form, illustrated with a hydraulic analogue, to discuss storage of energy by living organisms in terms of information theory. Appropriate terms can be added to Eq. (11.1) or Eq. (11.2) to cover animals producing milk, eggs, fleece etc.

The reduced forms of the power balance equations and their modifications are empirical with no theoretical foundations. Strunk (1971) has reported theoretical considerations of the various means whereby the heat of metabolism is dissipated to the environment via conduction, convection, etc., and the physical and geometrical properties of the thermal resistances to heat flow required by animals to produce and/or maintain internal temperature and temperature differences between themselves and their environments. He is justly critical of scientists who interpret their experimental data, involving flow of metabolic heat to the environment, as yielding measures of the physical thermal resistances of active tissues, and of the insulated surfaces separating the metabolically active volumes of the animals from the environment.

Animals are as the biologists receive them, there are neither blue prints showing the geometry of the animal's construction, nor bills of material showing the physical properties of the various tissues comprising the whole animal. Experimenters may therefore consider their whole animals as having lumped physical and thermal characteristics, the forms of which find justification in theoretical considerations such as Strunk's, or in sets of nested and adjacent compartments of tissues, carefully devised to make the physical and thermal properties more uniform in representing the whole animal. Milsum (1966) discusses the use of the various approaches to the whole animal in the context of biological control mechanisms. Equation (11.1) is an example of representing the power balance of whole animals with the lumped characteristics of energy intake, total energy storage and power dissipated in maintenance without regard to the physical parameters of the macro or micro systems integrated into the whole animal plus the food taken in. There are experiments reported in the literature, regarding the response of animals to different ambient temperatures, which cannot yield estimates of physical parameters, but which can give insight into operation of the whole animal provided the terms in equations like (11.1) or (11.2) are appropriately chosen. The experiment by Davis et al. (1973) is such an experiment on intake and utilisation of energy by mature laying hens in relation to ambient temperature.

## 11.3 Power Balance and Work

The power balance Eq. (11.1) reduces to Eq. (11.8) for an adult animal not storing energy and freely feeding in a constant environment to meet all of its energy needs.

$$\beta dF/dt = M. \tag{11.8}$$

The maintenance $M$ is composed of many mechanical, chemical, electrical, and thermal processes operating all the time over the entire body and its contact with the environment. These processes have components which are continuous, pulse type of high or low frequency, or almost constant. Chewing, walking, or a game of handball are examples of pulse type, high frequency mechanical energy dissipations. They are pulse type because they occupy relatively small amounts of wakeful time. Digestion is an example of a pulse type, low frequency chemical dissipation of energy. The flow of blood is almost a constant dissipation of energy. The losses through excretion are of a high frequency, pulse type but evaporative losses and the excitation of the nervous system are almost constant.

Thus there is a large number of different dissipations, each of which at any given time, contributes to the total dissipation. The spectrum of these contributions is highly variable from time to time. It is plausible that at any moment a demand that accentuates one or more components may modify the contributions of all the other processes to the total dissipation. Later this is shown to be more than a plausibility in the case of physical work demands.

We have seen, in the theory of feeding and growth (Chap. 7) that as animals mature DE's (7.1) and (7.2) for ad libitum feeding approach the controlled feeding DE (7.3). There we saw that the term $[(AB)/T_0]W$ represents the live weight "leakiness" of the system which was replenished by the live weight equivalent of the food intake, $(AB)dF/dt$. Analogously we can see that the energy "leakiness" can be written as $[(AB)/T_0]\Gamma W$, or its equal $k\Gamma W$, where $k$ is Brody's maturing constant, $\Gamma$ is the mature specific whole body energy, and $W$ is near the mature weight, $A$. This energy "leakiness" must be counter balanced by an equivalent energy intake as shown in Eq. (11.9),

$$\beta dF/dt = (k\Gamma)W. \tag{11.9}$$

This equation is applicable to any animal which is in a steady state of neither growing nor losing weight. Here it is seen that the maintenance term, $M$, in Eq. (11.8) is proportional to the weight, $W$, to the first power which is in line with the work of Thonney et al. (1976) re-evaluating the concept of metabolic size being proportional to $W^{0.75}$.

The caloric intake is of a pulse nature with ingestion occupying relatively small intervals with longer intervals of time between feeding periods. The caloric store dissipation and caloric intake vary almost incoherently from moment to moment so that an experimental power balance at instants of time is not possible. However, if the animal is in dynamic equilibrium with its environment, the mean variation of the difference between its dissipation of stored energy and its energy intake around zero will be of order zero over a sufficiently long time. Therefore Eq. (11.9) can be used to relate the rate of energy intake to the rate of energy output of the mature animal under steady conditions.

Mayer and Thomas (1967) discussed the control of feeding and concluded that there are three control systems, namely short, medium and long term which insure Eq. (11.9). Thus, the animal behaves like an integrator which minimises the errors of balance over time. Mayer et al. (1954) have used this knowledge of animal capabilities to design an experiment to investigate the effects of controlled work demands on the mechanical dissipation of energy by rats. Mayer et al. (1956) had also investigated the effects of different regimes of work on workers in a hemp mill in India.

This section puts the data of these studies by Mayer and his coworkers, of the effects of work on rats and men in a new perspective through a mathematical analysis based on Eq. (11.9) and draws several new conclusions. The results reported are accurate, but not precise, interpretations of the experiments. The raw data of these studies are not extant so the results are based on measurements from published graphs of the data shown in Fig. 11.9 and 11.10. The experiment on exercising rats will be described first.

A group of adult female rats, homogeneous with respect to age and weight, were fed a diet of fixed energy density and were exercised daily on a treadmill. The treadmill speed was constant at 1.61 km/h. The rats were pretreated by running them for at least 8 days for each period of exercise. This pretreatment was long enough for the long term feeding control to be operative and Eq. (11.9) was ex-

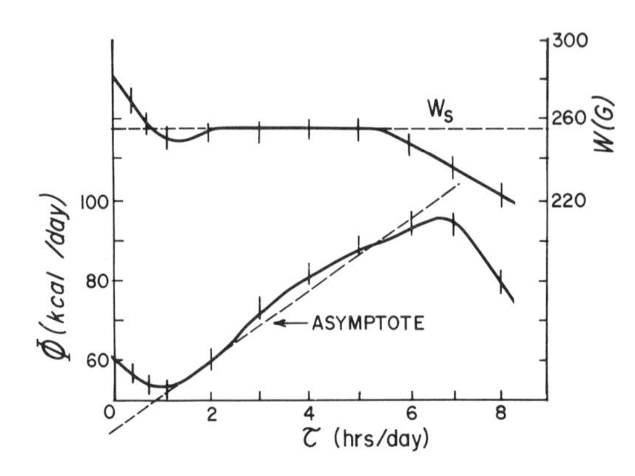


Fig. 11.9. Body weights, $W(\tau)$, and caloric intakes, $\Phi(\tau)$ versus time, $\tau$, hours per day enforced running on a treadmill at a speed of 1.61 km h$^{-1}$. (Reproduced with the permission of the American Journal of Physiology and Mayer et al. 1954)

pected to hold at a new equilibrium weight, $W$, and a new specific whole body combustion energy, $\Gamma$. The following six or seven days of exercise were used to obtain the stable weights and caloric intakes. The total days runs was at least fifteen at each level of daily exercise.

The data taken were average body weights and average daily caloric intakes per rat for 0, 1/3, 2/3, 1, 2, 3, 4, 5, 6, 7, and 8 h periods of exercise per day. These data are shown plotted in Fig. 11.9.

The experiment with the hemp mill workers is next described and followed by mathematical analysis of the two experiments in the same terms, because they seem to express the same phenomenon.

A sample was selected, from a group of 800 workers out of a total of 7,000 employees of a hemp mill in West Bengal, which provided data on height, weight, and type of work performed. The sample numbered about 25% of the group and was chosen to be homogeneous in regard to height with their types of work ranging from sedentary positions to heavy mill jobs.

The data on the individuals of the sample were average body weights and daily caloric intakes. These data were averaged by occupation and plotted by relative order of the daily work required in each occupation as shown in Fig. 11.10. The av-

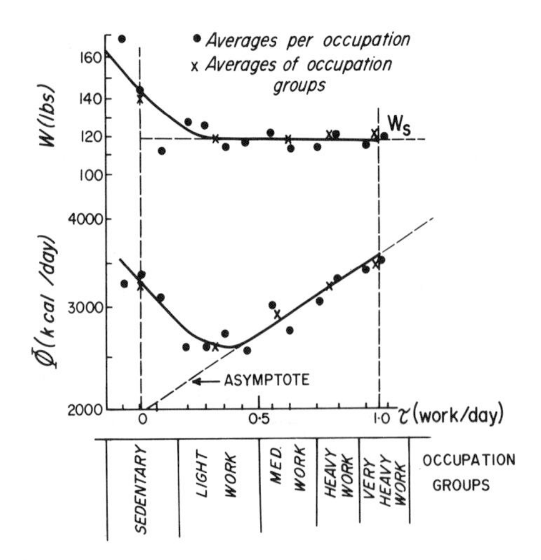


Fig. 11.10. Body weights, $W(\tau)$, and caloric intakes, $\Phi(\tau)$ of a group of Indian jute mill workers versus the physiological intensity of daily work, $\tau$, expressed from zero to one on a scale for occupations from sedentary work to heavy millweight work. (Reproduced with the permission of the American Journal of Clinical Nutrition and Mayer et al. 1956)

erage daily energy requirement for the work in each occupation was estimated from various physiological measurements made on individuals performing the work shown in the list along the abscissa of Fig. 11.10. Figures 11.9 and 11.10 show that in rats and men, the relations of body weights and daily caloric intakes to work reflect the same type of metabolic phenomenon. It is the purpose of this section to explore this type of metabolic behaviour from a quantitative point of view.

Equation (11.9) can be more conveniently written as

$$\phi = \gamma W, \tag{11.10}$$

where $\phi$ is $\beta dF/dt$ and $\gamma$ is $(k\Gamma)$. Equation (11.10) can be used to see what Gulick (1922) meant by such phrases as "spend thrift oxidation during over nourishment," and "economical oxidation during under nourishment" in describing his efforts to change his weight by changes in his caloric intake regimes. If the manner of living is unchanged and $W$ is constant, an increase of $\phi$ is reflected in an increase of $\gamma$ which signifies a lower efficiency of utilisation of the food. Similarly, a decrease in $\phi$ lowers $\gamma$ signifying a higher efficiency of utilisation.

Since the energy of work is shifting the body against internal resistances, it is reasonable to suppose that the loss rates per gram body weight due to physical work are lumped in a term denoted by $a$ and all other losses lumped into a term denoted by $\gamma_0$, so that $\gamma$ is $(\gamma_0 + a)$, thus giving

$$\phi = (\gamma_0 + a) W. \tag{11.11}$$

In exercising the rats, Mayer designed his experiments so that $a$ represented an imposed daily work which was proportional to the elapsed time of $\tau$ hours per day on the treadmill or

$$a = b\tau,$$

where $b$ is the constant of porportionality, dependent on the treadmill belt speed and having the units of kilocalories lost in physical work per gram body weight, per hour per day on the treadmill. Figure 11.9 and 11.10 show that $\phi$ and $W$ are functions of $\tau$ and it can also be assumed that $\gamma_0$ is a function of $\tau$ so that Eq. (11.11) can be more generally written as

$$\phi(\tau) = [\gamma_0(\tau) + b\tau] W(\tau). \tag{11.12}$$

What has been said here also applies to the Mayer et al. (1956) study of human body weight, daily caloric intake, and work. The units of $\tau$ and $b$ in this case are, of course, different, but the mathematical concepts are the same.

Figures 11.9 and 11.10 show that W($\tau$) decreases as $\tau$ increases and is clearly asymptotic to a stable weight $W_s$. The implication is that body weight greater than $W_s$ is sensitive to the energy losses due to work, defined as $a$. This sensitivity can be expressed as a general biological property in Eq. (11.13)

$$dW/da = -c(W - W_s), \tag{11.13}$$

where $c$ is the fractional decrease of $(W - W_s)$ per unit increase of $a$. Therefore $c$ is a measure of the sensitivity of body weight to work. Since $a = b\tau$, substitution in Eq. (11.12) gives

$$dW/d\tau = -bc[W(\tau) - W_s], \tag{11.14}$$

where $bc$ is now the fractional decrease of $[W(\tau)-W_s]$ per unit increase of daily elapsed time, $\tau$, on the treadmill. The figures also clearly show that the daily caloric intake, $\phi$, given by Eq. (11.12) is, for large $\tau$, asymptotic to a straight line the slope of which is $bW_s$ and the $\phi$ axis intercept of which is $\gamma_s W_s$, so that Eq. (11.15) applies along the asymptote

$$\phi(\tau)=(\gamma_s+b\tau)W_s. \tag{11.15}$$

The implication here is that $\gamma_0(\tau)$ in Eq. (11.12) is also sensitive to work and is asymptotic to $\gamma_s$ for large $a$ and the following relationship may be expected to apply;

$$d\gamma_0/da=-e(\gamma_0-\gamma_s), \tag{11.16}$$

where $e$ is the fractional decrease of $(\gamma_0-\gamma_s)$ per unit of increase of $a$. The coefficient $e$ therefore is a measure of the sensitivity to physical work of the energy losses other than work. If $b\tau$ is substituted for $a$ in Eq. (11.16) then the sensitivity of $\gamma_0$ to the amount of daily work $\tau$ is given by

$$d\gamma_0(\tau)/d\tau=-be[\gamma_0(\tau)-\gamma_s], \tag{11.17}$$

where $be$ is the fractional decrease of $[\gamma_0(\tau)-\gamma_s]$ per unit increase of $\tau$.

The DE's (11.14) and (11.17) can be solved using the initial conditions $\tau=0$, $W(\tau)=W(0)$ and $\gamma_0(\tau)=\gamma_0(0)$. The solutions are, respectively,

$$W(\tau)=[W(0)-W_s]\exp[-(bc)\tau]+W_s, \tag{11.18}$$

$$\gamma_0(\tau)=[\gamma_0(0)-\gamma_s]\exp[-(be)\tau]+\gamma_s. \tag{11.19}$$

Equation (11.18) shows how the body weight may be expected to decrease exponentially with decay constant, $(bc)$, as work increases. Equation (11.19) expresses a new relation not apparent from the graphs. It expresses how all the energy loss rates per gram of body weight, other than that due to physical work, can be expected to decrease exponentially with the decay constant, $(be)$, as work becomes heavier. Since $b$ is an experimental constant depending on the treadmill speed, the introduction of $c$ and $e$ as measures of the sensitivities of body weight and maintenance energy per unit body weight respectively to physical work demands are also new.

Equations (11.18) and (11.19) in conjunction with Eq. (11.12) define the caloric intake per day as a demand function of the daily exercise or work. They are reasonable representations of the data in Figs. 11.9 and 11.10 and can be used to estimate the values of the parameters $b$, $c$, $e$, $W_s$, and $\gamma_s$ from these data.

The parameter $W_s$ can be estimated from the figures and used in Eq. (11.15) with values of $\phi$ at various values of $\tau$ along the asymptotic straight line of caloric intake to estimate $b$ and $\gamma_s$. It remains to estimate $c$ and $e$. With $b$ known, parameter $c$ can be estimated using the following version of Eq. (11.18) on the $(\tau, W)$ data read from Figs. 11.9 and 11.10 and calculating a representative value of $c$ for each case.

$$c=(1/b\tau)\ln\{[W(0)-W_s]/[W(\tau)-W_s]\}.$$

Table 11.1. Estimated values $a$, $c$, $e$, $W_s$ and $\gamma_s$ and initial values of $W(0)$ and $\gamma(0)$ for studies of rats and exercise and of men and work (Mayer et al. 1954, 1956)

| | Rats | Men |
|---|---|---|
| $a$ | 0.04 kcal $gm^{-1}$ $hr^{-1}$ | 0.03 kcal $g^{-1}$ (unit of daily work)$^{-1}$ |
| $c$ | 69 g day $kcal^{-1}$ | 270 g fay $kcal^{-1}$ |
| $e$ | 42 g day $kcal^{-1}$ | 180 g day $kcal^{-1}$ |
| $W_s$ | 260 g | $52 \times 10^3$ g |
| $\gamma_s$ | 0.17 kcal $day^{-1}$ $g^{-1}$ | 0.036 kcal $day^{-1}$ $g^{-1}$ |
| $W(0)$ | 285 g | $63 \times 10^3$ g |
| $\gamma(0)$ | 0.20 kcal $day^{-1}$ $g^{-1}$ | 0.05 kcal $day^{-1}$ $g^{-1}$ |

The estimate of the parameter e for each case is similarly obtained from the following version of Eq. (11.19), namely

$$e = (1/b\tau)\ln\{[\gamma_0(0) - \gamma_s]/[\gamma_0(\tau) - \gamma_s]\},$$

and using values of $\gamma_0(\tau)$ calculated from the $W(\tau)$ and $\phi(\tau)$ data read from the figures substituted in the following version of Eq. (11.12), namely

$$\gamma_0(\tau) = \phi(\tau)/W(\tau) - b\tau.$$

This procedure was used to estimate the parameters $b$, $c$, $e$, $W_s$, and $\gamma_s$ from measurements made on the curves in Figs. 11.9 and 11.10 using dividers and a millimeter scale. The range of $\tau$ for the rat experiments was 0–5 h/day. For the study on men the range of $\tau$, as a measure of daily work, was arbitrarily taken to be unity, with zero being assigned to supervisors in the sedentary class of work and unity assigned to the average in the very heavy class of work. I am aware of the inaccuracies inherent in measuring graphs and claim nothing for the estimated values of the parameters other than order of magnitude (Table 11.1).

Table 11.1 shows the estimated values of the parameters $b$, $c$, $e$, $W_s$, and $\gamma_s$ and the initial values of $W$ and $\gamma_0$ for the studies of rats and men. Of these parameters $\gamma_0$, $W_s$, and $\gamma_s$ are more accurate than $c$ and $e$ because of the larger numbers of averaged data points in the straight line portions of the $W(t)$ and $\phi(t)$ curves. As a consequence, the % differences between $c$ and $e$ for rats (64) and men (50) probably are not significant. The difficulty of interpreting $c$ and $e$ as sensitivities to work is lessened by noting that the amounts of work required to reduce $(W - W_s)$ and $(\gamma_0 - \gamma_s)$ to one half their maximum values are also measures of the sensitivities of the body weight and maintenance energy to work. These values of work are $\ln 2/c$ and $\ln 2/e$, respectively. These formulae show more clearly that the higher the sensitivity, the less the work required to obtain a given reduction in weight or maintenance energy. On these bases the body weight sensitivities to work of rats and men are respectively 0.01 and 0.0026, and the sensitivities of the maintenance energy are respectively 0.017 and 0.0039.

Using Eqs. (11.12), (11.18), and (11.19) and the parameter values in Table 11.1, a computer program was written to compute $W(t)$ and $\phi(t)$ in a form suitable for a Calcomp plotter to produce the curves shown in Figs. 11.9 and 11.10. The result reproduced the main features of the rat curve for $0 < \tau \leqq 5$ and the man curve for

$0 \leqq \tau \leqq 1$, sufficiently closely to accept the values of the parameters in Table 11.1 as the best estimates I could get under the circumstances.

The equations proposed here, to describe quantitatively the functional relationships among the experimental data points of these studies by Mayer and associates, show some physical properties of the responses of weight and maintenance to work, which are not obvious in Figs. 11.9 and 11.10. These equations have no relation to the adaptability of animals to more or less instantaneous or short term demands for physical activity. They quantitatively emphasise the long term ability of the animal to control its caloric intake to keep its weight and maintenance at or above the levels $W_s$ and $\gamma_s$. It appears that in these studies Mayer et al. have uncovered a second long term control system which enables animals to maintain their overall energy balance to meet the demands of physical work.

The values of the parameters listed in Table 11.1 are not strictly comparable between rats and men across the table, because the data for men were not obtained from a treadmill type of experiment. Much work has been done with men exercising on treadmills but I know of none comparable to the exercising of rats by Mayer et al. However, because of the generality of the approach, some valid remarks can be made.

The maximum percent weight reduction, $100[W(0)-W_s]/W(0)$ achieved in exercising rats was 9% compared to 17% for working men. The maximum percent maintenance reduction, $100[\gamma_0(0)-\gamma_s]/\gamma_0(0)$, was 15% for rats and 25% for men. The amount of work to get half these maximum reductions shows that men have a wider physiological tolerance for work than rats by factors of about five (Table 11.1). The reduction in the maintenance coefficient, $\gamma_0$, illustrates how physical work can cause redistribution of the contributions of the various energy loss mechanisms to the total maintenance.

Figure 11.9 shows that the weight and caloric intakes of the rats could not be established for periods of exercise greater than 5 h/day. It appears that the greater periods of running on a treadmill at a speed of 1.61 km/h are outside the range of the adaptability of the rat to physical work. The 5 h/day of exercise can be taken as an estimate of the upper limit of this range of adaptability to work or exercise. However, these experiments do not estimate the lower limit of this range. The graphs and the equations for men and rats imply that in situations where the physical activity can be assigned negative values of $\tau$, the body weight and energy losses will increase and the caloric intake, $\phi$, will be adjusted upward to meet the new demands to maintain the new values of $W(0)$ and $\gamma_0$. The lower limit of adaptability might well be the response of $W$, $\gamma_0$, and $\phi$ to physical immobilisation of the animal.

It is interesting to note that in the case of the rats the asymptotic maintenance $\gamma_s$ contributes 46% to the total loss at the upper limit of adaptability ($\gamma_s + 5b$).

In the case of the men, the extreme of the range of adaptability to work was not approached. However, the percentage contribution of the maintenance $\gamma_s$ to the total ($\gamma_s + b$) for the very heavy workers is 55%. This is comparable to rats, but there remains the question of the upper limit of adaptability to work for men, which was not approached by the hemp mill workers. Thus, it appears that the metabolic efficiency of rats and men increases as work is increased towards some upper limit. When one recalls Gulick's remarks it appears, from the work of Mayer and coworkers, that body weight, fatness and metabolic efficiency are better regulated by enforced physical work than by voluntary caloric intake reduction.

Equations (11.12), (11.18), and (11.19) are suitable for use in nonlinear regression of the raw data from experiments such as these. The statistics of regression would show the adequacy of fit and yield estimates of the parameters. I shall next propose a theoretical basis for these equations.

### 11.3.1 A Theoretical Basis for the Effects of Work Routines on Mature Animals

The preceding section dealt with an empirical approach to the results of the experiments of Mayer et al. (1954) in which mature female rats were subjected to enforced daily exercise of increasing levels of severity. The approach yielded some new quantitative ideas in support of the discussion by Mayer et al. of the biology of the effects of routine work on the body weights and caloric intakes of their experimental animals. However I felt dissatisfied with the empirical approach because it did not focus on those quantities that defined the experiment, namely, the treadmill belt speed, $v$, the daily period of exercise, $\tau$, and the number of consecutive days, $n$, the rats were exercised for a given value of $\tau$. Intuitively I thought there must be some strong relation among these quantities which made possible the clear experimental results obtained.

In as much as all dynamic processes must operate against internal as well as external frictional resistances, it seemed reasonable to assume that an animal has a whole body frictional resistance which it must work against in order to move its body mass about in its living space. It is further assumed that this whole body friction is Newtonian; in other words the whole body force, resisting the shearing actions of tissues and bones resulting in moving the body about, is proportional to the speed of movement. In the c.g.s. (cm, gm, s) system of units a coefficient of Newtonian friction, $r$, has the units of dynes seconds per centimetre per gram of mass moved. It follows that the internal frictional force against which the animal must move its body weight, $W$, at a speed of $v$ cm/s is

$$Wrv \text{ dynes,}$$

and if it continues this speed for $\tau$ s/day the distance over which it works against the force is $v\tau$ cm/day, and therefore the internal energy generated is

$$Wrv^2\tau \text{ ergs/d.}$$

This energy appears as body heat in kilocalories, namely

$$JWrv^2\tau, \text{ kcal/d,}$$

where $J$ is the heat equivalent of work, namely $4.2 \times 10^{-7}$ kcal/erg. This mechanically generated body heat must be dissipated to the environment in addition to all the other types of body heat generated to maintain the animal as an ongoing biologically active entity.

In order to give the reader a better idea of the physical meaning of the Newtonian viscosity coefficient, $r$, as I use it here, I shall refer him to Eq. (7.4), Chap. 7, repeated here for convenience, but with all the terms multiplied by a mass m, and the right hand side modified as in the following equation.

$$m dv/dt + mrv = mg[1 - D(L)/D(m)]. \tag{11.20}$$

Here $g$ is the acceleration due to gravity in centimetres per second squared; $v$ is the speed of the body of mass, $m$ grams and density $D(m)$, in units of centimetres per second falling through liquid, $L$, having density $D(L)$ and coefficient of frictional force, $r$, acting against motion through it.

In the c.g.s. system of units mass in grams times acceleration in centimetres per second squared has the units of dynes of force. All the terms in Eq. (11.20) are forces and therefore have the units of dynes. It follows that the term, $mrv$, must have the units of dynes, so that the units of $r$ are dynes s/cm/g mass of the moving body. The factor $[1 - D(L)/D(m)]$ corrects the force of gravity, $mg$, for the bouyancy of the liquid acting on the body against the force of gravity.

Newton (d. 1727) wrote Eq. (11.20) as an aid to his experimental studies of bodies of various densities falling through liquids of various densities and varying in "thickness" or "thinness," for example treacle or water. He was able to show quantitatively that Aristotle was correct, in a limited way, in saying that heavier bodies (more dense) fall faster than lighter bodies (less dense) in the same liquid simply because the bouyancy correction factor $[1 - D(L)/D(m)]$ was nearer unity for the denser bodies. Newton also showed that if $D(L) = 0$ and $r = 0$ all bodies would fall with the same acceleration and speed regardless of their densities, in agreement with Galileo (d. 1642).

Furthermore, through his work with Eq. (11.20), Newton was able to integrate a great deal of common sense knowledge about the behaviour of bodies moving through various liquids. If $D(m) < D(L)$ then $[1 - D(L)/D(m)] < 0$ and the body will rise in the liquid at an acceleration and speed dependent on $r$. If $D(m) > D(L)$ and $r$ is very small, as in water or air, the bodies will fall with an acceleration close to $g$. If $D(m) > D(L)$ and $r$ is very large then $dv/dt$ will be small and the body will fall in the liquid at an almost constant speed $v = (g/r)[1 - D(L)/D(m)]$. Further if the liquid has such a high $r$ that it seems like a solid, for example tar, the same phenomenon will occur. The body will very slowly sink out of sight with a low constant speed.

I think the reader can understand that I prefer Newton's approach to Aristotle's "law" of falling objects rather than Galileo's logical argument because Newton's approach is more accurate in thought and practice and above all much wider in its intellectual reach into the world of practical activity. This small contribution by Newton, when the word "viscosity" was not in the language, was the seed from which grew the highly developed field of rheology today.

If the animal is in energetic balance, under the condition $\tau$ seconds per day of exercise at speed $v$, at body weight, $W(\tau)$, food intake $dF/dt = W(\tau)/T_0(\tau)$, the caloric intake as a function of daily exercise becomes

$$\Phi(\tau) = \beta dF/dt = [\gamma(\tau) + Jrv^2\tau]W(\tau), \text{ kcal/d}. \tag{11.21}$$

These equations are applicable to the experiment of Mayer et al. if $v$ is taken as the treadmill speed and $\tau$ is the period of time per day the rats are exercised on the treadmill.

Before the rats were subject to the exercise regimes, they had been raised under laboratory conditions to the steady mature weight $W_0$, food intake $dF/dt = W_0/T_0$ and energetic balance

$$\Phi = \gamma_0 W_0.$$

Subjecting these rats to exercise regimes of increasing severity reduced their weight, $W$, to a steady weight, $W_s$, which they maintained until they entered a state of exhaustion. The experiment demonstrated that exercising the rats for $\tau$ s/day over a long enough period of days, $n$ (about 10 days for the rats), a new steady weight, $W(\tau)$, and food intake $dF/dt = W(\tau)/T_0(\tau)$ and energetic balance given by Eq. (11.21) were attained. Here $W(\tau)$ is less than $W_0$ meaning that body mass $W_0 - W(\tau)$ was available to be catabolized to meet the extra demands of the enforced work.

Now let us suppose that the rats have reached a steady weight, $W(\tau)$, with body mass, $W(\tau) - W_s$, still available for use in meeting increased demands of daily exercise. If $\tau$ is increased by $\Delta\tau$ and the rats exercised for $n$ days to reach a steady weight of $W(\tau) - \Delta W$, then the fractional decrease of the available weight, namely $-\Delta W(\tau)/[W(\tau) - W_s]$, is $Jrv^2 n\Delta\tau/\lambda$. Here $\lambda$ is the energy obtained in catabolizing one gram of the available weight. Taking $\Delta\tau$ very small, namely $d\tau$, the preceding incremental equality becomes the following DE,

$$dW(\tau)/d\tau = -[W(\tau) - W_s]Jrv^2 n/\lambda.$$

The solution of this equation, with i.c. $W(\tau) = W_0$ when $\tau = 0$, is the steady weight as a function of the daily exercise period, $\tau$, given by

$$W(\tau) = (W_0 - W_s)\exp(-Jrv^2 n\tau/\lambda) + W_s. \tag{11.22}$$

When the body weight, $W(\tau)$, has been reduced from $W_0$ to $W(\tau)$ it is expected that the maintenance factor, $\gamma(\tau)$, for all those body processes not related to the imposed work will also be reduced from its original value, $\gamma_0$. The experiment showed that for $\tau$ large enough $\gamma(\tau)$ approached the minimum value of $\gamma_s$. Here again it can be seen that the fractional decrease of the available change in $\gamma(\tau)$, namely $-\Delta\gamma(\tau)/[\gamma(\tau)\gamma_s]$, is $Jrv^2 n\Delta\tau/u$. Here $u$ is the maintenance energy per gram associated with the decrease of body weight. Taking the increment of daily work, $\Delta\tau$, very small the preceding incremental quality becomes the DE,

$$d\gamma(\tau)/d\tau = -(\gamma(\tau) - \gamma_s)Jrv^2 n/u,$$

the solution of which, with i.c. $\gamma(\tau) = \gamma_0$ when $\tau = 0$, is

$$\gamma(\tau) = (\gamma_0 - \gamma_s)\exp(-Jrv^2 n/u) + \gamma_s. \tag{11.23}$$

Equations (11.21–11.23) are directly analogous to Eqs. (11.12), (11.18), and (11.19) and therefore show the same characteristics, namely for large $\tau$, $W(\tau)$ asymptotically approaches $W_s$, and the caloric intake $\Phi(\tau)$ approaches the straight line

$$\Phi(\tau) = \gamma_s W_s + (JW_s rv^2)\tau. \tag{11.24}$$

On the assumption that the animal has a whole body Newtonian viscosity coefficient, the theoretical equations show the relationship of the treadmill speed, $v$, the daily exercise period, $\tau$, and the number of days, $n$, at a given value of $\tau$ for the animal to reach a new steady state of weight and caloric intake. In Eqs. (11.22) and (11.23), if $v$ and $n$ are varied in such a manner that $(v^2 n)$ is constant, then the rates of decrease of $W(\tau)$ and $\Phi(\tau)$ would be unaffected, but the slope of the asymptotic straight line, Eq. (11.24) would vary as the square of the treadmill speed, $v^2$. This suggests a method of falsifying the theory presented here. Repeating this experiment with either an increased or decreased treadmill speed with n adjusted so that $(v^2 n)$ is the same as it was in this experiment, namely (1.61 km/h × 8 days) would critically test the assumption of a whole body Newtonian viscosity coefficient, $r$.

As the theoretical equations are directly related to the empirical ones, we can get for the rats estimates of the Newtonian viscosity coefficient, $r$, the energy of catabolizing a gram of available body weight, $\lambda$, and the maintenance energy per gram of body weight, $u$, associated with the loss of weight during exercise. Comparing the theoretical with the corresponding empirical equations we see from Table 11.1 for the rats,

1. the slope of the asymptotic straight line for $\phi(\tau)$ is $Jrv^2 = a/3{,}600$, from which
   $r = (0.04/3{,}600)/4.2 \times 10^{-7} \times (1.61 \times 10^5/3{,}600)^2$
   $= 0.013$ dynes sec. $g^{-1} cm^{-1}$.
2. the exponential decay constant for the steady weights, $W(\tau)$, is
   $Jrv^2 n/\lambda = ac/3{,}600$.
   From the preceding value of the slope
   $\lambda = n/c = 14/69$
   $= 0.20$ kcal $g^{-1}$ body weight catabolized.
3. the exponential decay constant for the maintenance factor, $\lambda(\tau)$, is
   $Jrv^2 n/u = ae/3{,}600$,
   so that
   $u = n/e = 14/42$
   $= 0.33$ kcal $g^{-1}$ maintenance energy per gram body weight corresponding to the weight catabolized.

These values of $r$, $\lambda$, and $u$ for the rats were used with the experimental values of $v$ and $n$ in Eqs. (11.22) and (11.21), with Eq. (11.23) substituted in (11.21), to calculate $W(\tau)$ and $\phi(\tau)$ for a range of values of $\tau$ from 0 to 5 h $d^{-1}$ exercise on the treadmill. The curves shown in Fig. 11.11 are graphs of the computed values $W(\tau)$ and $\phi(\tau)$ which show the main features of the experimental results illustrated in Fig. 11.9 where the vertical bars represent the errors of measurement of live weight, $W$, and caloric intake, $\phi$, at each period of daily exercise, $\tau$. The solid curves in Fig. 11.9 are the results of numerical analysis of the experimental data period by period and therefore must show some notable deviations from the smooth curves depicted in Fig. 11.11. These deviations may represent real phenomena not accounted for in the proposed theory, however I fail to understand biophysically, why $W(\tau)$ should undershoot $W_s$ in the neighbourhood of $\tau$ equal 1 h/day.

I had no previous ideas of the order of magnitudes of these apparently significant biological quantities. I had thought that $\lambda$ would be somewhere near 9 kcal/g in as much as the usual idea was that body fat was the primary tissue consumed in protracted work. However from the point of view of this theory burning off

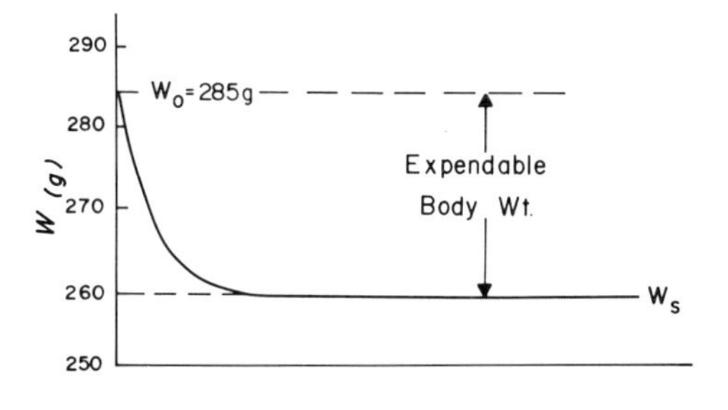


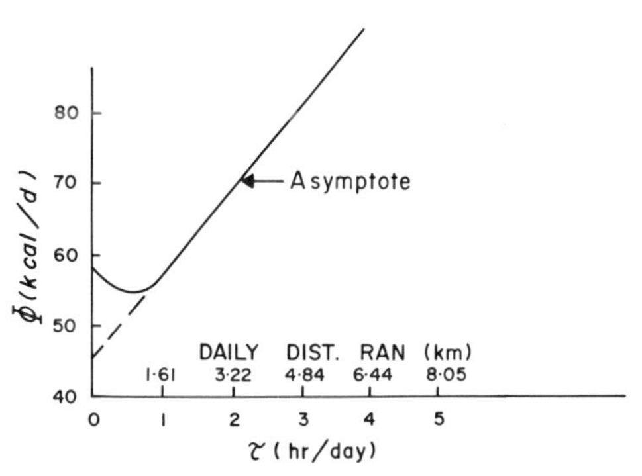


Fig. 11.11. Theoretical simulation of the forms of the live weight and caloric intakes functions, $W(\tau)$ and $\Phi(\tau)$ respectively, of rats exercised on a treadmill (Fig. 11.9) and men on various jobs in a jute mill (Fig. 11.10)

25 g, $W_0 - W_s$, of fat in working 1 h/day for 14 days at a speed of 44.7 cm/s seems an exaggeration (Fig. 11.11), especially since the amount of work the rats have done to lose the 25 g of body weight at most is

$$W_0 Jrv^2 n 3{,}600 = 156 \text{ kcals}$$

or 69% of heat required to burn off 25 g of fat at 9 kcal/g.

The very low value of the whole body friction coefficient, $r$, namely 0.013 dyne s $g^{-1}cm^{-1}$, indicates that these rats are indeed well oiled machines and can under ordinary conditions travel far and fast on their temporary stores of blood glucose and liver glycine in search of food and mates.

## References


Blaxter KL (1968) The effect of dietary energy supply on growth. In: Lodge GA, Lamming GE (eds) Growth and development of mammals. Plenum Press, New York

Brody S (1945) Bioenergetics and growth. First published: Reinhold, New York. (Reprinted: Hafner, New York, 1974)

Davis RH, Hassan, OEM, Sykes AW (1973) Energy utilization in the laying hen in relation to ambient temperature. J Agric Sci 81:173–177

Goodwin B (1963) Temporal organisation in the cell. A dynamic theory of cellular control processes. Academic Press, London New York

Gulick A (1922) A study of weight regulation in the adult human body during over nutrition. Am J Physiol 60:371–395

Haecker RW (1920) Investigations in beef production. Tech Bull 193 Minn Agric Exp Stn

Hilmi M (1975) The effect of growth on carcase composition of sheep. Unpubl thesis, Dep Vet Anat Univ Sydney

Joy MA, Mayer J (1968) Caloric expenditure in cold acclimating rats. Am J Physiol 215:757–761
Kleiber M (1961) Fire of life. Wiley, New York
Klotz JM (1957) Energetics in biochemical reactions. Academic Press, London New York
Mayer J (1955) The physiological basis of obesity and leaness. Nutr Abstr Rev Canberra: Commonw Bur Anim Nutr Tech Commun
Mayer J, Thomas D (1967) Regulation of food intake and obesity. Science 156:328–337
Mayer J, Vitale JJ (1957) Thermochemical efficiency of rats. Am J Physiol 189:9–42
Mayer J, Marshall NB, Vitale JJ, Christensen JH, Mashayekhi MB, Stare FJ (1954) Exercise, food intake and body weight in normal rats and genetically obese adult mice. Am J Physiol 177:544–548
Mayer J, Roy P, Mitra KP (1956) Relation between caloric intake, body weight and work. Am J Clin Nutr 4:169–176
McDonald P, Edwards RA, Greenhalgh JFD (1973) Animal nutrition, 2nd edn. Longman, New York
Milsum JH (1966) Biological control of systems analysis. McGraw, New York
Mitchell HH, Card LE, Hamilton TS (1926) The growth of white Plymouth Rock chickens. Ill Agric Exp Stn Bull 278
Mitchell HH, Card LE, Hamilton TS (1931) A technical study of the growth of White Leghorn chickens. Ill Agric Exp Stn Bull 367
Morowitz HJ (1968) Energy flow in biology. Academic Press, London New York
Parks JR (1970) Growth curves and the physiology of growth. III. Effects of dietary protein. Am J Physiol 219:840–843
Parks JR (1971) The effect of ambient temperature on the thermochemical efficiency of growth of cold acclimating rats. Am J Physiol
Reid JT and others (1968) Body composition in animals and man. Natl Acad Sci USA 1598
Scott D (1965) The determination and use of thermo-dynamic data in ecology. Ecology 46:673–680
Strunk TH (1971) Heat loss from a Newtonian animal. J Theor Biol 33:35–61
Thonney ML, Touchberry RW, Goodrich RD, Meiske JC (1976) Intraspecies relationship between fasting heat production and body weight; a reevaluation of $W^{0.73}$. J Anim Sci 43:692–703
Trincher KS (1965) Biology and information, elements of biological thermo-dynamics. Consultants Bureau, New York
Tulloh NM (1963) The carcase composition of sheep, cattle, and pigs. Symposium: carcase composition and appraisal of meat animals. Part 5, 15 pp. Ans Soc An Prod

# Appendix A Feeding and Growth Data

Table A-1. Class II data. Monthly weights and feed rates (lbs TDN/day) for Jersey and Holstein cattle. (Data of S. Brody 1945, p. 51)

| $i$ months | Jersey | | Holstein | |
|---|---|---|---|---|
| | $(dF/dt)_i$ lbs TDN/day | $W_i$ lbs | $(dF/dt)_i$ lbs TDN/day | $W_i$ lbs |
| 1 | 0.77 | 77 | 1.24 | 108 |
| 2 | 1.53 | 100 | 2.17 | 142 |
| 3 | 1.83 | 153 | 2.15 | 184 |
| 4 | 2.43 | 171 | 3.32 | 229 |
| 5 | 3.26 | 210 | 4.13 | 273 |
| 6 | 3.96 | 257 | 5.08 | 325 |
| 7 | 4.61 | 294 | 5.74 | 371 |
| 8 | 4.76 | 331 | 5.91 | 418 |
| 9 | 4.92 | 363 | 5.93 | 454 |
| 10 | 5.45 | 393 | 5.70 | 488 |
| 11 | 5.47 | 423 | 6.76 | 531 |
| 12 | 5.98 | 454 | 7.44 | 572 |
| 13 | 6.17 | 482 | 7.80 | 610 |
| 14 | 6.59 | 514 | 8.33 | 654 |
| 15 | 6.97 | 541 | 8.97 | 598 |
| 16 | 7.32 | 570 | 9.51 | 740 |
| 17 | 8.04 | 605 | 10.10 | 781 |
| 18 | 8.17 | 639 | 10.50 | 826 |
| 19 | 8.64 | 665 | 10.90 | 863 |
| 20 | 8.86 | 700 | 11.00 | 899 |
| 21 | 9.03 | 726 | 11.60 | 942 |

Table A-2. Class I data. Weight and food consumption of Jersey cattle from weaning age ($N=2$). (Unpublished data of Monteiro 1974)

| $i$ wks | $W_i$ kg | $f_i$ kg | $i$ wks | $W_i$ kg | $f_i$ kg |
|---|---|---|---|---|---|
| 0 | 58.38 | – | 42 | 240.38 | 104.77 |
| 2 | 65.33 | 35.32 | 44 | 245.86 | 104.64 |
| 4 | 72.66 | 43.40 | 46 | 256.19 | 108.08 |
| 6 | 78.63 | 44.61 | 48 | 264.19 | 106.76 |
| 8 | 86.61 | 48.18 | 50 | 272.25 | 110.99 |
| 10 | 95.19 | 55.99 | 52 | 280.08 | 109.70 |
| 12 | 104.05 | 56.44 | 54 | 286.97 | 112.40 |
| 14 | 113.27 | 65.49 | 56 | 294.66 | 110.42 |
| 16 | 122.63 | 69.99 | 58 | 303.16 | 114.38 |
| 18 | 131.75 | 73.67 | 60 | 310.86 | 110.28 |
| 20 | 141.94 | 76.39 | 62 | 317.41 | 111.51 |
| 22 | 151.00 | 78.92 | 64 | 322.50 | 110.14 |
| 24 | 160.33 | 86.42 | 66 | 329.41 | 108.55 |
| 26 | 172.66 | 91.79 | 68 | 335.72 | 104.82 |
| 28 | 180.77 | 92.40 | 70 | 344.00 | 107.61 |
| 30 | 189.72 | 92.11 | 72 | 352.63 | 104.23 |
| 32 | 199.75 | 98.45 | 74 | 358.55 | 109.83 |
| 34 | 208.00 | 99.89 | 76 | 366.63 | 110.87 |
| 36 | 216.50 | 100.43 | 78 | 372.86 | 111.30 |
| 38 | 224.69 | 102.72 | 80 | 381.05 | 110.52 |
| 40 | 232.22 | 100.14 | 82 | 388.22 | 113.04 |

Table A-3. Class I data. Weights and food consumption of Friesian cattle from weaning age ($N=2$). (Unpublished data of Montiero 1974)

| $i$ wks | $W_i$ kg | $f_i$ kg | $i$ wks | $W_i$ kg | $f_i$ kg |
|---|---|---|---|---|---|
| 0 | 89.63 | – | 42 | 357.36 | 135.35 |
| 2 | 100.42 | 58.91 | 44 | 367.92 | 137.39 |
| 4 | 111.60 | 64.43 | 46 | 379.00 | 133.39 |
| 6 | 122.68 | 68.42 | 48 | 389.97 | 135.39 |
| 8 | 134.50 | 75.56 | 50 | 400.42 | 139.19 |
| 10 | 147.15 | 81.79 | 52 | 414.04 | 137.22 |
| 12 | 161.28 | 83.98 | 54 | 424.73 | 138.48 |
| 14 | 174.97 | 89.61 | 56 | 435.31 | 141.42 |
| 16 | 188.65 | 95.27 | 58 | 443.55 | 141.45 |
| 18 | 203.02 | 98.31 | 60 | 455.89 | 145.24 |
| 20 | 216.18 | 100.04 | 62 | 464.94 | 147.74 |
| 22 | 233.55 | 113.62 | 64 | 477.60 | 146.81 |
| 24 | 245.13 | 113.90 | 66 | 486.63 | 147.58 |
| 26 | 259.26 | 117.58 | 68 | 496.81 | 145.36 |
| 28 | 273.42 | 118.33 | 70 | 507.02 | 146.58 |
| 30 | 285.97 | 123.87 | 72 | 518.92 | 144.64 |
| 32 | 297.52 | 128.17 | 74 | 527.89 | 145.73 |
| 34 | 310.13 | 123.98 | 76 | 540.94 | 144.43 |
| 36 | 323.52 | 133.86 | 78 | 551.13 | 141.21 |
| 38 | 333.10 | 130.77 | 80 | 558.63 | 139.66 |
| 40 | 346.86 | 133.93 | 82 | 566.02 | 134.00 |

Table A-4. Class I data. Periodic weights and weekly food consumption of broilertype females. (Unpublished data of Emmans 1974)

| $i$ wks | $f_i$ kg | $W_i$ kg |
|---|---|---|
| 0 | – | 0.0485 |
| 1 | 0.104 | 0.0920 |
| 2 | 0.192 | |
| 3 | 0.303 | 0.343 |
| 4 | 0.410 | |
| 5 | 0.536 | |
| 6 | 0.674 | 0.956 |
| 7 | 0.816 | |
| 8 | 0.898 | |
| 9 | 1.014 | 1.547 |
| 10 | 0.831 | |
| 11 | 1.049 | |
| 12 | 0.748 | 2.220 |
| 13 | 1.131 | |
| 14 | 1.025 | |
| 15 | 1.063 | 2.532 |
| 16 | 0.910 | |
| 17 | 0.929 | |
| 18 | 0.970 | 2.845 |
| 19 | 0.949 | |
| 20 | 0.883 | |
| 21 | 0.856 | |
| 22 | 0.859 | 3.155 |

Table A-5. Class I data. Weights and food consumption of swine from six weeks of age. (Data of Headley et al. 1961)

| $f_i$ lbs | $W_i$ lbs |
|---|---|
| – | 26.6 |
| 29.9 | 38.6 |
| 42.3 | 55.0 |
| 63.6 | 75.9 |
| 71.9 | 98.5 |
| 84.5 | 122.4 |
| 95.2 | 145.8 |
| 100.7 | 169.6 |
| 100.8 | 191.8 |
| 104.9 | 213.7 |

Table A-6. Class II data. Weights and ad libitum food intakes of male Great Dane dogs. (Unpublished data of Hedhammar 1972)

| $t$ wks | $dF/dt$ kg/d | $W$ kg | $t$ wks | $dF/dt$ kg/d | $W$ kg |
|---|---|---|---|---|---|
| 1 | 0.303 | 4.16 | 31 | 1.147 | 50.9 |
| 2 | 0.428 | 5.48 | 32 | 1.037 | 51.9 |
| 3 | 0.510 | 7.15 | 33 | 1.067 | 51.37 |
| 4 | 0.536 | 8.33 | 34 | 1.060 | 51.50 |
| 5 | 0.619 | 9.81 | 35 | 1.040 | 52.67 |
| 6 | 0.652 | 11.65 | 36 | 1.078 | 52.70 |
| 7 | 0.758 | 13.47 | 37 | 1.118 | 54.45 |
| 8 | 0.795 | 15.67 | 38 | 1.113 | 55.65 |
| 9 | 0.770 | 17.92 | 39 | 1.105 | 54.90 |
| 10 | 0.812 | 19.18 | 40 | 1.075 | 54.40 |
| 11 | 0.906 | 20.88 | 41 | 1.098 | 56.05 |
| 12 | 0.982 | 22.74 | 42 | 1.258 | 55.15 |
| 13 | 1.036 | 25.10 | 43 | 1.175 | 56.80 |
| 14 | 1.041 | 27.30 | 44 | 1.165 | 56.60 |
| 15 | 1.035 | 29.48 | 45 | 1.168 | 56.10 |
| 16 | 1.150 | 31.30 | 46 | 1.175 | 57.50 |
| 17 | 1.170 | 33.30 | 47 | 0.963 | 56.10 |
| 18 | 1.100 | 35.00 | 48 | 0.888 | 56.25 |
| 19 | 1.138 | 35.33 | 49 | 1.045 | 50.7 |
| 20 | 1.089 | 38.13 | 50 | 1.270 | 53.2 |
| 21 | 1.122 | 37.63 | 51 | 1.370 | 53.0 |
| 22 | 1.131 | 38.55 | 52 | 1.295 | 54.0 |
| 23 | 1.144 | 41.10 | 53 | 1.335 | 54.0 |
| 24 | 1.221 | 42.90 | 54 | 1.330 | 55.0 |
| 25 | 1.248 | 47.77 | 55 | 1.385 | 55.2 |
| 26 | 1.183 | 48.70 | 56 | 1.450 | 55.9 |
| 27 | 1.235 | 49.30 | 57 | 1.225 | 55.0 |
| 28 | 1.372 | 50.25 | 58 | 0.755 | 54.0 |
| 29 | 1.293 | 50.80 | 59 | 0.845 | 51.8 |
| 30 | 1.225 | 52.65 | 60 | 0.850 | 52.7 |

Table A-7. Class II data. Weights and 67% of ad libitum food intakes of male Great Dane dogs. (Unpublished data of Hedhammar 1972)

| $t$ wks | $dF/dt$ kg/d | $W$ kg | $t$ wks | $dF/dt$ kg/d | $W$ kg |
|---|---|---|---|---|---|
| 1 | 0.193 | 4.23 | 31 | 0.773 | 39.67 |
| 2 | 0.264 | 4.65 | 32 | 0.705 | 39.17 |
| 3 | 0.318 | 5.75 | 33 | 0.682 | 39.50 |
| 4 | 0.348 | 6.83 | 34 | 0.753 | 39.37 |
| 5 | 0.399 | 7.58 | 35 | 0.758 | 40.73 |
| 6 | 0.423 | 9.05 | 36 | 0.708 | 41.33 |
| 7 | 0.498 | 10.17 | 37 | 0.738 | 41.85 |
| 8 | 0.534 | 11.70 | 38 | 0.745 | 42.50 |
| 9 | 0.548 | 13.28 | 39 | 0.710 | 42.20 |
| 10 | 0.535 | 14.48 | 40 | 0.703 | 41.85 |
| 11 | 0.600 | 16.00 | 41 | 0.815 | 41.30 |
| 12 | 0.649 | 17.06 | 42 | 0.788 | 41.20 |
| 13 | 0.688 | 19.73 | 43 | 0.788 | 42.60 |

Table A-7 (continued)

| $t$ wks | $dF/dt$ kg/d | $W$ kg | $t$ wks | $dF/dt$ kg/d | $W$ kg |
|---|---|---|---|---|---|
| 14 | 0.680 | 20.75 | 44 | 0.755 | 43.90 |
| 15 | 0.681 | 22.33 | 45 | 0.763 | 42.20 |
| 16 | 0.721 | 23.90 | 46 | 0.743 | 43.60 |
| 17 | 0.766 | 24.85 | 47 | 0.613 | 42.75 |
| 18 | 0.749 | 26.45 | 48 | 0.695 | 42.60 |
| 19 | 0.758 | 26.05 | 49 | 0.895 | 41.10 |
| 20 | 0.740 | 28.65 | 50 | 0.945 | 44.10 |
| 21 | 0.731 | 28.10 | 51 | 0.850 | 44.80 |
| 22 | 0.749 | 28.85 | 52 | 0.895 | 44.50 |
| 23 | 0.752 | 30.50 | 53 | 0.905 | 43.60 |
| 24 | 0.816 | 31.07 | 54 | 0.920 | 45.00 |
| 25 | 0.825 | 34.95 | 55 | 1.015 | 45.00 |
| 26 | 0.825 | 35.60 | 56 | 0.880 | 45.70 |
| 27 | 0.840 | 36.35 | 57 | 0.685 | 46.80 |
| 28 | 0.903 | 37.15 | 58 | 0.915 | 46.80 |
| 29 | 0.858 | 38.65 | 59 | 0.915 | 46.80 |
| 30 | 0.786 | 39.65 | 60 | 0.915 | 46.10 |

Table A-8. Class III and IV data. Weight and cumulative food consumption of male chickens crossbred from Barred Plymouth Rock Females × Rhode Island Red males versus age with varying crude protein (CP) and nitrogen free extract (NFE). (Data of Hammond et al. 1939)

| | CP = 25% NFE = 49.20% | | CP = 23% NFE = 51.73% | | CP = 21% NFE = 53.20% | | CP = 19% NFE = 55.36% | |
|---|---|---|---|---|---|---|---|---|
| $t$ | $F$ kg | $W$ kg | $F$ kg | $W$ kg | $F$ kg | $W$ kg | $F$ kg | $W$ kg |
| 8 | 2.553 | 0.669 | 2.378 | 0.678 | 1.822 | 0.651 | 1.942 | 0.599 |
| 12 | 4.855 | 1.256 | 4.442 | 1.227 | 3.662 | 1.121 | 3.989 | 1.135 |
| 16 | 9.166 | 1.893 | 7.448 | 1.862 | 6.714 | 1.734 | 6.784 | 1.777 |
| 20 | 12.625 | 2.318 | 9.934 | 2.256 | 9.523 | 2.287 | 9.508 | 2.264 |
| 30 | 21.469 | 3.088 | 18.918 | 3.001 | 18.945 | 3.094 | 18.948 | 3.034 |
| 40 | 30.624 | 3.142 | 29.098 | 3.241 | 29.441 | 3.120 | 27.758 | 3.085 |
| 52 | 42.303 | 3.104 | 42.673 | 3.111 | 42.835 | 3.054 | 38.923 | 2.967 |

| | CP = 17% NFE = 57.18% | | CP = 15% NFE = 58.99% | | CP = 13% NFE = 60.80 | |
|---|---|---|---|---|---|---|
| $t$ | $F$ kg | $W$ kg | $F$ kg | $W$ kg | $F$ kg | $W$ kg |
| 8 | 2.057 | 0.665 | 1.874 | 0.559 | 1.803 | 0.467 |
| 12 | 4.424 | 1.193 | 4.162 | 1.092 | 3.281 | 0.773 |
| 16 | 7.088 | 1.692 | 6.959 | 1.637 | 5.878 | 1.160 |
| 20 | 9.706 | 2.010 | 9.570 | 2.041 | 8.718 | 1.672 |
| 30 | 18.715 | 3.067 | 18.251 | 3.004 | 18.112 | 2.724 |
| 40 | 28.318 | 3.105 | 27.331 | 3.141 | 28.105 | 3.110 |
| 52 | 40.597 | 3.004 | 38.125 | 3.000 | 40.741 | 3.024 |

Table A-9. Class III and IV data. Weight and cumulative food consumption of male chickens crossbred from Barred Plymouth Rock females × Rhode Island Red males versus age with varying crude protein (CP) and nitrogen free extract (NFE). The food consumed restricted to 70% of ad libitum. (Data of Hammond et al. 1939)

| | | CP=25% | CP=23% | CP=21% | CP=19% | CP=17% | CP=15% | CP=13% |
|---|---|---|---|---|---|---|---|---|
| *t* | *F* | *W* | *W* | *W* | *W* | *W* | *W* | *W* |
| wks | kg | kg | kg | kg | kg | kg | kg | kg |
| 8 | 1.463 | 0.600 | 0.570 | 0.573 | 0.535 | 0.570 | 0.490 | 0.390 |
| 12 | 3.097 | 1.161 | 1.074 | 1.078 | 0.986 | 1.010 | 0.917 | 0.710 |
| 16 | 5.170 | 1.709 | 1.608 | 1.611 | 1.470 | 1.480 | 1.403 | 1.089 |
| 20 | 7.495 | 2.038 | 1.973 | 1.896 | 1.815 | 1.885 | 1.775 | 1.484 |
| 30 | 13.788 | 2.574 | 2.428 | 2.409 | 2.304 | 2.377 | 2.338 | 2.244 |
| 40 | 20.340 | 2.837 | 2.571 | 2.657 | 2.544 | 2.631 | 2.711 | 2.536 |
| 52 | 28.224 | 2.761 | 2.743 | 2.639 | 2.584 | 2.639 | 2.740 | 2.707 |

Table A-10. Class III data. Weights and cumulative food consumed to attain these weights by crossbred male chickens. (Data of Card 1961)

| | Rhode Island Red X Barred Plymouth Rock | New Hampshire X Barred Plymouth Rock |
|---|---|---|
| *F* | *W* | *W*[a] |
| kg | kg | kg |
| 1 | 0.337 | 0.665 |
| 2 | 0.660 | 1.120 |
| 3 | 0.956 | 1.530 |
| 4 | 1.227 | 1.900 |
| 5 | 1.475 | 2.235 |
| 6 | 1.702 | 2.535 |
| 7 | 1.911 | 2.805 |
| 8 | 2.102 | |
| 9 | 2.277 | |
| 10 | 2.438 | |
| 11 | 2.585 | |
| 12 | 2.719 | |

[a] Diet typical of c. 1960

Table A-11. Class I data. Average weekly weights and food consumed per rat for four populations of 25 male Dan Rolfsmeyer Company strain of Sprague Dawley Albino rats each as affected by specific dietary energy. (Unpublished data of Parks 1968)

| Age | % RSE 100 | | 91.31 | | 82.64 | | 73.96 | |
|---|---|---|---|---|---|---|---|---|
| $i$ wks | $f_i$ kg | $W_i$ kg | $f_i$ kg | $W_i$ kg | $f_i$ kg | $W_i$ kg | $f_i$ kg | $W_i$ kg |
| 0 | – | 0.0596 | – | 0.0589 | – | 0.0611 | – | 0.0593 |
| 1 | 0.0764 | 0.0901 | 0.0783 | 0.0878 | 0.0777 | 0.0871 | 0.0863 | 0.0848 |
| 2 | 0.1092 | 0.1262 | 0.1154 | 0.1231 | 0.1279 | 0.1246 | 0.1300 | 0.1157 |
| 3 | 0.1350 | 0.1641 | 0.1465 | 0.1614 | 0.1623 | 0.1585 | 0.1545 | 0.1515 |
| 4 | 0.1519 | 0.2006 | 0.1606 | 0.1947 | 0.1807 | 0.1897 | 0.1912 | 0.1782 |
| 5 | 0.1512 | 0.2267 | 0.1671 | 0.2202 | 0.1836 | 0.2156 | 0.2014 | 0.2087 |
| 6 | 0.1604 | 0.2583 | 0.1745 | 0.2576 | 0.1870 | 0.2485 | 0.2052 | 0.2381 |
| 7 | 0.1725 | 0.2706 | 0.1908 | 0.2786 | 0.2017 | 0.2629 | 0.2291 | 0.2527 |
| 8 | 0.1547 | 0.2791 | 0.1807 | 0.2972 | 0.2010 | 0.2817 | 0.2219 | 0.2665 |
| 9 | 0.1695 | 0.3067 | 0.1861 | 0.3230 | 0.1996 | 0.2291 | 0.2199 | 0.2900 |
| 10 | 0.1768 | 0.3295 | 0.1798 | 0.3329 | 0.2006 | 0.3088 | 0.2145 | 0.2948 |
| 11 | 0.1646 | 0.3384 | 0.1811 | 0.3434 | 0.1889 | 0.3050 | 0.2308 | 0.3093 |
| 12 | 0.1729 | 0.3498 | 0.1867 | 0.3534 | 0.2063 | 0.3260 | 0.2348 | 0.3163 |
| 13 | 0.1776 | 0.3669 | 0.1966 | 0.3701 | 0.2155 | 0.3442 | 0.2328 | 0.3373 |
| 14 | 0.1832 | 0.3755 | 0.1999 | 0.3791 | 0.2093 | 0.3500 | 0.2486 | 0.3438 |
| 15 | 0.1755 | 0.3850 | 0.2035 | 0.3873 | 0.2587 | 0.3558 | 0.2378 | 0.3497 |
| 16 | 0.1796 | 0.3957 | 0.2002 | 0.3989 | 0.1762 | 0.3648 | 0.2509 | 0.3572 |
| 17 | 0.1752 | 0.4028 | 0.1950 | 0.4043 | 0.2141 | 0.3746 | 0.2380 | 0.3578 |
| 18 | 0.1713 | 0.4044 | 0.1964 | 0.4097 | 0.2092 | 0.3764 | 0.2322 | 0.3675 |
| 19 | 0.1648 | 0.4113 | 0.1880 | 0.4155 | 0.2043 | 0.3860 | 0.2238 | 0.3765 |
| 20 | 0.1799 | 0.4162 | 0.1949 | 0.4287 | 0.2041 | 0.3860 | 0.2325 | 0.3787 |

Table A-12. Class I data. Average weekly weights and foods consumed per rat for four populations of 25 male Younger Laboratory strain Sprague Dawley albino rats each as affected by dietary protein. The dietary metabolic energy level was 3.41 kcal/g. (Unpublished data of Parks 1968)

| Age | % CP 36.15 | | 25.20 | | 19.63 | | 14.05 | |
|---|---|---|---|---|---|---|---|---|
| $i$ wks | $f_i$ kg | $W_i$ kg | $f_i$ kg | $W_i$ kg | $f_i$ kg | $W_i$ kg | $f_i$ kg | $W_i$ kg |
| 0 | – | 0.0637 | – | 0.0630 | – | 0.0654 | – | 0.0616 |
| 1 | 0.0830 | 0.0966 | 0.0821 | 0.0924 | 0.0898 | 0.0961 | 0.0860 | 0.0858 |
| 2 | 0.1137 | 0.1347 | 0.1128 | 0.1324 | 0.1182 | 0.1370 | 0.1139 | 0.1211 |
| 3 | 0.1282 | 0.1759 | 0.1322 | 0.1742 | 0.1393 | 0.1804 | 0.1389 | 0.1591 |
| 4 | 0.1487 | 0.2199 | 0.1473 | 0.2221 | 0.1452 | 0.2209 | 0.1460 | 0.1981 |
| 5 | 0.1475 | 0.2532 | 0.1547 | 0.2545 | 0.1474 | 0.2514 | 0.1515 | 0.2277 |
| 6 | 0.1552 | 0.2802 | 0.1525 | 0.2845 | 0.1561 | 0.2815 | 0.1574 | 0.2543 |
| 7 | 0.1551 | 0.3069 | 0.1564 | 0.3108 | 0.1579 | 0.3071 | 0.1584 | 0.2797 |
| 8 | 0.1670 | 0.3249 | 0.1600 | 0.3299 | 0.1625 | 0.3250 | 0.1589 | 0.2951 |
| 9 | 0.1578 | 0.3424 | 0.1549 | 0.3472 | 0.1586 | 0.3440 | 0.1576 | 0.3128 |
| 10 | 0.1526 | 0.3558 | 0.1535 | 0.3615 | 0.1571 | 0.3571 | 0.1545 | 0.3276 |
| 11 | 0.1476 | 0.3647 | 0.1515 | 0.3724 | 0.1554 | 0.3687 | 0.1521 | 0.3423 |
| 12 | 0.1453 | 0.3724 | 0.1500 | 0.3837 | 0.1538 | 0.3804 | 0.1491 | 0.3544 |
| 13 | 0.1442 | 0.3834 | 0.1438 | 0.3920 | 0.1479 | 0.3889 | 0.1433 | 0.1364 |
| 14 | 0.1491 | 0.3921 | – | – | 0.1525 | 0.3967 | 0.1495 | 0.3738 |

Table A-13. Class I data. Average weekly weight and food consumption of five rats randomly from the population of 25 rats described by data in Table A-11. Percent RSE = 100

| | Rat E 31 | | Rat D 23 | | Rat D 10 | |
|---|---|---|---|---|---|---|
| $i$ wks | $f_i$ kg | $W_i$ kg | $f_i$ kg | $W_i$ kg | $f_i$ kg | $W_i$ g |
| 0 | – | 0.0385 | – | 0.0460 | – | 0.0890 |
| 1 | 0.0985 | 0.0720 | 0.0915 | 0.0740 | 0.0845 | 0.1280 |
| 2 | 0.1025 | 0.1065 | 0.0915 | 0.1030 | 0.1100 | 0.1305 |
| 3 | 0.1085 | 0.1315 | 0.0960 | 0.1380 | 0.1380 | 0.1945 |
| 4 | 0.1420 | 0.1670 | 0.1030 | 0.1645 | 0.1305 | 0.2225 |
| 5 | 0.1405 | 0.2069 | 0.1115 | 0.1950 | 0.1495 | 0.2695 |
| 6 | 0.1295 | 0.2305 | 0.1140 | 0.2185 | 0.1745 | 0.3065 |
| 7 | 0.1685 | 0.2530 | 0.1295 | 0.2480 | 0.1710 | 0.3400 |
| 8 | 0.1945 | 0.2650 | 0.1300 | 0.2635 | 0.1580 | 0.3485 |
| 9 | 0.1360 | 0.2845 | 0.1315 | 0.2795 | 0.1600 | 0.3680 |
| 10 | 0.1130 | 0.2830 | 0.1330 | 0.2935 | 0.1570 | 0.3790 |
| 11 | 0.1080 | 0.2885 | 0.1300 | 0.3005 | 0.1550 | 0.3855 |
| 12 | 0.1180 | 0.2970 | 0.1275 | 0.3105 | 0.1645 | 0.4015 |
| 13 | 0.0990 | 0.2985 | 0.1255 | 0.3185 | 0.1525 | 0.4100 |
| 14 | 0.1180 | 0.3060 | 0.1265 | 0.3240 | 0.1570 | 0.4205 |

| | Rat A 24 | | Rat A 22 | |
|---|---|---|---|---|
| $i$ wks | $f_i$ kg | $W_i$ kg | $F_i$ kg | $W_i$ kg |
| 0 | – | 0.0345 | – | 0.0325 |
| 1 | 0.0665 | 0.0565 | 0.0745 | 0.0615 |
| 2 | 0.0895 | 0.0915 | 0.1150 | 0.0955 |
| 3 | 0.1196 | 0.1400 | 0.1290 | 0.1380 |
| 4 | 0.1495 | 0.1860 | 0.1355 | 0.1765 |
| 5 | 0.1550 | 0.2305 | 0.1385 | 0.2155 |
| 6 | 0.1905 | 0.2635 | 0.1380 | 0.2565 |
| 7 | 0.1690 | 0.2910 | 0.1505 | 0.2755 |
| 8 | 0.1795 | 0.3160 | 0.1480 | 0.3085 |
| 9 | 0.1680 | 0.3345 | 0.1635 | 0.3310 |
| 10 | 0.1740 | 0.3575 | 0.1570 | 0.3500 |
| 11 | 0.1555 | 0.3615 | 0.1605 | 0.3615 |
| 12 | 0.1545 | 0.3700 | 0.1520 | 0.3735 |
| 13 | 0.1535 | 0.3825 | 0.1560 | 0.3820 |
| 14 | 0.1640 | 0.3910 | 0.1648 | 0.3990 |

Table A-14. Class IV data. Weight versus time since weaning of Hisaw-Harvard variety of Wistar Albino male rats on 10% and 60% protein diets. (Data by Mayer and Vitale 1957)

| | Percent protein | |
|---|---|---|
| | 10[a] | 60[b] |
| $i$ d | $W_i$ kg | $W_i$ kg |
| 0 | 0.038 | 0.038 |
| 6 | 0.052 | 0.068 |
| 12 | 0.072 | 0.102 |
| 18 | 0.101 | 0.136 |
| 24 | 0.125 | 0.178 |
| 30 | 0.146 | 0.213 |
| 60 | 0.266 | 0.339 |
| 90 | 0.327 | 0.377 |

[a] Composition of diet: 10% protein, 12% fat and 74.9% carbohydrate

[b] Composition of diet: 60% protein, 12% fat and 23.9% carbohydrate

*ME* of both diets, $\beta = 4.44$ kcal/g

Table A-15. Class II data. Post weaning weights and food intake of Hisaw-Harvard strain of Wistar albino male rats. Percent dietary protein = 25, and dietary energy level = 4.4 kcal/g. (Data of Mayer and Vitale 1957)

| $i$ d | $(dF/dt)_i$ kcal/d | $W_i$ kg |
|---|---|---|
| 0 | – | 0.038 |
| 3 | 42 | 0.048 |
| 6 | 48 | 0.065 |
| 9 | 55 | 0.081 |
| 12 | 62 | 0.099 |
| 18 | 69 | 0.139 |
| 24 | 75 | 0.179 |
| 30 | 79 | 0.211 |
| 45 | 79 | 0.289 |
| 60 | 78 | 0.332 |
| 90 | 79 | 0.372 |

Table A-16. Class III and IV data. Weights and cumulative food consumption of male and female mice versus time since weaning. (Unpublished data of Bateman and Slee 1974)

| | Male | | Female | |
|---|---|---|---|---|
| $t$ d | $F$ g | $W$ g | $F$ g | $W$ g |
| 0 | – | 11.3 | – | 10.8 |
| 2 | 6 | 12.3 | 6 | 11.5 |
| 9 | 37 | 21.6 | 35 | 18.8 |
| 12 | 55 | 25.6 | 50 | 20.3 |
| 14 | 68 | 26.9 | 61 | 21.8 |
| 16 | 79 | 28.6 | 71 | 23.3 |
| 23 | 122 | 31.7 | 108 | 25.3 |
| 30 | 165 | 33.7 | 145 | 26.3 |
| 37 | 208 | 35.2 | 183 | 28.1 |
| 44 | 255 | 36.5 | 226 | 29.9 |
| 51 | 298 | 37.5 | 266 | 30.4 |
| 58 | 342 | 38.1 | 307 | 31.1 |
| 65 | 385 | 38.4 | 347 | 31.3 |
| 67 | 396 | 39.2 | 357 | 32.0 |
| 72 | 426 | 39.2 | 386 | 32.5 |
| 79 | 468 | 39.4 | 427 | 32.4 |
| 86 | 508 | 40.3 | 465 | 32.8 |
| 93 | 546 | 41.4 | 503 | 34.1 |

Table A-17. Class V data. Live weights and ad libitum food intakes versus age after weaning of selected (HM) mice read from the HM curves in Fig. 4.7a, b. (Data of Timon and Eisen 1970)

| Age − 22 d | *dF/dt* g/d | Age − 21 d | *W* g |
|---|---|---|---|
| 0.0 | 3.19 | 0.0 | 12.8 |
| 3.2 | 4.49 | 3.7 | 15.5 |
| 6.4 | 5.33 | 7.4 | 20.5 |
| 9.6 | 5.97 | 11.1 | 24.5 |
| 12.8 | 6.36 | 16.0 | 26.7 |
| 19.2 | 6.65 | 18.5 | 30.3 |
| 22.4 | 6.59 | 22.1 | 32.2 |
| 25.6 | 6.49 | 25.8 | 33.1 |
| 28.8 | 6.36 | 29.5 | 33.9 |
| 32.0 | 6.24 | 33.2 | 34.8 |

Table A-18. Class V data. As Table A-17 except from unselected (control) male mice

| Age − 22 d | *dF/dt* g/d | Age − 21 d | *W* g |
|---|---|---|---|
| – | – | 0.0 | 12.82 |
| 3.2 | 3.97 | 3.69 | 15.46 |
| 6.4 | 4.75 | 7.38 | 19.05 |
| 9.6 | 5.33 | 11.07 | 22.64 |
| 12.8 | 5.72 | 14.76 | 24.82 |
| 16.0 | 5.99 | 18.45 | 26.85 |
| 19.2 | 6.08 | 22.14 | 28.56 |
| 22.4 | 6.14 | 25.83 | 29.66 |
| 25.6 | 6.12 | 29.52 | 30.44 |
| 28.8 | 6.10 | 33.21 | 31.22 |
| 32.0 | 6.10 | | |

Table A-19. Class V data. As Table A-17 except from selected (high) female mice

| Age − 22 d | *dF/dt* g/d | Age − 21 d | *W* g |
|---|---|---|---|
| – | – | 0.0 | 12,82 |
| 3.2 | 4.16 | 3.69 | 14.68 |
| 6.4 | 4.04 | 7.38 | 18.58 |
| 9.6 | 5.64 | 11.07 | 22.01 |
| 12.8 | 6.01 | 14.76 | 24.82 |
| 16.0 | 6.18 | 18.45 | 26.84 |
| 19.2 | 6.20 | 22.14 | 28.41 |
| 22.4 | 6.14 | 25.83 | 28.72 |
| 25.6 | 6.01 | 29.52 | 28.72 |
| 28.8 | 5.87 | 33.21 | 29.50 |
| 32.0 | 5.83 | | |

Table A-20. Class V data. Same as Table A-17 except from unselected (control) female mice

| Age − 22 d | *dF/dt* g/d | Age − 21 d | *W* g |
|---|---|---|---|
| – | – | 0.0 | 12.82 |
| 3.2 | 4.55 | 3.69 | 14.06 |
| 6.4 | 5.13 | 7.38 | 16.55 |
| 9.6 | 5.52 | 11.07 | 19.36 |
| 12.8 | 5.75 | 14.76 | 21.23 |
| 16.0 | 5.85 | 18.45 | 23.26 |
| 19.2 | 5.81 | 22.14 | 24.35 |
| 22.4 | 5.71 | 25.83 | 24.82 |
| 25.6 | 5.62 | 29.52 | 25.13 |
| 28.8 | 5.50 | 33.21 | 25.60 |

Table A-21. Class IV data. Weights versus age of a group of white mice reared at North Carolina State University at Raleigh. (Unpublished data of J. Rutledge 1971)

| $t$<br>d | $W$<br>g | $t$<br>d | $W$<br>g |
|---|---|---|---|
| 0 | 1.61 | 33 | 25.03 |
| 3 | 2.38 | 36 | 26.81 |
| 6 | 4.09 | 39 | 28.15 |
| 9 | 6.16 | 42 | 29.39 |
| 12 | 7.95 | 49 | 30.97 |
| 15 | 9.11 | 56 | 32.57 |
| 18 | 10.11 | 63 | 34.04 |
| 21 | 12.87 | 70 | 34.44 |
| 24 | 15.98 | 77 | 35.22 |
| 27 | 19.03 | 84 | 35.56 |
| 30 | 22.31 | | |

Table A-23. Class V data. Points selected graphically from a smooth curve drawn by eye through the broken curve in Fig. 6.3 which depicts the detailed response of one of the Taylor Ayrshire cattle fed on constant food intake of 77 lbs/wk. These points are plotted in Fig. 6.4

| | $t$<br>yrs | $W$<br>lbs |
|---|---|---|
| Initial condition | 0.58 | 306 |
| | 0.98 | 490 |
| | 1.37 | 580 |
| | 1.76 | 630 |
| | 2.15 | 630 |
| | 2.45 | 670 |
| | 2.93 | 670 |
| | 3.32 | 690 |
| | 3.71 | 720 |
| | 4.10 | 740 |
| | 4.87 | 750 |
| | 5.67 | 760 |
| | 6.45 | 760 |

Table A-22. Class IV data. Weights versus age of male and female Rhode Island Red and White Leghorn chickens. (Unpublished data of Grossman 1971)

| | Rhode Island Red | | White Leghorn | |
|---|---|---|---|---|
| Age<br>wks | Male<br>$W$<br>kg | Female<br>$W$<br>kg | Male<br>$W$<br>kg | Female<br>$W$<br>kg |
| 0 | 0.0388 | 0.0387 | 0.0379 | 0.0381 |
| 3 | 0.1590 | 0.1465 | 0.1268 | 0.1250 |
| 6 | 0.4068 | 0.3607 | 0.3010 | 0.2693 |
| 9 | 0.7880 | 0.6676 | 0.5770 | 0.5015 |
| 12 | 1.1938 | 0.9542 | 0.8367 | 0.6866 |
| 15 | 1.7808 | 1.3273 | 1.1947 | 0.9361 |
| 18 | 2.2498 | 1.6081 | 1.5268 | 1.1123 |
| 21 | 2.6020 | 1.7728 | 1.7849 | 1.2204 |
| 24 | 2.9215 | 2.0228 | 1.9471 | 1.3327 |
| 27 | 3.0806 | 2.2035 | 2.0138 | 1.4992 |
| 30 | 3.1292 | 2.2164 | 2.0762 | 1.6408 |
| 33 | 3.1793 | 2.2379 | 2.1178 | 1.6719 |
| 36 | 3.1884 | 2.2702 | 2.1287 | 1.7119 |
| 39 | 3.1583 | 2.2766 | 2.1213 | 1.7242 |
| 42 | 3.1317 | 2.2742 | 2.1120 | 1.7273 |
| 45 | 3.1183 | 2.3200 | 2.1162 | 1.7130 |

Table A-24. Class I data. BG 54. Treatment ALF (ad libitum). Weekly foods consumed and live weight versus age from hatching of Ross Ltd. male fryer chickens. (Data of Wilson et al. 1975)

| Age wks | $f_i$ kg | $W_i$ kg | Age wks | $f_i$ kg | $W_i$ kg |
|---|---|---|---|---|---|
| 1 | 0.151 | 0.127 | 16 | 1.771 | 4.618 |
| 2 | 0.241 | 0.288 | 17 | 1.950 | 4.867 |
| 3 | 0.397 | 0.522 | 18 | 1.938 | 4.912 |
| 4 | 0.616 | 0.828 | 19 | 1.530 | 5.043 |
| 5 | 0.748 | 1.153 | 20 | 1.443 | 5.068 |
| 6 | 0.956 | 1.478 | 21 | 1.322 | 5.036 |
| 7 | 0.817 | 1.776 | 22 | 1.490 | 5.119 |
| 8 | 1.093 | 2.151 | 23 | 1.366 | 5.209 |
| 9 | 1.070 | 2.615 | 24 | 1.668 | 5.419 |
| 10 | 1.250 | 2.898 | 25 | 1.473 | 5.425 |
| 11 | 1.502 | 3.279 | 26 | 1.499 | 5.467 |
| 12 | 1.313 | 3.577 | 27 | 1.266 | 5.407 |
| 13 | 1.638 | 3.817 | 28 | 1.484 | 5.448 |
| 14 | 1.411 | 4.031 | 29 | 1.378 | 5.517 |
| 15 | 1.705 | 4.443 | 30 | 1.557 | 5.482 |

At hatching $W_0 = 0.045$ kg

Table A-25. BG 54. Ross Ltd. fryer males. Treatment CF1. Live weights versus age. Constant weekly food consumed $q_1(t) = 0.575$ kg/wk beginning at age $T_{01} = 4.5$ wks and $W_{01} = 0.912$ kg

| $t$ wks | $W(t)$ kg |
|---|---|
| 5 | 1.011 |
| 6 | 1.146 |
| 7 | 1.331 |
| 8 | 1.482 |
| 9 | 1.621 |
| 10 | 1.760 |
| 11 | 1.875 |
| 12 | 2.023 |
| 13 | 2.043 |
| 14 | 2.102 |
| 15 | 2.152 |
| 16 | 2.219 |
| 17 | 2.286 |
| 18 | 2.302 |
| 19 | 2.311 |
| 20 | 2.367 |
| 21 | 2.431 |
| 22 | 2.447 |
| 23 | 2.464 |
| 24 | 2.400 |
| 25 | 2.537 |
| 26 | 2.568 |
| 27 | 2.625 |
| 28 | 2.461 |
| 29 | 2.624 |
| 30 | 2.671 |

Table A-26. BG 54. Ross Ltd. fryer males. Treatment LF1. Weekly food consumed and live weights versus age under linear feeding beginning at age $T_{02} = 6.4$ wks, $q_{02} = 0.575$ kg/wk and $W_{02} = 1.237$ kg

| $t$ wks | $f(t)$[a] kg | $W(t)$ kg |
|---|---|---|
| 7 | 0.590 | 1.373 |
| 8 | 0.635 | 1.549 |
| 9 | 0.681 | 1.741 |
| 10 | 0.727 | 1.766 |
| 11 | 0.774 | 2.145 |
| 12 | 0.820 | 2.364 |
| 13 | 0.866 | 2.461 |
| 14 | 0.913 | 2.609 |
| 15 | 0.959 | 2.833 |
| 16 | 1.005 | 3.001 |
| 17 | 1.051 | 3.211 |
| 18 | 1.098 | 3.351 |
| 19 | 1.144 | 3.522 |
| 20 | 1.190 | 3.666 |
| 21 | 1.237 | 3.818 |
| 22 | 1.283 | 4.106 |
| 23 | 1.329 | 4.254 |
| 24 | 1.375 | 4.235 |
| 25 | 1.422 | 4.562 |
| 26 | 1.468 | 4.777 |
| 27 | 1.514 | 4.959 |
| 28 | 1.561 | 4.720 |
| 29 | 1.607 | 4.914 |
| 30 | 1.653 | 4.900 |

[a] $q_2(t) = 0.575 + 0.0457\,(t - T_{02})$

Table A-27. BG54. Ross Ltd. fryer males. Treatment LF2. Weekly food consumed and live weights versus age under linear feeding beginning at age $T_{03} = 11$ wks, $q_{03} = 0.575$ kg/wk and $W_{03} = 1.843$ kg

| $t$ wks | $f(t)$[a] kg | $W(t)$ kg |
|---|---|---|
| 12 | 0.594 | 1.933 |
| 13 | 0.615 | 1.993 |
| 14 | 0.636 | 2.086 |
| 15 | 0.657 | 2.163 |
| 16 | 0.678 | 2.279 |
| 17 | 0.699 | 2.322 |
| 18 | 0.721 | 2.405 |
| 19 | 0.742 | 2.486 |
| 20 | 0.763 | 2.522 |
| 21 | 0.784 | 2.604 |
| 22 | 0.805 | 2.722 |
| 23 | 0.827 | 2.803 |
| 24 | 0.848 | 2.840 |
| 25 | 0.869 | 2.999 |
| 26 | 0.890 | 3.118 |
| 27 | 0.911 | 3.244 |
| 28 | 0.932 | 3.202 |
| 29 | 0.954 | 3.441 |
| 30 | 0.975 | 3.485 |

[a] $q_3(t) = 0.575 + 0.0211\,(t - T_{03})$

Table A-28. BG54. Ross Ltd. fryer males. Treatment CF2. Live weights versus age. Constant weekly food consumed at age $q_4(t) = 0.893$ kg/wk beginning at age $T_{04} = 13.6$ wks and $W_{04} = 2.529$ kg

| $t$ wks | $W(t)$ kg | $W(t)$ kg |
|---|---|---|
| 14 | 2.574 | |
| 15 | 2.671 | |
| 16 | 2.823 | |
| 17 | 2.980 | |
| 18 | 3.042 | |
| 19 | 3.061 | |
| 20 | 3.111 | |
| 21 | 3.106 | |
| 22 | 3.237 | |
| 23 | 3.221 | |
| 24 | 3.124 | |
| 25 | 3.318 | |
| 26 | 3.357 | |
| 27 | 3.456 | |
| 28 | 3.248 | |
| 29 | 3.554 | |
| 30 | 3.629 | |

Table A-29. Class I data. BG54. Ross Ltd. fryer males. Treatment RF. Random weekly foods consumed and the corresponding weekly live weights beginning at age $T_{05}=6$ wks, $q_{05}=0.575$ kg/wk and $W_{05}=1.146$ kg

| $t$ wks | $f(t)$ kg | $W(t)$ kg |
|---|---|---|
| 7 | 0.584 | 1.336 |
| 8 | 0.597 | 1.491 |
| 9 | 0.621 | 1.598 |
| 10 | 0.662 | 1.793 |
| 11 | 0.637 | 1.940 |
| 12 | 0.641 | 2.048 |
| 13 | 0.792 | 2.125 |
| 14 | 0.686 | 2.312 |
| 15 | 0.780 | 2.529 |
| 16 | 0.828 | 2.641 |
| 17 | 0.719 | 2.716 |
| 18 | 0.900 | 2.747 |
| 19 | 0.833 | 2.997 |
| 20 | 0.631 | 2.879 |
| 21 | 0.666 | 2.910 |
| 22 | 0.730 | 3.024 |
| 23 | 0.785 | 3.050 |
| 24 | 0,666 | 2.858 |
| 25 | 1.106 | 3.193 |
| 26 | 1.199 | 3.442 |
| 27 | 0.984 | 3.835 |
| 28 | 1.179 | 3.627 |
| 29 | 0.893 | 3.803 |
| 30 | 1.171 | 4.152 |

Table A-30. Class I data. BG54. Ross Ltd. fryer females. Treatment ALF (ad libitum). Weekly foods consumed and live weights versus age from hatching. (Data of Wilsons et al. 1975)

| $t$ wks | $f(t)$ kg | $W(t)$ kg |
|---|---|---|
| 1 | 0.148 | 0.121 |
| 2 | 0.277 | 0.266 |
| 3 | 0.365 | 0.465 |
| 4 | 0.529 | 0.728 |
| 5 | 0.689 | 1.012 |
| 6 | 0.748 | 1.326 |
| 7 | 0.792 | 1.547 |
| 8 | 0.916 | 1.793 |
| 9 | 0.979 | 2.075 |
| 10 | 1.116 | 2.305 |
| 11 | 1.211 | 2.586 |
| 12 | 1.158 | 2.764 |
| 13 | 1.274 | 2.908 |
| 14 | 1.112 | 3.043 |
| 15 | 1.306 | 3.282 |
| 16 | 1.406 | 3.473 |
| 17 | 1.493 | 3.681 |
| 18 | 1.350 | 3.845 |
| 19 | 1.291 | 4.021 |
| 20 | 1.223 | 4.202 |
| 21 | 1.240 | 4.185 |

At hatching $W_0=0.045$ kg

Table A-31. Class I data. Ross Ltd. fryer females. Treatment CF1. Live weights versus age. Constant weekly food consumed $q_1(t)=0.496$ kg/wk beginning at $T_{01}=3.8$ wks and $W_{01}=0.663$ kg

| $t$ wks | $W(t)$ kg |
|---|---|
| 4 | 0.713 |
| 5 | 0.902 |
| 6 | 1.014 |
| 7 | 1.155 |
| 8 | 1.273 |
| 9 | 1.378 |
| 10 | 1.493 |
| 11 | 1.525 |
| 12 | 1.592 |
| 13 | 1.640 |
| 14 | 1.670 |
| 15 | 1.691 |
| 16 | 1.760 |
| 17 | 1.817 |
| 18 | 1.818 |
| 19 | 1.812 |
| 20 | 1.829 |
| 21 | 1.838 |

Table A-32. Class I data. BG54. Ross Ltd. fryer females. Treatment LF1. Weekly foods consumed and live weights versus age under linear feeding beginning at age $T_{02}=6.7$ wks, $q_{02}=0.496$ kg/wk and $W_{02}=1.130$ kg

| $t$ wks | $f(t)$[a] kg | $W(t)$ kg |
|---|---|---|
| 7 | 0.509 | 1.180 |
| 8 | 0.547 | 1.351 |
| 9 | 0.586 | 1.466 |
| 10 | 0.626 | 1.653 |
| 11 | 0.665 | 1.788 |
| 12 | 0.704 | 1.953 |
| 13 | 0.744 | 2.028 |
| 14 | 0.783 | 2.193 |
| 15 | 0.822 | 2.318 |
| 16 | 0.862 | 2.502 |
| 17 | 0.901 | 2.639 |
| 18 | 0.941 | 2.778 |
| 19 | 0.980 | 2.893 |
| 20 | 1.019 | 3.039 |
| 21 | 1.059 | 3.173 |

[a] $q_2(t)=0.496+0.0393\,(t-T_{02})$

Table A-33. Class I data. BG54. Ross Ltd. fryer females. Treatment LF2. Weekly foods consumed and live weight versus age under linear feeding beginning at age $T_{03}=11$ wks, $q_{03}=0.496$ kg/wk and $W_{03}=1.57$kg

| $t$ wks | $f(t)$[a] kg | $W(t)$ kg |
|---|---|---|
| 12 | 0.513 | 1.682 |
| 13 | 0.532 | 1.740 |
| 14 | 0.551 | 1.807 |
| 15 | 0.570 | 1.891 |
| 16 | 0.589 | 1.995 |
| 17 | 0.609 | 2.098 |
| 18 | 0.628 | 2.172 |
| 19 | 0.647 | 2.243 |
| 20 | 0.666 | 2.302 |
| 21 | 0.685 | 2.362 |

[a] $q_3(t)=0.496+0.0189\,(t-T_{03})$

Table A-34. Class I data. BG54. Ross Ltd. fryer females. Treatment CF2. Live weights versus age. Constant weekly food consumed $q_4(t)=0.496$ kg/wk beginning at $T_{04}=13$ wks and $W_{04}=2.04$ kg

| $t$ wks | $W(t)$ kg |
|---|---|
| 14 | 2.174 |
| 15 | 2.271 |
| 16 | 2.398 |
| 17 | 2.474 |
| 18 | 2.578 |
| 19 | 2.643 |
| 20 | 2.714 |
| 21 | 2.767 |

Table A-35. Class I data. BG54. Ross Ltd. fryer females. Treatment RF. Random weekly foods consumed and the corresponding weekly live weights beginning at age $T_{05}=6$ wks, $q_{05}=0.496$ kg/wk and $W_{05}=1.014$ kg

| $t$ wks | $f(t)$ kg | $W(t)$ kg |
|---|---|---|
| 7 | 0.503 | 1.181 |
| 8 | 0.515 | 1.336 |
| 9 | 0.535 | 1.418 |
| 10 | 0.570 | 1.614 |
| 11 | 0.548 | 1.714 |
| 12 | 0.552 | 1.758 |
| 13 | 0.681 | 1.829 |
| 14 | 0.586 | 1.916 |
| 15 | 0.666 | 2.064 |
| 16 | 0.709 | 2.189 |
| 17 | 0.617 | 2.247 |
| 18 | 0.767 | 2.274 |
| 19 | 0.711 | 2.484 |
| 20 | 0.542 | 2.378 |
| 21 | 0.568 | 2.410 |

Table A-36. Class I data. BG54. Apollo males. Treatment ALF (ad libitum). Weekly food consumed and live weight versus age from hatching. At hatching $W_0=0.035$ kg

| $t$ wks | $f(t)$ kg | $W(t)$ kg | $t$ wks | $f(t)$ kg | $W(t)$ kg |
|---|---|---|---|---|---|
| 1 | 0.060 | 0.083 | 16 | 0.848 | 1.724 |
| 2 | 0.132 | 0.148 | 17 | 0.851 | 1.776 |
| 3 | 0.184 | 0.232 | 18 | 1.087 | 1.869 |
| 4 | 0.259 | 0.356 | 19 | 0.950 | 1.896 |
| 5 | 0.349 | 0.479 | 20 | 1.021 | 1.925 |
| 6 | 0.455 | 0.625 | 21 | 1.051 | 1.958 |
| 7 | 0.474 | 0.759 | 22 | 0.853 | 1.938 |
| 8 | 0.550 | 0.863 | 23 | 0.866 | 2.041 |
| 9 | 0.592 | 1.011 | 24 | 0.937 | 1.965 |
| 10 | 0.649 | 1.170 | 25 | 0.932 | 2.010 |
| 11 | 0.782 | 1.206 | 26 | 0.993 | 2.037 |
| 12 | 0.874 | 1.394 | 27 | 0.898 | 2.046 |
| 13 | 0.796 | 1.482 | 28 | 0.955 | 2.058 |
| 14 | 0.774 | 1.564 | 29 | 0.963 | 2.076 |
| 15 | 0.791 | 1.644 | 30 | 0.898 | 2.153 |

Table A-37. Class I data. BG54. Apollo males. Treatment CF1. Live weights versus age. Constant weekly food consumed $q_1(t) = 0.315$ kg/wk beginning at $T_{01} = 4.6$ wks and $W_{01} = 0.422$ kg

| $t$ wks | $W(t)$ kg |
|---|---|
| 5 | 0.466 |
| 6 | 0.522 |
| 7 | 0.623 |
| 8 | 0.660 |
| 9 | 0.721 |
| 10 | 0.739 |
| 11 | 0.791 |
| 12 | 0.783 |
| 13 | 0.830 |
| 14 | 0.918 |
| 15 | 0.907 |
| 16 | 0.896 |
| 17 | 0.887 |
| 18 | 0.896 |
| 19 | 0.934 |
| 20 | 0.917 |
| 21 | 0,927 |
| 22 | 0.980 |
| 23 | 0.972 |
| 24 | 0.933 |
| 25 | 1.006 |
| 26 | 1.084 |
| 27 | 1.063 |
| 28 | 0.931 |
| 29 | 1.085 |
| 30 | 1.150 |

Table A-38. Class I data. BG54. Apollo males. Treatment LF1. Weekly foods consumed and live weights versus age under linear feeding beginning at $T_{02} = 7.6$ wks, $q_{02} = 0.315$ kg/wk and $W_{02} = 0.643$ kg

| $t$ wks | $f(t)$[a] kg | $W(t)$ kg |
|---|---|---|
| 8 | 0.321 | 0.660 |
| 9 | 0.344 | 0.758 |
| 10 | 0.367 | 0.786 |
| 11 | 0.390 | 0.863 |
| 12 | 0.414 | 0.909 |
| 13 | 0.437 | 0.989 |
| 14 | 0.460 | 1.025 |
| 15 | 0.483 | 1.108 |
| 16 | 0.506 | 1.143 |
| 17 | 0.529 | 1.172 |
| 18 | 0.553 | 1.236 |
| 19 | 0.576 | 1.276 |
| 20 | 0.599 | 1.256 |
| 21 | 0.622 | 1.347 |
| 22 | 0.645 | 1.475 |
| 23 | 0.668 | 1.496 |
| 24 | 0.691 | 1.448 |
| 25 | 0.715 | 1.547 |
| 26 | 0.738 | 1.580 |
| 27 | 0.761 | 1.774 |
| 28 | 0.784 | 1.656 |
| 29 | 0.807 | 1.835 |
| 30 | 0.830 | 1.842 |

[a] $q_2(t) = 0.315 + 0.0230\,(t - T_{02})$

Table A-39. Class I data. BG 54. Apollo males. Treatment LF 2. Weekly foods consumed and live weights versus age under linear feeding beginning at $T_{03}=12.2$ wks, $q_{03}=0.315$ kg/wk and $W_{03}=0.780$ kg

| $t$ wks | $f(t)$[a] kg | $W(t)$ kg |
|---|---|---|
| 13 | 0.322 | 0.767 |
| 14 | 0.334 | 0.773 |
| 15 | 0.345 | 0.810 |
| 16 | 0.356 | 0.807 |
| 17 | 0.368 | 0.841 |
| 18 | 0.379 | 0.854 |
| 19 | 0.390 | 0.864 |
| 20 | 0.402 | 0.888 |
| 21 | 0.413 | 0.891 |
| 22 | 0.424 | 0.999 |
| 23 | 0.436 | 1.003 |
| 24 | 0.447 | 0.940 |
| 25 | 0.459 | 1.045 |
| 26 | 0.470 | 1.154 |
| 27 | 0.481 | 1.175 |
| 28 | 0.493 | 1.078 |
| 29 | 0.504 | 1.293 |
| 30 | 0.515 | 1.364 |

[a] $q_3(t)=0.315+0.0112\,(t-T_{03})$

Table A-40. Class I data. BG 54. Apollo males. Treatment CF 2. Live weights versus age. Constant weekly food consumed $q_{04}(t)=0.427$ kg/wk beginning at $T_{04}=12.6$ wks and $W_{04}=0.952$ kg

| $t$ wks | $W(t)$ kg |
|---|---|
| 13 | 0.980 |
| 14 | 1.000 |
| 15 | 1.039 |
| 16 | 1.083 |
| 17 | 1.101 |
| 18 | 1.116 |
| 19 | 1.099 |
| 20 | 1.091 |
| 21 | 1.040 |
| 22 | 1.097 |
| 23 | 1.086 |
| 24 | 1.065 |
| 25 | 1.092 |
| 26 | 1.178 |
| 27 | 1.180 |
| 28 | 1.035 |
| 29 | 1.267 |
| 30 | 1.247 |

Table A-41. Class I data. Bg54. Apollo males. Treatment RF. Random weekly foods consumed and the corresponding live weights beginning at $T_{05}=8$ wks, $q_{05}=0.319$ kg/wk and $W_{05}=0.663$ kg

| $t$ wks | $f(t)$ kg | $W(t)$ kg |
|---|---|---|
| 9 | 0.327 | 0.723 |
| 10 | 0.345 | 0.742 |
| 11 | 0.339 | 0.825 |
| 12 | 0.340 | 0.801 |
| 13 | 0.403 | 0.860 |
| 14 | 0.353 | 0.869 |
| 15 | 0.397 | 1.000 |
| 16 | 0.417 | 0.981 |
| 17 | 0.369 | 0.983 |
| 18 | 0.448 | 1.015 |
| 19 | 0.424 | 1.151 |
| 20 | 0.341 | 1.040 |
| 21 | 0.347 | 1.019 |
| 22 | 0.377 | 1.081 |
| 23 | 0.398 | 1.095 |
| 24 | 0.347 | 1.009 |
| 25 | 0.568 | 1.106 |
| 26 | 0.438 | 1.306 |
| 27 | 0.493 | 1.465 |
| 28 | 0.580 | 1.288 |
| 29 | 0.427 | 1.381 |
| 30 | 0.607 | 1.557 |

Table A-42. Class I data. BG54. Apollo females. Treatment ALF (ad libitum). Weekly foods consumed and live weights versus age from hatching. (Data of Wilson et al. 1975)

| $t$ wks | $f(t)$ kg | $W(t)$ kg |
|---|---|---|
| 1 | 0.065 | 0.080 |
| 2 | 0.125 | 0.135 |
| 3 | 0.177 | 0.207 |
| 4 | 0.224 | 0.291 |
| 5 | 0.288 | 0.399 |
| 6 | 0.358 | 0.512 |
| 7 | 0.445 | 0.608 |
| 8 | 0.434 | 0.719 |
| 9 | 0.441 | 0.783 |
| 10 | 0.500 | 0.896 |
| 11 | 0.520 | 0.963 |
| 12 | 0.559 | 1.063 |
| 13 | 0.600 | 1.135 |
| 14 | 0.635 | 1.175 |
| 15 | 0.734 | 1.242 |
| 16 | 0.704 | 1.274 |
| 17 | 0.599 | 1.344 |
| 18 | 0.697 | 1.405 |
| 19 | 0.701 | 1.459 |
| 20 | 0.712 | 1.543 |
| 21 | 0.839 | 1.609 |

At hatching $W_0=0.035$ kg

Table A-43. Class I data. BG 54. Apollo females. Treatment CF 1. Live weights versus age. Constant weekly food consumed $q_1(t)=0.261$ kg/wk beginning at $T_{01}=4.6$ wks, $q_{01}=0.261$ kg/wk and $W_{01}=0.350$ kg

| $t$ wks | $W(t)$ kg |
|---|---|
| 5 | 0.390 |
| 6 | 0.448 |
| 7 | 0.522 |
| 8 | 0.557 |
| 9 | 0.603 |
| 10 | 0.614 |
| 11 | 0.658 |
| 12 | 0.681 |
| 13 | 0.650 |
| 14 | 0.647 |
| 15 | 0.679 |
| 16 | 0.658 |
| 17 | 0.690 |
| 18 | 0.683 |
| 19 | 0.707 |
| 20 | 0.713 |
| 21 | 0.670 |

Table A-44. Class I data. BG 54. Apollo females. Treatment LF 1. Weekly foods consumed and live weights versus age under linear feeding beginning at $T_{02}=7.6$ wks, $q_{02}=0.261$ kg/wk and $W_{02}=0.537$ kg

| $t$ wks | $f(t)$[a] kg | $W(t)$ kg |
|---|---|---|
| 8 | 0.264 | 0.548 |
| 9 | 0.280 | 0.609 |
| 10 | 0.296 | 0.647 |
| 11 | 0.312 | 0.701 |
| 12 | 0.329 | 0.768 |
| 13 | 0.345 | 0.832 |
| 14 | 0.362 | 0.808 |
| 15 | 0.378 | 0.868 |
| 16 | 0.394 | 0.882 |
| 17 | 0.411 | 0.928 |
| 18 | 0.427 | 0.965 |
| 19 | 0.443 | 0.975 |
| 20 | 0.460 | 0.993 |
| 21 | 0.476 | 1.012 |

[a] $q_2(t)=0.261+0.0160\,(t-T_{02})$

Table A-45. Class I data. BG 54. Apollo females. Treatment LF2. Weekly foods consumed and live weight versus age under linear feeding beginning at $T_{03} = 13.6$ wks, $q(t) = 0.261$ kg/wk and $W_{03} = 0.677$ kg

| $t$ wks | $f(t)$[a] kg | $W(t)$ kg |
|---|---|---|
| 14 | 0.274 | 0.691 |
| 15 | 0.282 | 0.737 |
| 16 | 0.290 | 0.740 |
| 17 | 0.299 | 0.792 |
| 18 | 0.307 | 0.762 |
| 19 | 0.315 | 0.791 |
| 20 | 0.323 | 0.846 |
| 21 | 0.332 | 0.811 |

[a] $q_3(t) = 0.261 + 0.00969\,(t - T_{03})$

Table A-46. Class I data. BG 54. Appolo females. Treatment CF2. Live weight versus age. Constant weekly food consumed $q_4(t) = 0.336$ kg/wk beginning at $T_{04} = 12.2$ wks and $W_{04} = 0.780$ kg

| $t$ wks | $W(t)$ kg |
|---|---|
| 13 | 0.829 |
| 14 | 0.806 |
| 15 | 0.833 |
| 16 | 0.813 |
| 17 | 0.832 |
| 18 | 0.842 |
| 19 | 0.855 |
| 20 | 0.862 |
| 21 | 0.830 |

Table A-47. Class I data. BG 54. Apollo females. Treatment RF. Random weekly foods consumed and the corresponding live weights beginning at $T_{05} = 8$ wks, $q_{05} = 0.261$ kg/wk and $W_{05} = 0.554$ kg

| $t$ wks | $f(t)$ kg | $W(t)$ kg |
|---|---|---|
| 9 | 0.269 | 0.596 |
| 10 | 0.281 | 0.634 |
| 11 | 0.277 | 0.676 |
| 12 | 0.279 | 0.663 |
| 13 | 0.322 | 0.701 |
| 14 | 0.287 | 0.708 |
| 15 | 0.317 | 0.800 |
| 16 | 0.331 | 0.764 |
| 17 | 0.299 | 0.771 |
| 18 | 0.353 | 0.773 |
| 19 | 0.336 | 0.918 |
| 20 | 0.280 | 0.790 |
| 21 | 0.282 | 0.787 |

Table A-48. Class IV data. Weekly weights versus age of Rhode Island Red Male chickens with and without skim milk available with diet. (Data of Titus and Jull 1928)

| | Without skim milk | With skim milk |
|---|---|---|
| Age wks | $W$ kg | $W$ kg |
| 0 | 0.04040 | 0.03982 |
| 1 | 0.06395 | 0.06299 |
| 2 | 0.08578 | 0.09922 |
| 3 | 0.11027 | 0.16333 |
| 4 | 0.13184 | 0.21833 |
| 5 | 0.17396 | 0.31762 |
| 6 | 0.24581 | 0.44950 |
| 7 | 0.31742 | 0.54237 |
| 8 | 0.42252 | 0.66550 |
| 9 | 0.52145 | 0.81537 |
| 10 | 0.59984 | 0.97337 |
| 11 | 0.71339 | 1.54750 |
| 12 | 0.87984 | 1.32112 |
| 13 | 1.05290 | 1.47432 |
| 14 | 1.23484 | 1.62850 |
| 15 | 1.39339 | 1.74412 |
| 16 | 1.49758 | 1.83437 |
| 17 | 1.63080 | 1.96181 |
| 18 | 1.76403 | 2.08925 |
| 20 | 2.11355 | 2.30012 |
| 22 | 2.27419 | 2.47225 |
| 24 | 2.55984 | 2.65000 |
| 26 | 2.70661 | 2.79962 |
| 28 | 2.93193 | 2.92050 |
| 30 | 3.05081 | 3.08137 |
| 32 | 3.25403 | 3.18875 |
| 34 | 3.32258 | 3.43700 |

Table A-49. Class V data. Read from Ricard's graph (Fig. 10.1) with a divider and a millimeter scale. The data are weight by age of his base population X-38 and the four populations selected from it for high and low body weights at 8 and 36 weeks respectively over nine generations. Estimated error = 0.028 kg. $W_0 = 0.035$ kg

| Age wks | X-44 kg | X.22 kg | Base Pop. kg | X-11 kg | X-33 kg |
|---|---|---|---|---|---|
| 2.02 | 0.08 | 0.15 | 0.11 | 0.08 | 0.16 |
| 4.04 | 0.16 | 0.28 | 0.21 | 0.17 | 0.32 |
| 6.06 | 0.27 | 0.52 | 0.41 | 0.31 | 0.59 |
| 8.08 | 0.41 | 0.73 | 0.59 | 0.50 | 0.85 |
| 10.10 | 0.55 | 0.98 | 0.82 | 0.69 | 1.13 |
| 12.12 | 0.71 | 1.20 | 1.07 | 0.97 | 1.45 |
| 14.14 | 0.89 | 1.36 | 1.26 | 1.23 | 1.72 |
| 16.16 | 1.06 | 1.51 | 1.45 | 1.51 | 2.02 |
| 18.18 | 1.24 | 1.64 | 1.64 | 1.77 | 2.26 |
| 20.20 | 1.37 | 1.73 | 1.76 | 1.97 | 2.45 |
| 22.22 | 1.50 | 1.82 | 1.91 | 2.21 | 2.64 |
| 24.24 | 1.59 | 1.88 | 2.00 | 2.40 | 2.78 |
| 26.26 | 1.64 | 1.93 | 2.06 | 2.56 | 2.90 |
| 28.28 | 1.68 | 1.96 | 2.12 | 2.74 | 3.01 |
| 30.30 | 1.71 | 1.99 | 2.17 | 2.86 | 3.09 |
| 32.32 | 1.73 | 2.01 | 2.20 | 2.99 | 3.16 |
| 34.34 | 1.74 | 2.02 | 2.22 | 3.11 | 3.23 |
| 36.36 | 1.75 | 2.03 | 2.24 | 3.18 | 3.28 |
| 38.38 | 1.76 | 2.05 | 2.27 | 3.26 | 3.32 |
| 40.40 | 1.76 | 2.06 | 2.27 | 3.26 | 3.33 |

# Appendix B Standard Deviations of Life Weight Data

Table B-1. The standard deviation $s(w)$ of each mean weight in Table 2.2. (Data of Jull and Titus 1928)

| | Female | | Male | |
|---|---|---|---|---|
| Age wks | $s(W)$ Lot 1 | $s(W)$ Lot 2 | $s(W)$ Lot 3 | $s(W)$ Lot 4 |
| 0 | 2.02 | 2.32 | 2.15 | 2.16 |
| 2 | 16.00 | 12.56 | 15.21 | 14.06 |
| 4 | 35.26 | 33.44 | 41.04 | 42.73 |
| 6 | 56.25 | 68.19 | 94.77 | 75.92 |
| 8 | 77.68 | 85.86 | 143.05 | 114.70 |
| 10 | 105.39 | 106.55 | 184.47 | 145.74 |
| 12 | 122.91 | 109.73 | 183.26 | 160.31 |
| 14 | 145.36 | 118.03 | 232.93 | 183.03 |
| 16 | 163.37 | 131.67 | 233.64 | 178.71 |
| 18 | 170.38 | 132.10 | 238.68 | 189.72 |
| 20 | 178.96 | 186.00 | 255.00 | 191.76 |
| 22 | 218.54 | 218.88 | 284.16 | 241.83 |
| 24 | 269.01 | 249.14 | 295.31 | 258.35 |

$s(W)$ is in g

Table B-2. Standard deviation $s(W)$ of live weight versus age and cumulative food consumed of two groups of Sprague Dawley male Albino rats fed diets of 100% and 90% relative specific energies (RSE) respectively

| | 100% | | 90% | |
|---|---|---|---|---|
| $t$ wks | $F$ kg | $s(W)$ kg | $F$ kg | $s(W)$ kg |
| 1 | 0.0766 | 0.00918 | 0.0782 | 0.01277 |
| 2 | 0.1897 | 0.01482 | 0.1941 | 0.01579 |
| 3 | 0.3239 | 0.01512 | 0.3352 | 0.01771 |
| 4 | 0.4725 | 0.01553 | 0.4933 | 0.01875 |
| 5 | 0.6304 | 0.02162 | 0.6629 | 0.02547 |
| 6 | 0.7943 | 0.02376 | 0.8401 | 0.02284 |
| 7 | 0.9622 | 0.03184 | 1.0230 | 0.02665 |
| 8 | 1.1330 | 0.03304 | 1.2080 | 0.03041 |
| 9 | 1.3050 | 0.04435 | 1.3970 | 0.03661 |
| 10 | 1.4780 | 0.03501 | 1.5860 | 0.03137 |
| 11 | 1.6510 | 0.03279 | 1.7770 | 0.03117 |
| 12 | 1.8260 | 0.03697 | 1.9690 | 0.03235 |
| 13 | 2.0000 | 0.03772 | 2.1610 | 0.03452 |
| 14 | 2.1750 | 0.04100 | 2.3530 | 0.03651 |
| 15 | 2.3490 | 0.04243 | 2.5460 | 0.03655 |
| 16 | 2.5240 | 0.04116 | 2.7380 | 0.04039 |
| 17 | 2.6990 | 0.03882 | 2.9310 | 0.04108 |
| 18 | 2.8740 | 0.03892 | 3.1240 | 0.04164 |
| 19 | 3.0490 | 0.04079 | 3.3170 | 0.04046 |
| 20 | 3.2240 | 0.04267 | 3.5100 | 0.04418 |

Table B-3. Standard deviation $s(W)$ of live weight versus age and cumulative food consumed of Sprague Dawley male albino rats fed diets of 80% and 70% relative specific energies (RSE) respectively

| | 80% | | 70% | |
|---|---|---|---|---|
| $t$ wks | $F$ kg | $s(W)$ kg | $F$ kg | $s(W)$ kg |
| 1 | 0.078 | 0.01096 | 0.085 | 0.00874 |
| 2 | 0.206 | 0.01248 | 0.215 | 0.00994 |
| 3 | 0.364 | 0.01490 | 0.375 | 0.01134 |
| 4 | 0.541 | 0.02289 | 0.558 | 0.01589 |
| 5 | 0.729 | 0.03850 | 0.756 | 0.02438 |
| 6 | 0.924 | 0.02648 | 0.966 | 0.02317 |
| 7 | 1.123 | 0.03148 | 1.183 | 0.02273 |
| 8 | 1.324 | 0.03742 | 1.406 | 0.02706 |
| 9 | 1.527 | 0.04236 | 1.632 | 0.02563 |
| 10 | 1.730 | 0.03194 | 1.862 | 0.02737 |
| 11 | 1.934 | 0.02494 | 2.093 | 0.02860 |
| 12 | 2.139 | 0.03223 | 2.326 | 0.02997 |
| 13 | 2.343 | 0.03520 | 2.559 | 0.02733 |
| 14 | 2.548 | 0.03743 | 2.794 | 0.02904 |
| 15 | 2.753 | 0.03656 | 3.028 | 0.03056 |
| 16 | 2.958 | 0.04001 | 3.264 | 0.03238 |
| 17 | 3.163 | 0.03948 | 3.499 | 0.03256 |
| 18 | 3.368 | 0.03791 | 3.735 | 0.03567 |
| 19 | 3.573 | 0.02825 | 2.970 | 0.03580 |
| 20 | 3.778 | 0.03715 | 4.206 | 0.03670 |

Table B-4. Standard deviations $s(W)$ of live weight versus age of male chickens fed diets with and without access to sour skim milk. (Data of Titus and Jull 1928)

| $t$ wks | No milk $s(W)$ kg | With milk $s(W)$ kg |
|---|---|---|
| 1 | 0.00958 | 0.01117 |
| 2 | 0.01901 | 0.02005 |
| 3 | 0.02599 | 0.03413 |
| 4 | 0.03239 | 0.04301 |
| 5 | 0.04497 | 0.05849 |
| 6 | 0.07769 | 0.08067 |
| 7 | 0.10740 | 0.08375 |
| 8 | 0.12958 | 0.09706 |
| 9 | 0.14703 | 0.11577 |
| 10 | 0.16058 | 0.12772 |
| 11 | 0.17656 | 0.14858 |
| 12 | 0.22891 | 0.17984 |
| 13 | 0.24657 | 0.15796 |
| 14 | 0.26658 | 0.18396 |
| 18 | 0.32127 | 0.22185 |
| 22 | 0.31114 | 0.23830 |
| 26 | 0.36163 | 0.25500 |
| 30 | 0.40024 | 0.29817 |
| 34 | 0.38401 | 0.29446 |

Table B-5. Standard deviations $s(W)$ of live weight versus age of Rhode Island Red male chickens. (Unpublished data of Grossman 1969)

| $t$ wks | $s(W)$ kg | $t$ wks | $s(W)$ kg |
|---|---|---|---|
| 1 | 0.00711 | 24 | 0.35667 |
| 2 | 0.01689 | 25 | 0.36214 |
| 3 | 0.03084 | 26 | 0.36109 |
| 4 | 0.04364 | 27 | 0.36133 |
| 5 | 0.05788 | 28 | 0.35423 |
| 6 | 0.07467 | 29 | 0.35299 |
| 7 | 0.09128 | 30 | 0.35475 |
| 8 | 0.10992 | 31 | 0.35320 |
| 9 | 0.12704 | 32 | 0.35475 |
| 10 | 0.14358 | 33 | 0.35249 |
| 11 | 0.15801 | 34 | 0.35597 |
| 12 | 0.17514 | 35 | 0.35196 |
| 13 | 0.19133 | 36 | 0.34758 |
| 14 | 0.20885 | 37 | 0.34928 |
| 15 | 0.22754 | 38 | 0.34631 |
| 16 | 0.24700 | 39 | 0.34200 |
| 17 | 0.26144 | 40 | 0.33892 |
| 18 | 0.27708 | 41 | 0.33925 |
| 19 | 0.28579 | 42 | 0.33672 |
| 20 | 0.29718 | 43 | 0.34242 |
| 21 | 0.31091 | 44 | 0.34571 |
| 22 | 0.32342 | 45 | 0.33868 |
| 23 | 0.34289 | | |

Table B-6. Standard deviations $s(W)$ of live weight versus age of Rhode Island Red female chickens. (Unpublished data of Grossman 1969)

| $t$ wks | $s(W)$ kg | $t$ wks | $s(W)$ kg |
|---|---|---|---|
| 1 | 0.00734 | 24 | 0.25309 |
| 2 | 0.01542 | 25 | 0.25380 |
| 3 | 0.02568 | 26 | 0.25005 |
| 4 | 0.03749 | 27 | 0.24805 |
| 5 | 0.05132 | 28 | 0.24446 |
| 6 | 0.06670 | 29 | 0.25479 |
| 7 | 0.08028 | 30 | 0.24869 |
| 8 | 0.09882 | 31 | 0.25096 |
| 9 | 0.11199 | 32 | 0.25247 |
| 10 | 0.12368 | 33 | 0.25411 |
| 11 | 0.12646 | 34 | 0.26160 |
| 12 | 0.13505 | 35 | 0.26376 |
| 13 | 0.14375 | 36 | 0.26665 |
| 14 | 0.15463 | 37 | 0.26968 |
| 15 | 0.15946 | 38 | 0.27511 |
| 16 | 0.16713 | 39 | 0.27422 |
| 17 | 0.17094 | 40 | 0.27999 |
| 18 | 0.17404 | 41 | 0.27490 |
| 19 | 0.18080 | 42 | 0.27797 |
| 20 | 0.18875 | 43 | 0.28085 |
| 21 | 0.19882 | 44 | 0.28698 |
| 22 | 0.21833 | 45 | 0.29229 |
| 23 | 0.24376 | | |

Table B-7. Standard deviations $s(W)$ of live weight versus age of White Leghorn male chickens. (Unpublished data of Grossman 1969)

| $t$ wks | $s(W)$ kg | $t$ wks | $s(W)$ kg |
|---|---|---|---|
| 1 | 0.00660 | 24 | 0.22252 |
| 2 | 0.01520 | 25 | 0.22722 |
| 3 | 0.02601 | 26 | 0.23272 |
| 4 | 0.03856 | 27 | 0.23477 |
| 5 | 0.05283 | 28 | 0.22890 |
| 6 | 0.06821 | 29 | 0.22830 |
| 7 | 0.08321 | 30 | 0.22991 |
| 8 | 0.10052 | 31 | 0.23348 |
| 9 | 0.11255 | 32 | 0.23574 |
| 10 | 0.12223 | 33 | 0.23299 |
| 11 | 0.12552 | 34 | 0.23365 |
| 12 | 0.12837 | 35 | 0.23610 |
| 13 | 0.13491 | 36 | 0.23429 |
| 14 | 0.14176 | 37 | 0.23566 |
| 15 | 0.15315 | 38 | 0.23752 |
| 16 | 0.16533 | 39 | 0.23346 |
| 17 | 0.16632 | 40 | 0.23457 |
| 18 | 0.18135 | 41 | 0.23290 |
| 19 | 0.19292 | 42 | 0.23481 |
| 20 | 0.19777 | 43 | 0.23468 |
| 21 | 0.20330 | 44 | 0.23366 |
| 22 | 0.21003 | 45 | 0.23610 |
| 23 | 0.21933 | | |

Table B-8. Standard deviations $s(W)$ of live weight versus age of White Leghorn female chickens. (Unpublished data of Grossman 1969)

| $t$ wks | $s(W)$ kg | $t$ wks | $s(W)$ kg |
|---|---|---|---|
| 1 | 0.00668 | 24 | 0.15686 |
| 2 | 0.01645 | 25 | 0.17364 |
| 3 | 0.02780 | 26 | 0.19187 |
| 4 | 0.03980 | 27 | 0.20314 |
| 5 | 0.05396 | 28 | 0.20347 |
| 6 | 0.06835 | 29 | 0.20861 |
| 7 | 0.08326 | 30 | 0.21099 |
| 8 | 0.09695 | 31 | 0.20918 |
| 9 | 0.10869 | 32 | 0.20649 |
| 10 | 0.11918 | 33 | 0.21112 |
| 11 | 0.11577 | 34 | 0.21550 |
| 12 | 0.11620 | 35 | 0.21678 |
| 13 | 0.11984 | 36 | 0.22308 |
| 14 | 0.11968 | 37 | 0.22358 |
| 15 | 0.12554 | 38 | 0.22645 |
| 16 | 0.12846 | 39 | 0.22426 |
| 17 | 0.12835 | 40 | 0.22587 |
| 18 | 0.12980 | 41 | 0.22901 |
| 19 | 0.13096 | 42 | 0.22688 |
| 20 | 0.13340 | 43 | 0.23070 |
| 21 | 0.13479 | 44 | 0.22735 |
| 22 | 0.13728 | 45 | 0.21984 |
| 23 | 0.14661 | | |

Table B-9. Standard deviations $s(W)$ versus age for mice reared in the Animal Science Department of North Carolina State University at Rayleigh. (Unpublished data of Rutledge 1971)

| $t$ d | $s(W)$ g |
|---|---|
| 0 | 0.117 |
| 3 | 0.360 |
| 6 | 0.578 |
| 9 | 0.748 |
| 12 | 0.900 |
| 15 | 1.000 |
| 18 | 1.330 |
| 21 | 1.620 |
| 24 | 2.200 |
| 27 | 2.560 |
| 30 | 2.640 |
| 33 | 2.470 |
| 36 | 2.310 |
| 39 | 2.240 |
| 42 | 2.230 |
| 49 | 2.280 |
| 56 | 2.540 |
| 63 | 2.710 |
| 70 | 2.650 |
| 77 | 2.730 |
| 84 | 2.800 |

# Appendix C Partial and Complete Starvation Data

Table C-1. Starvation of *Cassiopea Xamachana*. Average weight of 6 animals starved in the dark. (Data of Mayer 1914)

| *t*<br>d | *W*<br>g | *t*<br>d | *W*<br>g |
|---|---|---|---|
| 0 | 85.59 | 14 | 27.23 |
| 1 | 79.57 | 15 | 27.12 |
| 2 | 75.00 | 16 | 26.49 |
| 3 | 70.90 | 17 | 24.87 |
| 4 | 63.35 | 18 | 23.24 |
| 5 | 56.20 | 19 | 20.97 |
| 6 | 52.60 | 20 | 18.85 |
| 7 | 51.71 | 21 | 17.56 |
| 8 | 42.09 | 22 | 15.31 |
| 9 | 41.03 | 23 | 14.38 |
| 10 | 40.96 | 24 | 13.07 |
| 11 | 39.40 | 25 | 10.78 |
| 12 | 33.54 | 26 | 8.99 |
| 13 | 28.51 | | |

Table C-2. Starvation of *Cassiopea Xamachana*. Average weight of 10 animals versus time in diffuse light. (Data of Mayer 1914)

| *t*<br>d | *W*<br>g | *t*<br>d | *W*<br>g |
|---|---|---|---|
| 0 | 130.75 | 13 | 72.31 |
| 1 | 128.96 | 14 | 69.27 |
| 2 | 119.58 | 15 | 66.81 |
| 3 | 115.34 | 17 | 61.24 |
| 4 | 108.97 | 18 | 56.79 |
| 5 | 106.58 | 19 | 53.45 |
| 6 | 103.55 | 20 | 51.81 |
| 7 | 97.60 | 21 | 48.20 |
| 8 | 97.22 | 22 | 25.84 |
| 9 | 92.14 | 23 | 44.71 |
| 10 | 88.71 | 24 | 42.78 |
| 11 | 86.24 | 25 | 39.70 |
| 12 | 79.54 | | |

Table C-3. Starvation of dog Oscar. Pecent of original weight versus days since starvation began. (Data of Kleiber 1961)

| *t*<br>d | *W*<br>% |
|---|---|
| 0 | 100 |
| 10 | 90 |
| 20 | 82 |
| 30 | 74 |
| 40 | 68 |
| 50 | 63 |
| 60 | 61 |
| 70 | 54 |
| 80 | 48 |
| 90 | 46 |
| 100 | 42 |
| 110 | 40 |

Table C-4. Starvation of man Succi. Average weight at Florence and Naples versus time during 30-day tests. (Data of Benedict 1907)

| *t*<br>d | *W*<br>kg | *t*<br>d | *W*<br>kg |
|---|---|---|---|
| 0 | 63.45 | 16 | 54.51 |
| 1 | 62.55 | 17 | 54.08 |
| 2 | 61.07 | 18 | 53.55 |
| 3 | 59.87 | 19 | 53.08 |
| 4 | 59.67 | 20 | 52.63 |
| 5 | 59.02 | 21 | 52.23 |
| 6 | 58.24 | 22 | 51.88 |
| 7 | 57.54 | 23 | 51.48 |
| 8 | 57.14 | 24 | 51.08 |
| 9 | 56.84 | 25 | 51.13 |
| 10 | 56.71 | 26 | 50.93 |
| 11 | 56.13 | 27 | 50.88 |
| 12 | 55.50 | 28 | 50.68 |
| 13 | 55.12 | 29 | 50.08 |
| 14 | 54.87 | 30 | 50.83 |
| 15 | 54.69 | | |

Table C-5. Forty day fast of Succi. (Data from Table 2, p. 912 of Sect. 4 of Handbook of Physiology 1964)

| $t$ d | $W$ kg |
|---|---|
| 0 | 55.9 |
| 14 | 48.8 |
| 16 | 48.2 |
| 20 | 46.9 |
| 29 | 44.4 |
| 30 | 44.3 |
| 31 | 44.3 |
| 40 | 41.8 |

Table C-6. Average weekly weights of 32 men during preconditioning period prior to semi-starvation. (Data of Keys et al. 1950)

| $t$ wks | $W$ kg |
|---|---|
| 1 | 70.19 |
| 2 | 70.02 |
| 3 | 69.95 |
| 4 | 69.88 |
| 5 | 69.82 |
| 6 | 69.71 |
| 7 | 69.64 |
| 8 | 69.51 |
| 9 | 69.39 |
| | Mean $W = 69.79$ |

Mean caloric intake = 3492.5 kcal/d per man

Table C-7. Average weekly weights of the 32 men in Table C-6, except the group was placed on a semi-starvation regime equivalent to 44.94% of the mean pre-experimental caloric intake per man

| $t$ wks | $W$ kg | $t$ wks | $W$ kg |
|---|---|---|---|
| 1 | 68.35 | 13 | 56.60 |
| 2 | 66.80 | 14 | 56.16 |
| 3 | 65.76 | 15 | 55.69 |
| 4 | 64.29 | 16 | 54.70 |
| 5 | 63.33 | 17 | 54.28 |
| 6 | 62.16 | 18 | 54.08 |
| 7 | 61.11 | 19 | 53.51 |
| 8 | 60.31 | 20 | 53.18 |
| 9 | 59.56 | 21 | 52.99 |
| 10 | 58.71 | 22 | 52.90 |
| 11 | 58.14 | 23 | 52.83 |
| 12 | 57.28 | 24 | 52.57 |

# Appendix D Whole Body Combustion Data

Table D-1. Empty body weight $W$, body composition and specific combustion energy $\Gamma$ for Haecker's steers. (Data of Kleiber 1961)

| $W$ kg | Fat % | Protein % | Water % | Ash % | $\Gamma$ kcal/g |
|---|---|---|---|---|---|
| 45.4 | 4.0 | 19.9 | 71.8 | 4.3 | 1.50 |
| 91.0 | 6.3 | 19.6 | 69.5 | 4.6 | 1.70 |
| 181.0 | 10.6 | 19.3 | 65.7 | 4.4 | 2.11 |
| 273.0 | 14.0 | 19.2 | 62.2 | 4.6 | 2.42 |
| 364.0 | 19.2 | 18.7 | 57.9 | 4.2 | 2.89 |
| 454.0 | 25.5 | 17.6 | 53.1 | 3.8 | 3.43 |
| 545.0 | 31.1 | 15.6 | 48.6 | 3.7 | 3.84 |

Table D-2. Body weight $W$, energy intake $\beta dF/dt$, whole body combustion energy $\Gamma W$, specific combustion energy $\Gamma$ versus time since weaning for rats fed a 25% protein diet. (Data of Mayer and Vitale 1957)

| $t$ d | $W$ g | $\beta dF/dt$ kcal/d | $\Gamma W$ kcal | $\Gamma$ kcal/g |
|---|---|---|---|---|
| 0 | 38 | 33 | 57 | 1.48 |
| 3 | 48 | 42 | 64 | 1.36 |
| 6 | 65 | 48 | 101 | 1.56 |
| 9 | 81 | 55 | 127 | 1.56 |
| 12 | 99 | 62 | 167 | 1.58 |
| 18 | 139 | 69 | 252 | 1.81 |
| 24 | 179 | 75 | 379 | 2.13 |
| 30 | 211 | 79 | 472 | 2.24 |
| 45 | 289 | 79 | 695 | 2.39 |
| 60 | 332 | 78 | 965 | 2.89 |
| 90 | 372 | 79 | 1099 | 2.92 |

Dietary metabolizble energy $\beta = 4.44$ kcal/g
Mature weight $A = 462$ g

Table D-3. Fasting body weight $W$, food intake $dF/dt$, and whole body consumption energy $\Gamma W$ versus time for 7 individual rats housed at 25 °C. $ME$ of diet $\beta = 3.6$ kcal/g. (Data of Joy and Mayer 1968)

| $t$ d | $W$ g | $dF/dt$ g/d | $\Gamma W$ kcal | $t$ d | $W$ g | $dF/dt$ g/d | $\Gamma W$ kcal |
|---|---|---|---|---|---|---|---|
| 0 | 199 | 20.9 | 374.0 | 22 | 352 | 27.3 | 737.8 |
| | 185 | 20.0 | 266.8 | | 287 | 21.4 | 488.1 |
| | 186 | 19.9 | 292.0 | | 344 | 21.4 | 706.4 |
| | 197 | 22.6 | 260.0 | | 373 | 26.7 | 766.0 |
| | 198 | 21.9 | 280.0 | | 396 | 25.6 | 637.3 |
| | 184 | 20.1 | 295.3 | | 380 | 24.4 | 809.7 |
| | 186 | 19.4 | 257.5 | | 364 | 26.4 | 657.1 |
| 4 | 196 | 22.4 | 332.4 | 36 | 422 | 29.0 | 782.2 |
| | 197 | 23.3 | 295.2 | | 442 | 25.3 | 1017.0 |
| | 213 | 24.6 | 290.1 | | 449 | 28.3 | 1019.3 |
| | 165 | 17.0 | 216.0 | | 462 | 28.1 | 995.1 |
| | 189 | 17.0 | 298.2 | | 412 | 30.0 | 804.0 |
| | 238 | – | 365.7 | | 493 | 27.4 | 1151.2 |
| | 227 | 17.0 | 313.5 | | 499 | 25.1 | 1201.6 |
| 8 | 241 | – | 390.8 | 50 | 457 | 24.4 | 1104.9 |
| | 200 | 17.4 | 396.6 | | 412 | 22.6 | 1275.8 |
| | 249 | 22.3 | 389.8 | | 441 | 24.1 | 1069.0 |
| | 270 | 26.1 | 423.3 | | 522 | 29.9 | 1166.6 |
| | 275 | 24.4 | 529.4 | | 486 | 23.6 | 1213.2 |
| | 247 | 22.1 | 360.2 | | 552 | 27.0 | 1516.7 |
| | 237 | 20.9 | 363.5 | | 557 | 27.4 | 1700.1 |
| 15 | 291 | 22.0 | 464.2 | 64 | 549 | 25.1 | 1585.8 |
| | 300 | 23.0 | 552.0 | | 541 | 23.7 | 1488.1 |
| | 302 | 22.3 | 534.1 | | 640 | 29.9 | 1921.8 |
| | 226 | 23.0 | 444.7 | | 690 | 30.7 | 2032.0 |
| | 347 | 29.7 | 671.8 | | 655 | 26.9 | 1955.2 |
| | 275 | 15.9 | 425.5 | | 634 | 27.0 | 2275.6 |
| | 337 | 21.6 | 688.8 | | 581 | 25.6 | 1543.6 |

Table D-4. Fasting body weight $W$, food intake $dF/dt$, and whole body combustion energy $\Gamma W$ versus time of five rats housed at 5 °C. (Data of Joy and Mayer 1968)

| $t$<br>d | $W$<br>g | $dF/dt$<br>g/d | $\Gamma W$<br>kcal |
|---|---|---|---|
| 0 | 241 | 20.5 | 390.8 |
| | 227 | 17.4 | 313.5 |
| | 200 | 17.6 | 396.6 |
| | 196 | 22.4 | 319.3 |
| | 199 | 20.9 | 357.6 |
| 4 | 223 | 28.3 | 304.3 |
| | 204 | 24.8 | 294.0 |
| | 202 | 25.8 | 317.2 |
| | 201 | 26.5 | 323.3 |
| | 212 | 32.3 | 337.0 |
| 8 | 200 | 28.4 | 297.0 |
| | 204 | 29.0 | 249.9 |
| | 221 | 31.3 | 280.6 |
| | 196 | 29.8 | 279.6 |
| | 199 | 29.0 | 291.3 |
| 15 | 225 | 30.6 | 328.8 |
| | 228 | 38.9 | 368.8 |
| | 240 | 35.0 | 330.7 |
| | 213 | 34.7 | 259.5 |
| | 262 | 37.1 | 477.5 |
| 22 | 262 | 42.3 | 356.2 |
| | 212 | 28.3 | 315.1 |
| | 274 | 34.7 | 383.9 |
| | 223 | 33.9 | 325.3 |
| | 278 | 25.1 | 456.2 |
| 36 | 260 | 35.0 | 369.7 |
| | 291 | 39.6 | 495.0 |
| | 275 | 42.4 | 389.3 |
| | 306 | 39.4 | 486.4 |
| | 327 | 41.4 | 735.7 |
| 50 | 323 | 43.9 | 547.2 |
| | 251 | 24.3 | 321.1 |
| | 282 | 38.4 | 443.4 |
| | 338 | 41.6 | 481.5 |
| | 293 | 37.7 | 532.5 |
| 64 | 367 | 41.7 | 593.8 |
| | 365 | 41.0 | 705.6 |
| | 408 | 46.7 | 869.9 |
| | 280 | 39.9 | 485.9 |
| | 301 | 33.4 | 626.8 |

# Glossary of Mathematical Symbols

| Symbol | Units | Meaning |
|---|---|---|
| *Growth and Feeding Data* | | |
| $W_0$ or $W(0)$ | kg or g | Birth weight or weaning weight or weight in units of kilograms or grams at beginning of the experiment |
| $t$ | mo., wk., d | Age or time elapsed since beginning of experiment in units of months, weeks or days |
| $W$ or $W(t)$ | kg or g | Live weight at age $t$ or at time $t$ since experiment began |
| $F$ or $F(t)$ | kg or g | Cumulative food consumed since experiment began |
| $\Delta t$ | mo., wk., d | A short time period |
| $\Delta W$ | kg or g | Live weight gain in time period $\Delta t$ |
| $\Delta F$ | kg or g | Food consumed in time period $\Delta t$ |
| *Quantities Derived from Data* | | |
| $\Delta W/\Delta t$ | kg or g/mo, wk or d; kg or g, $mo^{-1}$, $wk^{-1}$ or $d^{-1}$ | Estimated growth rate |
| $\Delta W/W\Delta t$ | per mo., wk or d $mo^{-1}$, $wk^{-1}$ or $d^{-1}$ | Estimated relative growth rate |
| $\Delta W/\Delta F$ | unitless | Estimated growth efficiency |
| $\Delta F/\Delta t$ | kg or g/mo, wk, d kg or g $mo^{-1}$, $wk^{-1}$, $d^{-1}$ | Estimated food intake |
| *Diet Composition Data* | | |
| Proximate analysis of 1 g of diet | | |
| $p$ | fraction | Crude protein |
| $c$ | | Nitrogen-free extract or NFE |
| $f$ | | Ether extract, fat or oil |
| $b$ | | Fibre |
| $a$ | | Ash or mineral content |
| $w$ | | Water |
| Balance equation | | $p+c+f+b+a+w=1$ |

| Symbol | Units | Meaning |
|---|---|---|
| *Derived Diet Composition* | | |
| $\eta$ | unitless | Gram of nutrients per gram of diet |
| $\pi$ | | Crude protein per gram of nutrient |
| $\phi$ | | Ether extract, fat or oil per gram of nutrients. Also used in Chap. 11 to represent metabolisable or net energy intake in units $\mathrm{kcal d^{-1}}$ |
| $\kappa$ | | Carbohydrate as NFE, glucose or starch per gram of nutrients |
| Balance equation | | $\pi+\phi+\kappa=1$ |
| *Ad Libitum Feeding and Growth Parameters* | | |
| $D$ | kg or g/mo, wk, d kg or g $\mathrm{mo^{-1}}$, $\mathrm{wk^{-1}}$, $\mathrm{d^{-1}}$ | Initial food intake or feeding rate at $t=0$ |
| $C(=A/T_0)$ | kg or g/mo, wk, d kg or g $\mathrm{mo^{-1}}$, $\mathrm{wk^{-1}}$, $\mathrm{d^{-1}}$ | Mature food intake |
| $t^*$ | mo, wk, d | Is the time required for the animal to increase its food intake to 63% of its mature food intake, $C$, from that at birth. The internal resistance to build up of appetite. Brody's $t^*$. |
| $A$ | kg or g | Mature live weight |
| $B$ | $\mathrm{kg^{-1}}$ or $\mathrm{g^{-1}}$ | It is the experimental decay constant of growth. $Ln2/B$ is the amount of food the animal must consume to build its body weight from $W(t)$ to $[W(t)+A]/2$. $B$ is also the probability per unit of food consumed the animal will grow |
| $(AB)$ | unitless | Growth efficiency factor. This is the product of $A$ and $B$ put in parentheses to indicate the product is a parameter |
| $T_0$ | mo, wk, d | The Taylor time constant. It is the slope of the diagonal of the Growth Phase Plane (GPP). $T_0=A/C$ |
| *Some Quantities Derived from the Feeding and Growth Parameters* | | |
| $k$ | $\mathrm{mo^{-1}}$, $\mathrm{wk^{-1}}$, $\mathrm{d^{-1}}$ | Brody's maturing rate constant $k=(AB)C/A=(AB)/T_0$ |
| $\alpha$ | unitless | The ad libitum feeding and growth discriminant; $\alpha\leqq 1$ feeding is ad libitum, $\alpha>1$ feeding is controlled. $\alpha=kt^*=(AB)t^*/T_0=(AB)Ct^*/A$ |
| $C/t^*=At^*/T_0$ | kg or g $\mathrm{Mo^{-2}}$, $\mathrm{wk^{-2}}$, $\mathrm{d^{-2}}$ | Appetency. Internal drive to increase appetite to $C$ |

| Symbol | Units | Meaning |
|---|---|---|
| *Derived Feeding and Growth Unitless Variables* | | |
| $u$ | unitless | Degree of maturity, $u = W(t)/A$ |
| $T$ | | Normalized age or time, $t$. It can be $kt$, $t/t^*$ or $t/t_{0.95}$ where $t_{0.95}$ is the age at which $W(t) = 0.95A$ |
| $Z(t)$ | | $Z(t) = (AB)F(t)/A$ which is the body weight equivalent of the cumulative food consumed in units of the mature weight $A$ |
| $Q(t)$ | | Normalized feeding rate at age or time $t$. $Q(t) = (dF/dt)/C = T_0(dF/dt)/A$ |
| *Quantities Expressed as Derivatives of Continuous Feeding and Growth Functions* | | |
| $dW/dt$ | kg or g $mo^{-1}$, $wk^{-1}$, $d^{-1}$ | Growth rate |
| $dF/dt$ | kg or g $mo^{-1}$, $wk^{-1}$, $d^{-1}$ | Appetite or feeding rate |
| $dW/dF$ | unitless | Growth efficiency |
| $g^*(t) = dF/dt$ | | Ad libitum feeding rate |
| $g(t)$ | | Feeding rate other than ad libitum |
| $dq^*/dt = d^2F/dt^2$ | kg or g $mo^{-2}$, $wk^{-2}$, $d^{-2}$ kcal/$g^{-1}$ | Ad libitum feeding acceleration |
| *Specific energies* | | |
| $E_c$ | | Combustion energy per gram of diet |
| $\beta$ | | Metabolisable or net energy of diet |
| ME | | Metabolisable energy |
| $\Gamma$ | | Specific whole body combustion energy |

*Constants in Equations*

$a$, $b$, $c$, and so on

*Variables*

$x$, $y$, $z$, and so on

The units and meanings of constants and variables are appropriate to the equations in which they appear, unless otherwise specified.

# Glossary of Words and Phrases

| Symbol | Units | Meaning |
| --- | --- | --- |
| Appetency | $C/t^*$ or $A/T_0t^*$ | Inner drive to increase appetite towards the maximum $C$ |
| Appetite | $dF/dt$ or $q^*(t)$ | Rate of change of the cumulative food consumed. Rate of feeding |
| Biotrace | $u=u(T, Z)$ | The three-dimensional Euclidean space curve generated by any growing animal. The axes of the space are the dimensionless quantities, $u$, $T$ and $Z$ |
| Differential equation | DE | A general statement of the fundamental dynamic character of a class of systems expressed as a function of the rates of change of some property common to all the members of the class and a set of quantities referred to as structural parameters. For example in the equation $\sum_{i,j}^{N,n} a_i d^jy/dx^j=0$, the $d^jy/dx^j$ are the rates of change of $y$ relative to change of $x$, and the quantities $a_i$ are the structural parameters |
| Factor | $(AB)$, $C$ etc. | A quantity used as a multiplier in the arithmetic sense, given a name or meaning depending on the context in which it is used |
| Function (mathematical) | $y=y(x)$ or $y=f(x)$ | Here the function is considered continuous, differentiable and single valued in some specified range of $x$ |
| Growth curve | $W=W(t)$ | The two dimensional Euclidean space curve generated by any growing animal. The axes of the space are $W$ and $t$ |
| Growth phase plane | GPP | A rectangle of height $A$, the mature live weight, and width $C$, the mature food intake, which contains all the possible $[q(t), W(t)]$ data points for an animal |
| Growth promoting ability | GPA ($x$) | The ability of some environmental variable, $x$, to increase the growth of an animal |
| Relative | GPA ($x$)/GPA ($r$) | Here $r$ names some reference environmental variable of type $x$ |
| Hyperspace | | A space of more than three dimensions |

| Symbol | Units | Meaning |
|---|---|---|
| Initial conditions | i.c. | All the values of $y$ and its derivatives with respect to $x$ at $x$ equal to zero which permit solution of a given DE for a particular function, $y=f(x)$, describing a particular system in the class of systems to which the DE is appropriate. The total number of parameters appearing in the solution, $y=f(x)$, will be the sum of the number of structural parameters and the number of initial conditions |
| Normalize | Dimensionless | Reduce a variable quantity measured in some unit to a dimensionless number by dividing the variable by a constant expressed in the same units, such as $W/A$, $t/t^*$ etc |
| Space growth curve | | The curve generated by a growing animal in a three dimensional Euclidian space the axes of which are $W$, $F$ and $t$. Synonymous with trace |
| Taylor diagonal | | The diagonal of the GPP drawn from the point (0, 0) to the point $(C, A)$ |
| Taylor time constant | $T_0$ | The slope of the Taylor diagonal of the GPP, namely $A/C$ |
| Trace | | A synonym for 'space growth curce' |
| Length of | $s$ | The distance along a trace from a given point on the trace to some chosen point on the trace |

# Subject Index

**Volume 7**
J. K. Matsushima

## Feeding Beef Cattle

1979. 31 figures, 23 tables. IX, 128 pages
ISBN 3-540-09198-X

"...The book is written as a practical handbook on a theoretically qualified basis... It is systematically composed, richly provided with tables and practical directions, and written in an intelligible language – without excessive use of difficult technical words and complicated explanations. In a simple way, Matsushima's little handbook explains the character of many feed items and their suitability in various ratios in compound feeds and for different cattle groups. ...The relevant reference may be found quickly and the necessary, practical knowledge may easily be acquired. Although written for the American feed-lot farmer, this book has a message for several groups other than the large-scale beef producers. It will be **an excellent tool for the trained cattle husbandry researcher,** wherever he might be, and it will also be **a good handbook** for new beef cattle systems in technical poor or undeveloped countries. No matter the feed items are different or the technology missing, Matsushima's book may be useful. ...."

*Animal Feed Sci. and Technol.*

**Volume 8**
R. J. Hanks, G. L. Ashcroft

## Applied Soil Physics

**Soil Water and Temperature Applications**
1980. 55 figures, 19 tables. VI, 159 pages
ISBN 3-540-09457-1

Developed from a series of lectures by two leading experts in agriculture and soil science, *Applied Soil Physics* thoroughly explores all aspects of the physical properties of soils. Special emphasis is on quantity potentials and flow of water in soils, relations between soils, plants and atmosphere, and temperature and heat flow in soils.
Teachers, students, and researchers in agronomy and related fields will find this book a useful and important tool.

**Volume 9**
J. Palti

## Cultural Practices and Infectious Crop Diseases

1981. 43 figures. XVI, 243 pages
ISBN 3-540-11047-X

The prevention of plant disease is generally not a major consideration in crop planning, sowing, fertilization and harvesting. Although these and other agricultural practices have long been recognized as a major component in integrated disease control, far more research has been invested in resistance breeding and the application of chemical agents, leaving growers to rely on experience and tradition rather than on sound scientific principles in the conduct of successful antipathogenic agriculture.
**Cultural Practices and Infectious Crop Diseases** is the first monograph to deal comprehensively with the effects of agricultural practices on plant health. Following the effects of background factors (climate, soil, stress, and crop age), the author describes the impact of the major farming operations on the development of crop diseases, among them crop planning and alternation, multiple cropping, tillage, fertilization, moisture management, sowing, harvesting, and sanitation. The author concludes with a consideration of integrated disease control combining agricultural practices, resistance breeding and the use of chemical agents.
This volume will prove an invaluable aid to agricultural advisors, growers, researchers, students and teachers in appreciating the importance of appropriate practices management in the prevention of crop disease.

**Volume 10**
E. Bresler, B. L. McNeal

## Saline and Sodic Soils

**Principles – Dynamics – Modeling**
With a Contribution by D. L. Carter
1982. 78 figures, 23 tables. Approx. 280 pages
ISBN 3-540-11120-4

*Saline and Sodic Soils* is a comprehensive exposition of the principles and processes involved in the genesis, formation and reclamation of saline and sodic soils. The coverage includes critical interpretations of models characterizing the physical and chemical behavior of salt concentration and composition and their effects on soils and plants, as well as practical suggestions for the control of soil salinity to improve economic potential.
Each topic in this volume is clearly explained and readily accessible for all students and professionals in the agricultural and environmental sciences with a general background in mathematics, physics, chemistry and biology.

**Volume 12**
J. Hagin, B. B. Tucker

## Fertilization of Dryland and Irrigated Soils

1982. Approx. 9 figures, approx. 7 tables.
Approx. 250 pages
ISBN 3-540-11121-2
In preparation

Springer-Verlag
Berlin Heidelberg New York